HERBIVORY

The Dynamics of Animal–Plant Interactions

STUDIES IN ECOLOGY

GENERAL EDITORS

D. J. ANDERSON BSc, PhD
Department of Botany
University of New South Wales
Sydney

P. GREIG-SMITH MA, ScD
School of Plant Biology
University College of North Wales
Bangor

and

FRANK A. PITELKA PhD
Department of Zoology
University of California
Berkeley

STUDIES IN ECOLOGY VOLUME 10

HERBIVORY
The Dynamics of Animal–Plant Interactions

by MICHAEL J. CRAWLEY

Department of Pure and Applied Biology
Imperial College, London

UNIVERSITY OF CALIFORNIA PRESS

BERKELEY AND LOS ANGELES 1983

UNIVERSITY OF
CALIFORNIA PRESS

Library of Congress
Catalog Card No 82–45903

Printed in Great Britain

CONTENTS

PREFACE

This book is about the population dynamics of plants and the animals that feed on them. It aims to show how plant numbers, sizes, chemical composition and spatial distribution affect the birth, death and dispersal rates of their herbivores, and how the timing, intensity, selectivity and spatial pattern of animal feeding affect the establishment, growth and seed set of plants.

I use the word herbivore to mean an animal that eats living plant tissues. Herbivores differ from the other major group of plant-feeding animals, the detritivores, in that they have the potential to influence the rate at which their food resources are produced. My definition includes grazers like sheep and horses, browsers like goats and deer, leaf-chewing insects (the so-called phytophages), root-feeding species, plant parasites (the sap-sucking in-vertebrates like aphids and spider-mites), seed-feeding (granivorous) insects and vertebrates, flower and fruit feeders (frugivores), and animals like bark beetles or wood wasps which act as vectors of debilitating fungal and viral plant diseases. The definition is stretched to include this last class of animals because the abundance of the vectors affects the death rate of the plant population and, therefore, has a direct impact on plant population dynamics. Also, many vectors only attack living plants, so the animals' feeding activity affects the future availability of their own food resources.

The subject matter of herbivory is extremely broad, and crosses the traditional divide between botany and zoology. The study of plant–animal interaction draws on fields ranging from plant biochemistry to the ecology of seed dispersal, and from animal behaviour to nutritional physiology. The fact that a review of herbivory has not yet been attempted is doubtless due in no small part to this daunting array of subject matter.

Whereas classical ecological studies deal with four complementary sets of factors (climatic, physiographic, edaphic and biotic), this book is concerned almost solely with biotic interactions. This is not because climate, physiography or soil are any less important in the dynamics of animals and plants, but simply because it is impossible to accommodate the four sets of factors in one coherent framework. Soils and physiography change much too slowly compared with the characteristic time scales of population dynamics, and weather changes are too fast and too unpredictable to form an integral part of plant–herbivore theory. The abiotic factors figure in the present account

ix

in so far as climate is one of the main factors introducing randomness (stochasticity) in time, and physiography and soils are important determinants of the spatial heterogeneity which can have such a profound effect on both the abundance and stability of populations.

The study of plant–herbivore dynamics is still in an embryonic state. There has been a woeful lack of theoretical models against which to sharpen plausible hypothesis prior to testing them in the field, and a tendency for field workers to observe animals and plants rather than to experiment with them. Short-term funding of research has led to a proliferation of two-year studies and a dearth of detailed, long-term work. I hope that, despite these difficulties, the book will provide a focus for debate and will stimulate experimental field studies aimed at elucidating the effects of herbivorous animals on the demography of plants.

I am indebted to a great many friends and colleagues including Roy Anderson, John Beddington, Annette Duggan, Charles Godfray, Mike Hassell, Richard Hill, John Lawton, Bob May, George McGavin, Stuart McNeill, Imanuel Noy Meir, Jill Packham, Frank Pitelka, Dick Root, Jorge Soberon, Jeff Waage, Nadia Waloff, Mike Way, Mark Westoby, Imogen Yates, all the students of ecology at the University of Bradford, and the proprietor of the Jai Alai.

Finally, warmest thanks are due to John Harper who started it all by rekindling the spirit of Darwinism in plant ecology.

M. J. Crawley
Imperial College

CHAPTER 1

INTRODUCTION

1.1   POPULATION REGULATION OF HERBIVORES

One of the most persistent controversies in population ecology concerns the 'food limitation hypothesis'. Its proponents believe that herbivorous animals regulate the abundance of plant populations (Brues 1946) or even control the function of whole ecosystems (Chew 1974, Mattson & Addy 1975), and that the numbers of most herbivore populations are food limited. The opponents argue that herbivorous animals are typically scarce in relation to their food and have little impact on plants because their numbers are controlled by the depredations of natural enemies (Hairston *et al*. 1960) or the weather (Andrewartha & Birch 1954). A few even consider that herbivore feeding is generally beneficial to plants (Owen 1980).

These extreme views have been expressed about herbivorous insects as in the following examples. After an extensive study of grasshoppers, Richards and Waloff (1954) conclude that 'food is of little importance in the choice of habitats; moreover the food supply is never a limiting factor under the conditions of the British Isles'. In contrast, the American entomologist Brues (1946), generalizing about herbivorous insects, writes that 'Under natural conditions insects are a prime factor in regulating the abundance of all plants, particularly the flowering plants as the latter are especially prone to insect attack'.

In a brave attempt to bring order to this chaos of contradictory information, Hairston *et al*. (1960) proposed the following sequence of logic. Cases of obvious depletion of green plants by herbivores or by meteorological catastrophe are exceptions, therefore plants must be resource limited (by competition for light, water or nutrients). Since plants are not regulated by herbivores, the usual condition is for populations of herbivores not to be food limited. All populations are regulated by something, so herbivores must be regulated by natural enemies.

The spirit of this argument has survived despite bitter and sometimes scathing criticism (Murdoch 1966, Ehrlich & Birch 1967). There are, however, two principal flaws in the assumptions. First, the world is not always green, and herbivore population regulation may well occur during these non-green periods when plant food is actually very scarce (in winter or in drought; after fire or flood). Second, all that is green is not food. Vast

1

differences in food quality, in digestibility and in the levels of nitrogen, toxins and repellants, mean that most green matter is simply inedible to most herbivores (Sinclair 1975).

The regulation of abundance need not be reciprocal. It is entirely plausible for a herbivore to be food limited and yet its host plant not to be herbivore limited. This may come about because plant recruitment is limited by microsite availability and plant competition, or because herbivore numbers are determined by food quality rather than by plant abundance.

The view one holds about the veracity of the food limitation hypothesis is likely to be determined by the kind of animal and plant one studies. Workers in the field of biological weed control are likely to believe that herbivores can be food limited; to believe otherwise would be heresy! If herbivores were not food limited, it is most unlikely that they could increase to sufficiently high densities to bring about the required dramatic reductions in the numbers of weeds. However, an entomologist studying the leaf-mining Diptera of bracken fern, where less than 1 pinnule in 50 is mined, and the flies suffer 80% parasitism, is unlikely to believe that his animals are food limited or that they have much impact on the abundance of the fern.

In fact, there is a vast range of examples, from almost certain food limitation to almost certain natural enemy limitation. Between these extremes lie the bulk of cases, where population regulation is brought about by a complex of interacting non-linear processes which defies simple one-factor explanations.

The numerous examples of successful weed control using introduced herbivorous insects provide strong support for the argument that plants can be herbivore regulated. Conversely, the outbreak of previously unnoticed herbivorous insects to pest status following the application of broad-spectrum insecticides, suggests strongly that the insects had previously been regulated by natural enemies.

1.2   PLANT AND ANIMAL DEMOGRAPHY

This book is concerned with the effects of animals on plants and of plants on animals in so far as they affect one another's rates of birth, death, immigration and emigration. These four demographic parameters determine not only whether a population will increase or decrease, but also how stable it will be and how rapidly it will respond to perturbations. The factors affecting these rates, whether they act immediately or after a time-lag, and the extent to which they are influenced by the population densities of the animals and plants will form the subject matter of the first two sections.

The rate of birth is measured by fecundity: the number of seeds produced, eggs laid or young born to an individual per unit time. Fecundity is defined as being age dependent, size dependent, density dependent or time dependent in different applications. Where the organisms are separate sexed (dioecious) as in most herbivores and a few plants, fecundity is measured as young produced per female, and a correction is made for the sex ratio at birth (usually assumed to be 50% female).

The rate of death is described by two related parameters. The mortality rate (at a specified age, population density or time) is the fraction of individuals that dies per unit time. The survivorship curve is a description of the fraction of the young born (or seeds produced) in a cohort at time zero, that survives to a specific age. If the annual death rate is constant at 50%, the survivorship curve will be 1, 0.5, 0.25, 0.125 and so on. Survivorship curves tend to be used as a broad summary of the species' life history, independent of the details of density dependence or time dependence in the death rate.

The rates of immigration and emigration are generally specified as time dependent and density dependent. They can apply to animals of any age but tend only to be applied to the seeds or spores of plants. They are measured as the rate of arrival or departure of individuals per unit area.

A population that is scarce in relation to its resources (but sufficiently common that reproduction is not depressed because of difficulties of mate location and similar factors) will tend to grow exponentially. Once the population has attained a stable age distribution, the rate of exponential growth is described by the slope of a graph of the natural logarithm of population density plotted against time. The slope of this graph is called the intrinsic rate of increase and is denoted by r_{max}. The value of r_{max} is important because it measures the rate at which the population can recover from low densities and thus profoundly affects the stability properties of the population equilibrium.

The actual rate of increase of a population (r) is found by solving the integral

$$\int_{x=0}^{\infty} e^{-rx} l_x m_x = 1 \tag{1.1}$$

for species which reproduce more or less continously, or the summation

$$\sum_{x=1}^{\infty} e^{-rx} l_x m_x = 1 \tag{1.2}$$

for those cases where reproduction is discrete. The age of the organism is x; l_x is the age-specific survivorship and m_x is the age-specific fecundity (for details see Southwood 1976). Note that the summation begins from $x = 1$ and *not* $x = 0$ as in many of the standard references (cf. Krebs 1978).

A good approximation for r can be obtained by dividing the natural logarithm of the expected number of female offspring produced by the average female over her lifetime (R) by the average generation time (T):

$$r \approx \frac{\ln R}{T}. \tag{1.3}$$

R is simply $\sum l_x m_x$ and T is $\sum x l_x m_x / R$ (May 1976). Thus it is plain that r increases with the expected number of offspring, and declines as the average generation time increases.

For a population in equilibrium r will be zero. Increasing populations have positive values for r and declining populations have negative values. In populations that cycle, the value of r fluctuates between positive and negative values. Thus r is merely a summary of the current balance between loss and recruitment under the particular conditions that the l_x and m_x data were gathered; it is a variable while r_{max} is a constant.

1.2.1   Herbivore feeding and plant demography

Plant survivorship is affected in different ways by different kinds of herbivores. Leaf-feeding species like cows grazing on pasture grasses, or moth caterpillars defoliating deciduous forest trees, typically have little effect on plant death rate. Their feeding is essentially parasitic rather than predatory; they reduce plant growth rate but tend not to kill the plants. The impact of leaf feeding on survivorship is critically dependent on the timing of attack relative to plant development and on the history of defoliation experienced by the plants (see section 2.2).

Sap-sucking species may kill the plants by draining their carbohydrate reserves but more frequently they increase the death rate indirectly by transmitting viral diseases. Bark-feeding vertebrates may kill trees directly by ring barking but most bark-feeding invertebrates rely on the fungal diseases they transmit to kill the host plants. Sucking and bark-feeding insect vectors can have a dramatic effect in reducing plant survivorship (see section 2.2).

Seed-feeding animals have the most obvious, direct effect on plant survival rate. Seeds are killed in very large numbers by vertebrates like finches and squirrels and by invertebrates like weevils and moths. We do not know for certain, however, how seed feeding affects the full survivorship curve of long-lived plants like trees. It is possible, for example, that reduced sapling mortality may compensate for high levels of seed feeding (see section 2.7), or that plant recruitment is limited by the availability of germination microsites rather than by seed availability (see section 2.6). Some kinds of seed feeding may actually increase plant fitness as with seed-hoarding animals like squirrels when the benefits of dispersal of those seeds that survive in undiscovered caches outweigh the costs of predation on the remaining seeds (see section 2.5.2).

Plant fecundity is more dramatically affected by herbivore feeding than is survivorship. Leaf-feeding animals reduce plant growth rate and alter plant shape. Less carbohydrate is available for seed production, and there may be fewer sites for flower initiation (see section 2.7.5). Defoliation often delays flowering so that the plant's rate of increase suffers doubly; fewer seeds are produced and development time is increased (see equation 1.3).

Sap-sucking animals can reduce vegetative growth and seed production by tapping the plant's carbohydrate supply lines directly. There is some scope, however, for the plant to compensate for this kind of feeding if its rate of photosynthesis is limited by carbohydrate build-up (see section 2.7.2).

Flower- and fruit-feeding animals have the most obvious effects in reducing plant fecundity (see section 2.4.2). On the other hand, many plant species produce far more flowers and fruits than they could ever fill with carbohydrate and protein-rich seeds. Thus there is considerable scope for compensation for such feeding by a reduction in the rate of natural flower and fruit abortion (see section 2.7.5).

1.2.2   Plant attributes and animal demography

Animal survivorship is affected by both the quantity and quality of the plant food available. One of the most obvious differences between predator–prey and plant–herbivore systems is the increased importance for herbivores of changes in food quality. The chemical composition of different animal tissues is relatively uniform and a high proportion of their mass is digestible. Plant tissues, however, are largely comprised of indigestible substances like cellulose, lignins and so on, and contain widely variable concentrations of nutrients like amino acids, and of attractant, repellant and toxic chemicals. Thus the survival rate of herbivores may be very low at certain seasons because of low food quality despite an apparent abundance of suitable food;

starvation in the midst of plenty (see sections 3.2.4 and 3.5).

Shortages of food can have a great effect on the death rate, but mass starvation of wild herbivores is a relatively rare phenomenon (see section 3.2.4). Most mobile herbivores respond to reductions in food availability or declining food quality by emigrating in search of pastures new. The losses during dispersal may be dramatic, especially when the distances between patches of suitable food are large, or when the density of herbivores is already high in all the patches (see section 3.3).

Typically, an increase in food availability leads to an increase in feeding rate by the herbivores (the so-called functional response; see section 3.1.2) and hence to improved survival. In many grasslands, however, increases in plant biomass only occur at the expense of reductions in food quality, because average leaf age is higher and the proportion of dead and non-nutritious tissues is larger. For herbivores like sheep, therefore, that feed at a slower rate on poor quality food, increases in total plant biomass availability may actually lead to reductions in the rate of feeding (see section 3.1.3).

Herbivore fecundity is critically dependent upon the plane of nutrition obtained. Low food quality may cause delayed sexual maturation, reduced ovulation and increased rates of abortion in vertebrates (see section 3.2.2). Insect herbivores grow less quickly on poor diets and lay fewer eggs. Thus the potential exists for population regulation by food availability, without there ever being evident starvation or mass emigration, when the number of births is determined by the feeding success of the adult and maturing animals.

The numerical response of herbivorous animals, therefore, depends not only upon how much food is available, but on its quality and on the ability of the herbivore population to exploit the food efficiently (see section 3.1.4).

I shall say little about 'coevolution' but the interested reader will find numerous references in Ehrlich and Raven (1965) and Gilbert (1979).

1.2.3   Plant–herbivore dynamics

In analysing the dynamics of plant–herbivore interactions, we shall attempt to establish the conditions under which population equilibria occur, and what kinds of factors affect their local and global stability. Instability tends to be due to over-exploitation of the plant population by the animals, and typically leads to the demise of one or both of the species. Stable systems will usually display one or more mechanisms that prevent over-exploitation;

these may take the form of refuges for the plant (ungrazable reserves of carbohydrate, inaccessible foliage, persistent banks of seed in the soil, immigration of seed from outside, etc.). Alternatively, the herbivores' fecundity or survival may decline so rapidly with reduced food availability that the plant's persistence is never jeopardized. Stability may also be enhanced when the herbivores switch to alternative plants, or emigrate in increasing numbers as food supplies decline.

We shall be particularly concerned to establish how closely the dynamics of the plant and the herbivore are coupled. Do herbivores merely track changes in plant abundance, for example, or do they cause the changes?

When the herbivore is maintained at low density by natural enemies or by shortage of suitable breeding sites, it may have little effect on plant fitness. Similarly, when feeding occurs for only part of the year (either because the herbivore enters diapause, migrates or switches to alternative food plants), the plant may compensate during the herbivore-free period for any losses due to earlier feeding. Alternatively, plants that succumb to herbivore attack may be those that are, in any case, near to death from other causes; again, herbivore feeding will have little impact on plant fitness.

Conversely, herbivores that have few natural enemies and are not limited by shortage of other resources are likely to have a profound effect on plant fitness. There are several possibilities. First, there may be no stable equilibrium; the herbivores may eat the plant to extinction and then either become extinct themselves or switch to less attractive foods. Second, if the survivorship or the fecundity of the herbivores is closely dependent upon food availability, the plant and herbivore may coexist in stable equilibrium. The extent to which herbivore feeding depresses the plant population below the level it would attain in the absence of feeding ($q = 1 - \overset{*}{N}/K$, see section 6.1.3) is determined by a complex of factors including the demographic parameters of the plant and herbivore populations, the searching efficiency of the herbivores, levels of herbivore density dependence and the degree of spatial heterogeneity in attack (Beddington *et al.* 1978; see also section 4.3).

Finally, the populations may undergo regular cycles in abundance that are driven by environmental periodicity or by internally generated demographic processes. These might be due to time-lags in the response of herbivore numbers to plant density, or by non-linearities in either the response of the plants to attack or of the animals to changes in the quantity or quality of food (see sections 4.2.4 and 4.3.6).

The importance of particular biological processes on population dynamics is investigated by exploring the behaviour of simple mathematical models that are structured to include different processes in turn (see section 4.3). As stressed by May (1981a), 'We are relatively uninterested in the algebraic details of any one particular formula, but are instead interested in questions of the form: which factors determine the numerical magnitude of the population; which parameters determine the time scale on which it will respond to natural or man-made disturbances; will the system track environmental variations, or will it average over them?'

1.2.4 Community dynamics

Plant–herbivore systems do not exist in isolation but as components of ecosystems. Plants share their trophic level with other species, many of which are likely to be potential or actual competitors for germination sites, light, water, nutrients, pollinators, dispersal agents and so on. Herbivores, too, may have competitors sharing one or several of their food-plant species (see sections 5.1 and 5.5).

The plant community affects the abundance and species of the herbivores in two main ways. First, the more plant species there are, the greater the potential to support specialist herbivores (although, as we shall see in section 5.5 this correlation is often poor). Second, the more structurally diverse the vegetation, the greater will be both herbivore numbers and species richness. Plant community structure also affects the natural enemies of the herbivores, providing them with the supplementary foods they need, and with places to lie in wait for the herbivores (see section 5.8).

The impact of herbivores on the structure and species richness of the plant community can be dramatic; for example, feeding by vertebrates can prevent the successional change of grasslands into woodlands (see section 5.3). Conversely, prolonged heavy grazing by domestic livestock can turn semi-arid grasslands into barren, thorny scrub (see section 5.2.1). The principal effect of herbivores on plant species richness acts not through the animals eating plants to extinction (although this can happen), but through their feeding modifying the competitive abilities of one plant species with another (see section 5.1.2).

Spatial heterogeneity plays a central role in the dynamics of all manner of relationships between plants and their herbivores, between competing herbivore species and between herbivores and their natural enemies (see sections 3.3, 4.3 and 5.1). It may allow, for example, the coexistence of herbivores which would otherwise be impossible in a homogeneous environment by providing a refuge for the inferior competitor. It may allow a natural enemy to regulate herbivore numbers at such a low, stable equilibrium that no impact of herbivore feeding is felt by the plant.

Temporal heterogeneity can have similar effects; for instance, differences in weather conditions from year to year may favour first this, then another species, cancelling out any competitive superiority that otherwise would be exhibited in a constant environment. Population densities are likely to be lower, and species richnesses higher, in heterogeneous than in uniform, constant environments.

The assemblages of plants and animals that we observe in the field may not represent equilibrium communities structured by competition. Instead, spatial and temporal heterogeneity may create mosaics of non-equilibrium patches in which competition is infrequent or weak, and where populations are essentially ephemeral (see sections 5.6 and 5.7).

1.3   THE TYPES OF HERBIVOROUS ANIMALS

The animal groups that spend most of their lives as herbivores are shown in Table 1.1. Groups that live by filtering dead organic matter from sediments have been omitted from the table, although obviously they may take living plants from time to time.

There are numerous bizarre feeding habits that confound any attempt at a complete classification. For example, there are the purely herbivorous Carnivora, like the panda and the giant panda; there is the palm nut vulture, a strange, but locally abundant herbivore in west Africa which is the only herbivorous member of the Falconiformes.

No plant tissue escapes the attention of herbivores (Table 1.2) although young leaves are attacked by many more kinds of animals than eat healthy wood. Generally, the number of herbivores feeding on a tissue is proportional to its quality as food; its digestibility, nitrogen content and so on.

1.4   HERBIVORE IMPACT IN NATURAL ECOSYSTEMS

All the carbon fixed by plants in photosynthesis is destined to be eaten. Some of it will be eaten when the tissue is still alive by herbivores, the rest, once it has died, by detritivores and saprophytes. It is of some interest to know what fraction of net primary production is consumed by herbivores. The danger is that once a figure has been obtained (usually after great effort), it is likely to be taken too seriously. The amount eaten by herbivores varies from place to place and from year to year, and it is in many ways more interesting to know

Table 1.1(a) Invertebrate herbivores.

Phylum	Order	Frequency of herbivores within Order	Name	Example	Example of tissue eaten	Mode of feeding
PROTOZOA	Several	Many	Amoeba	*Amoeba dubia*	Diatoms	Cytoplasmic engulfing
NEMATODA	Several	Many	Nematodes	*Tylenchus tritici*	Wheat flower	Cell contents
ECHINODERMATA	ECHINOIDEA	Many	Sea-urchins	*Echinus esculentus*	Seaweeds	Rasping
MOLLUSCA	GASTEROPODA	Most	Slugs, snails	*Deroceras reticulum*	Leaves, fruits	Rasping
TARDIGRADA		All	Bear animalcules	*Macrobiotus macronyx*	Sphagnum moss	Sucking cell contents
ARTHROPODA Class	Subclass					
CRUSTACEA	BRANCHIOPODA	Many	Water fleas	*Daphnia pulex*	Diatoms	Selective filtering
	COPEPODA	Most	Copepods	*Calanus*	Phytoplankton	Selective filtering
	MALACOSTRACA Order					
	ISOPODA	Some	Slaters	*Ligia oceanica*	Seaweed	Chewing
	AMPHIPODA	Some	Gammarids	*Gammarus neglectus*	Phytoplankton	Selective filtering
	EUPHAUSIACEA	Most	Krill	*Euphausia superba*	Phytoplankton	Selective filtering
	DECAPODA	Few	Crabs, lobsters	*Birgus latro*	Fruits, etc.	Chewing

	Subclass					
MYRIAPODA	DIPLOPODA	Few	Millipedes	*Colobognathus*	Sap	Sucking
INSECTA	COLLEMBOLA	Few	Springtails	*Sminthurus viridis*	Legume leaves	Chewing
	EPHEMEROPTERA	Few	Mayflies	*Ephemera danica*	Algae	Chewing
	ORTHOPTERA	All	Grasshoppers	*Locusta migratoria*	Leaf	Chewing
	DERMAPTERA	Few	Earwigs	*Forficula tomis*	Tobacco leaves, etc.	Chewing
	PSOCOPTERA	Most	Booklice		Algae, lichen	Sucking
	PLECOPTERA	Few	Stoneflies	*Nemoura variegata*	Algae	Filtering
	ISOPTERA	All	Termites	*Kalotermes capicola*	Wood	Chewing
	THYSANOPTERA	Most	Thrips	*Kakothrips robustus*	Pea pod cells, etc.	Sucking
	HOMOPTERA	All	Aphids, hoppers	*Aphis fabae*	Phloem sap	Sucking
	HETEROPTERA	Many	Bugs	*Leptopterna dolobrata*	Grass cells	Sucking
	HYMENOPTERA	Many	Sawflies, ants	*Prolasius pallidus*	Eucalypt seeds	Chewing
	COLEOPTERA	Many	Weevils, chrysomelids, etc.	*Anthonomus grandis*	Cotton bolls	Chewing
	TRICHOPTERA	Some	Caddis flies	*Limnephilus rhombicus*	Epiphytic algae	Scraping
	LEPIDOPTERA	All	Moths, butterflies	*Anthocharis cardamines*	Immature fruits	Chewing
	DIPTERA	Many	Flies	*Phytomyza ilicis*	Holly leaves	Miner
ARACHNIDA	ACARINA	Many	Mites	*Tetranychus urticae*	Tomato leaves	Cell sucking

Table 1.1 (b) Vertebrate herbivores.

Class	Order	Frequency of herbivores within Order	Name	Example	Example of tissue eaten
ACTINOPTERYGII		Some	Bony fishes	Carp	Water weed
CHOANICHTHYES		Few	Lung fishes	Tilapia	Stoneworts
AMPHIBIA		Few	Frogs, toads	Rana tadpoles	Water weed
REPTILIA	CHELONIA	Some	Tortoises, turtles	Giant tortoise	Grasses, sedges
	SQUAMATA	Few	Snakes, lizards	Giant iguana	Seaweeds
AVES	ANSERIFORMES	Many	Ducks, geese	Greylag goose	Roots, leaves, fruits
	FALCONIFORMES	Few	Eagles, hawks	Palm nut vultures	Palm nuts
	GALLIFORMES	Most	Game birds	Red grouse	Heather shoots, insects
	COLUMBIFORMES	All	Pigeons	Woodpigeon	Legume leaves, seeds, etc.
	PSITTACIFORMES	Most	Parrots	Macaw	Fruits, seeds
	MICROPODIFORMES	Some	Humming-birds, swifts	Trochus	Nectar, fruit
	PASSERIFORMES	Many	Perching birds	Finches	Seeds, buds, fruits
MAMMALIA	MARSUPALIA	Many	Koala, kangaroo, etc.	Wombat	Roots, leaves, fruits
	CHIROPTERA	Few	Bats	Pteropus	Fruit, nectar
	DERMOPTERA	All	Flying lemurs	Cynocephalus	Leaves, fruit
	EDENTATA	Some	Sloths, armadilloes	Tree sloth	Foliage, fruit
	PRIMATES	Most	Apes, monkeys, lemurs	Man	All but wood
	RODENTIA	Most	Voles, mice, squirrels	Lemmings	Sedges, grasses
	LAGOMORPHA	All	Hares, rabbits	Snowshoe hare	Grasses, twigs, etc.
	CARNIVORA	Few	Dogs, cats, bears	Giant panda	Bamboo
	HYRACOIDEA	All	Conies	Procavia	Grass
	PROBOSCIDEA	All	Elephants	African elephant	Grass, leaves, twigs
	SIRENIA	All	Sea cows	Manatee	Water weed
	PERISSODACTYLA	All	Odd-toed ungulates	Horse	Grass, forbs
	ARTIODACTYLA	All	Even-toed ungulates	Goats	All

Table 1.2 Plant tissues and the herbivores which feed on them.

Tissue	Mode of feeding	Examples of feeders
Leaves	Clipping	Ungulates, slugs, sawflies, butterflies, etc.
	Skeletonizing	Beetles, sawflies, capsid bugs
	Holing	Moths, weevils, pigeons, slugs, etc.
	Rolling	Microlepidoptera, aphids
	Spinning	Lepidoptera, sawflies
	Mining	Microlepidoptera, Diptera
	Rasping	Slugs, snails
	Sucking	Aphids, psyllids, hoppers, whitefly, mites, etc.
Buds	Removal	Finches, browsing ungulates
	Boring	Hymenoptera, Lepidoptera, Diptera
	Deforming	Aphids, moths
Herbaceous stems	Removal	Ungulates, sawflies, etc.
	Boring	Weevils, flies, moths
	Sucking	Aphids, scales, cochineals, bugs
Bark	Tunnelling	Beetles, wasps
	Stripping	Squirrels, deer, goats, voles
	Sucking	Scales, bark lice
Wood	Felling	Beavers, large ungulates
	Tunnelling	Beetles, wasps
	Chewing	Termites
Flowers	Nectar drinking	Bats, humming-birds, butterflies, etc.
	Pollen eating	Bees, butterflies, mice
	Receptacle eating	Diptera, microlepidoptera, thrips
	Spinning	Microlepidoptera
Fruits	Beneficial	Monkey, thrushes, ungulates, elephants
	Destructive	Wasps, moths, rodents, finches, flies, etc.
Seeds	Predation	Deer, squirrels, mice, finches, pigeons
	Boring	Weevils, moths, bruchids
	Sucking	Lygaeid bugs
Sap	Phloem	Aphids, whitefly, hoppers
	Xylem	Spittlebugs, cicadas
	Cell contents	Bugs, hoppers, mites, tardigrades, etc.

Table 1.2 (*cont.*)

Tissue	Mode of feeding	Examples of feeders
Roots	Clipping	Beetles, flies, rodents, ungulates, etc.
	Tunnelling	Nematodes, flies
	Sucking	Aphids, cicadas, nematodes, etc.
Galls	Leaves	Hymenoptera, Diptera, aphids, mites
	Fruits	Hymenoptera
	Stems	Hymenoptera, Diptera
	Roots	Aphids, weevils, Hymenoptera

why the amount changes, than to have a precise estimate of the value for one field in one year.

The major ecosystems of the world differ by more than two orders of magnitude in their mean net primary productivity (from less than 30 to over 3000 g dry matter m^{-2} yr^{-1}; Whittaker 1975) and there are vast differences in the total amount of carbohydrate potentially available to herbivores. Net production can be predicted with considerable precision from a knowledge of actual evapotranspiration (Rosenweig 1968), and herbivore biomass (at least for vertebrates) is closely correlated with net production (Coe *et al.* 1976).

What little evidence there is suggests no obvious relationship between the amount of primary production and the fraction of primary production that is eaten by herbivores as a group. There does, however, appear to be a rough correlation between primary productivity and the fraction of productivity that is consumed by invertebrate herbivores; insects appear to take a greater proportion of plant production in more productive environments (Andrzejewska & Gyllenberg 1980). Comparative data for vertebrates are scarce, but very high rates of consumption occur occasionally in low-productivity environments (e.g. tundra lemmings) and regularly in some productive natural grasslands (e.g. Serengeti ungulates; see Table 1.3).

On average, about 10% of net primary productivity seems to be taken by herbivores and about 90% by decomposers in most natural ecosystems.

However, these figures mask large and important variations in the feeding of individual herbivore species and in the kind of tissues taken and the timing of their removal. Forest insects like larch budmoth may take less than 2% of net production in most years but may occasionally take 100%. Animals that defoliate evergreen trees can sometimes consume more than

Table 1.3 The fraction of primary production eaten by different herbivores.

Plant	Herbivore	Percentage of primary production eaten	Source
Grasses and herbs			
Grassland	Invertebrates	9.6	van Hook et al. (1970)
Meadow plants	Invertebrates	14.0	Andrzejewska & Wojcik (1970)
Sedges	Invertebrates	8.0	Andrzejewska & Wojcik (1970)
Grassland	Nematodes	30–60	Andrzejewska & Gyllenberg (1980)
Spartina	Invertebrates	7.0	Smalley (1959)
Spartina	Invertebrates	4.6	Teal (1962)
Tundra graminoids	Lemming	20–90	Schultz (1969)
All plants	Giant tortoise	0.7–11.3	Coe et al. (1979)
Serengeti (all)	Vertebrates and invertebrates	23–43	Sinclair & Norton Griffiths (1979)
East African grasslands	Ungulates	30–60	Wiegert & Evans (1967)
Aristolochia	Swallowtail butterfly	45	Rausher & Feeny (1980)
Ryegrass/clover	Sheep	50–60	Frame (1966)
Pasture	Dairy cattle	25	Campbell (1966)
Grassland	Voles	0.1–1.0	Grodzinski et al. (1966)
Grass/forbs	Invertebrates	4–20	Odum et al. (1962)
Grass/forbs	Invertebrates	0.5	Wiegert (1965)
Alfalfa	Invertebrates	2.5	Wiegert (1965)
Texan rangelands	Scale insect	30.0	Schuster et al. (1971)
Pasture grass	Invertebrates	15.0	Clements (1978)
African field crops	Insect pests	10–20	Bullen (1970)
Tallgrass prairie	Granivorous birds	0.17	Risser (1972)
Desert seeds	Kangaroo-rat	6.9	Scholt (1973)

Table 1.3 *(cont.)*

Plant	Herbivore	Percentage of primary production eaten	Source
Mountain grassland	Grasshoppers	2.4	White E.G. (1974)
Prairie	Grasshoppers	2.0	Bailey & Riegert (1973)
Echinochloa grassland	Froghoppers	16.0	Andrzejewska (1967)
Holcus grassland	Heteroptera	0.15	McNeill (1971)
Salt marsh	Snow geese	58.0	Smith & Odum (1981)
Trees and Shrubs			
Tundra shrubs	Invertebrates	1–10	Haukioja (1981)
Mangrove	All herbivores	1.5	Golley et al. (1962)
Oak	All herbivores	1.7	Rothacher et al. (1954)
Elm/ash	All herbivores	2.0	Lindquist (1938)
Beech	Invertebrates	8.0	Bukovskii (1936)
Oak	Invertebrates	10.6	Bray (1964)
Maple/beech	Invertebrates	6.6	Bray (1964)
Tulip/poplar	Invertebrates	2.6	Reichle et al. (1973)
Creosote bush	Grasshopper	1.5	Mispagel (1978)
Heather	Grouse	1.5–2.5	Savory (1978)
Sagebrush	Kangaroo-rat	1.95	Chew & Chew (1970)
Pine/oak/alder	Invertebrates	3.4–4.0	Kaczmarek (1967)
Pine	Invertebrates	9.2	Kaczmarek (1967)

100% of annual shoot production (Rafes 1970). Any uniformity in consumption at ecosystem level masks a rich variety of counterbalancing changes in the amounts taken by individual species.

The difficulty of relating real grazing impact to energetics measures (such as percentage biomass consumption) is exemplified by work on grasshopper feeding on the marsh black needle rush, *Juncus roemerianus* (Parsons & de la Cruz 1980). The insects were found to consume only 0.33% of the annual net production of the rush and yet, by the end of the season, 56% of the *Juncus* leaves had been attacked. Also, because of the grasshopper's habit of feeding in mid-leaf, twice the amount eaten by the animals was clipped off and added to the litter layer.

It should be remembered in all discussions of harvesting and production that grazing itself affects the rate of primary production. Productivity in a Russian meadow was greatest at vole densities of $100 \, \text{ha}^{-1}$ when about 20% of the plant growth was being harvested (Coupland 1979). Grass production in the Serengeti is increased by moderate levels of defoliation as grazing-intolerant grasses are replaced by more productive, grazing-tolerant species (McNaughton 1979a).

Taking ecosystems as a whole, about 33% of all the species they contain (plant and animal) is made up of herbivores, and the range for 40 food webs catalogued by Briand (see Yodzis 1981) is from 14%, in Arctic and Antarctic seas, to 56%, in a mangrove swamp. Great caution must be exercised with these figures, however; the most complex community in Briand's list, for example, had only 9 plant species in it! Also, simple ecosystems are bound to be vastly over-represented in the set of ecosystems that have been studied in detail.

1.5 INDIRECT EFFECTS OF HERBIVORES IN ECOSYSTEMS

1.5.1 Habitat structure

By altering the physical structure of the plant community by regular defoliation (as in grasslands) or irregular defoliation (as in forests), herbivores affect the microclimate experienced by all the other species. Reduced plant height and increased plant patchiness affect air flow, altering the seed set of wind-pollinated plants and affecting the dispersal of small flying insects such as aphids.

Altered temperature regimes affect the development rates of invertebrates and may lead to a lack of synchrony between the emergence of an insect and its parasite or between an insect and the plant tissues on which it feeds.

Defoliation alters the water balance of the whole community, reducing the area of transpiring leaves and exposing areas of the soil surface to the drying influence of sun and wind.

Changes in the physical structure of the plant canopy have a profound effect on the abundance and distribution of predators and parasites; defoliation by one species of herbivorous insect can alter radically the rate of parasitism suffered by another.

1.5.2   Nutrient cycling

Outbreaks of herbivorous insects in nutrient-poor habitats can increase the rate of nutrient turnover (if only for a short time). Evergreen coniferous trees that keep their leaves for up to four years tie up a substantial fraction of the ecosystem's nutrients in their foliage. In years of peak caterpillar abundance the trees may be entirely defoliated, suffering a loss of foliage well in excess of annual net primary production. The rain of frass, cast skins and larval corpses constitutes a large input of readily available nutrients which may be sufficient to allow rapid growth in the following year. Increased ring widths in trees in the year following defoliation have been noted frequently and have often been attributed to this nutrient release (Owen 1980). The effect has yet to receive experimental verification.

Defoliation by insect herbivores can also increase the rate of nutrient loss from ecosystems; for example, chronic defoliation of mixed hardwood forests by the autumn canker-worm, *Alsophila pometaria*, led to a substantial increase in stream export of soluble nitrate nitrogen (Swank *et al.* 1981). However, as we shall see (see section 3.2.5) the probability of insect outbreak may itself be a function of nutrient levels. Increased nutrient availability may increase the likelihood of outbreak in some cases and decrease it in others (see section 3.6.5). It is by no means obvious, therefore, whether increased nutrient outflow represents a net drain on the ecosystem, or provides a mechanism whereby equilibrium nutrient levels are maintained.

Increased soil surface temperatures in severely defoliated communities may speed the decomposition of organic litter (Collins 1961). When defoliation stimulates regrowth from stored reserves, or increases the rate of leaf fall during the growing season because of wastage or induced abscission, herbivores may increase the rate of nutrient input to the soil. They increase the availability of organic matter for the decomposer organisms at the time of year when conditions for litter breakdown are ideal.

The importance of dung and urine from vertebrate herbivores hardly needs to be stressed. As Harper (1977) so graphically puts it, to be covered by a cow pat is a disaster for most plants; if they are not killed by the darkness or by the concentrated nutrients, they may have their living space usurped by creeping plants, which grow over the top of the dung. A cow pat has a long-lasting impact on the vegetation both because fouled areas are avoided by cattle, and because certain plants thrive in the nutrient-rich conditions and pinpoint the position of the pat long after the faeces have decomposed (Marsh & Campling 1970).

Herbivore faeces may be more, or less, readily decomposed than the plants of which it is composed. Usually, insect frass or vertebrate faeces is rapidly broken down. Occasionally, however, the local decomposer fauna may be unable to cope with the faeces of introduced animals, as happened on certain cow pastures in Australia. Dung beetles introduced from Africa increased the rate of disappearance of cow pats, and may also have been instrumental in reducing problems from cattle pests like bushfly, *Musca vetustissima*, and buffalo fly, *Haematobia irritans*, which spend their larval period in dung (Hughes 1975).

Herbivores also affect the spatial pattern of nutrient distribution. Animals that feed over a wide area, yet defecate over a small area can have a substantial effect on pasture productivity. Rabbits, for example, defecate in defined latrines, and sheep congregate in camps at night or for shade, so that about 35% of their faeces is distributed over less than 5% of the area of the paddocks, an effect known as nutrient dislocation (Spedding 1971). Similar effects occur with wild animals. Hippopotamus graze within 3 km of water, removing up to 18 kg dry matter per animal per day from short grasslands, but since they defecate in the water they cause a substantial nutrient drain on the terrestrial community (Lock 1972).

Rodent cycles affect nutrient cycles (and may even be affected by them; Schultz 1969, and see section 4.2.4). At Point Barrow in Alaska, nutrients become progressively less available to the plants at low lemming densities because dead leaves accumulate, there is a reduction in the exploitable root volume (due to reduced depth of thaw caused by the build-up of an insulating mat of litter), and most of the dead roots freeze rather than decompose. In a winter of peak lemming numbers, almost all the mosses are consumed along with many of the shoot bases, so that 40% of the nitrogen and 50% of the phosphorus in the above-ground biomass is ingested. Much of this is returned to the soil nutrient pool almost immediately as faeces, or later as lemming corpses, predator faeces, and so on. The insulating layer is removed, so the ground thaws to a greater depth, creating an increased exploitable soil volume and allowing the decomposition of previously frozen roots (Dowding *et al.* 1981).

1.5.3   Soil structure

Large herbivores do much more than merely eat. They trample, scratch at, lie on and otherwise abuse the vegetation. They leave footprints and bare areas of soil where germination conditions are quite different to those in the unbroken sward. They alter the physical structure of the soil by their trampling; it is compacted along paths and trails or puddled in the wet places near drinking holes and gateways. Compaction and puddling both lead to the development of characteristic plant communities quite distinct from the predominant vegetation (Duffey *et al.* 1974).

Heavy grazing of arid grasslands may alter the surface properties of the soil to such an extent that the vertical profile of soil water distribution is altered in favour of deep-rooting plants, so that grasses are replaced by woody scrub (Walker *et al.* 1981).

Fig. 1.1 shows how infiltration rate declines on heavily grazed pastures in Australia as the cover of pasture species is reduced by Merino sheep (Willoughby & Davidson 1979).

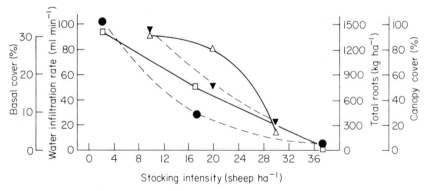

Fig. 1.1 Herbivores and soil properties. Increasing the stocking rate of Merino sheep from 4 to 36 ha^{-1} leads to a reduction in water infiltration rate (●) from 100 to less than 10 ml min^{-1}. This is associated with parallel reductions in total canopy cover (□), basal cover of sown species (△) and total root (▼) (after Willoughby & Davidson 1979).

CHAPTER 2
PLANT POPULATIONS

2.1   PLANT DEMOGRAPHY

There is no formal difference of approach between studies of plant and animal populations dynamics, although zoologists are extremely envious of the fact that plants 'stand still' to be counted. The aim in both fields is to account for changes in the number of individuals per unit area by studying the rates of birth, death, immigration and emigration. However, because individual plant size is so variable, and so many demographic parameters are size dependent rather than age dependent, the study of plant demography is considerably more difficult than might at first appear. Whereas in traditional animal population dynamics it is usually assumed that change in numbers alone is an adequate description of population behaviour, the plant ecologist must consider changes in mean plant size and changes in the relative abundance of large and small plants in addition to changes in the number of genetic individuals (Harper 1977). Furthermore, there is a class of plants for which traditional methods seem completely inappropriate; these are the patch-forming rhizomatous or stoloniferous species in which the identification of genetic individuals is always difficult and frequently impossible. In communities where these plants are abundant (e.g. grasslands) the plant ecologist resorts to describing the dynamics of such modular units as tillers, shoots or rooted nodes.

Six generalizations form the core of plant demographic theory; I shall outline each of them briefly, and then consider how each is affected by herbivore feeding.

(1) Plant fecundity is strongly density dependent. Cereal farmers have long known that increasing the density of seeds sown does not increase the yield of seeds harvested. Fig. 2.1 shows a typical yield/seed density curve and highlights that once a threshold (and usually rather low) density of seeds has been sown, the rate of seed production per unit area is more or less constant (Willey & Heath 1969, Busch & Ergun 1973).

This constancy is due to density-dependent reductions in plant size that lead to either a reduction in the mean number of seeds produced per plant or to density-dependent mortality. It means that, in a monoculture, a large number of plant deaths can be accommodated without a loss in reproductive output.

21

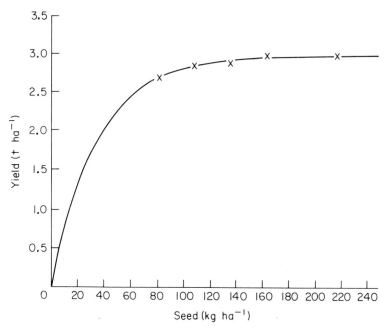

Fig. 2.1 Density-dependent plant fecundity. Increasing the density of seed sown has little effect on the amount of seed produced because of reductions in the number of seeds produced per plant and in the number of plants reaching maturity.

(2) Population density and soil fertility. Plant population densities are lower on fertile soils than on poor soils. This initially surprising result has long been familiar to foresters. Gause (1934) quotes the Russian Sukatschev, who found 1300 'predominant trunks' and 640 'oppressed trunks' in a 60-year-old fir stand on good soil, but 2780, more than twice as many predominant trunks with 760 oppressed trunks on a poor soil. Sukatschev also experimented with scentless mayweed, *Matricaria inodora*, on fertilized and unfertilized soils. He found a 6% death rate on unfertilized soil but a 25% death rate on fertilized soil. When he reduced plant density (by increasing spacing from 3 to 10 cm) the effect disappeared; there was no difference in death rate on fertilized and unfertilized soils.

Plant deaths due to intraspecific competition are caused mainly by shading. Plants only grow big enough to shade one another on fertile soils and the death rate only increases when they grow sufficiently close to one another that small neighbours can be suppressed.

(3) Mortality rate is size dependent. Deaths of established plants occur

mainly in the smallest size classes, whether these are young seedlings or suppressed older individuals (Harper & White 1974, White 1980).

(4) Yoda's rule. In dense plant populations there is an inverse relationship between population density and mean plant size. Individual plants only grow in size at the expense of the death of others. This is known as 'self-thinning'.

Population density is measured as numbers per unit area, N, and has the dimensions L^{-2}. Plant weight is a function of plant volume and has the dimensions L^3. So plant weight as a function of plant density should scale as $(N^{-1/2})^3$ which is $N^{-3/2}$. A large body of empirical evidence backs up this relationship between plant size and population density first noted by Yoda *et al.* (1963). Graphs of the logarithm of mean plant weight against the logarithm of the density of surviving plants have slopes that are rarely significantly different from -1.5.

Despite the wide application of the $-3/2$ thinning rule the details of the simple derivation are not accurately descriptive of the shapes of real plants, for example, shoot weight in trees scales as trunk diameter to the power 2.5 rather than as diameter cubed, and plant shapes change as they develop (White 1981). Also, the shape of the survivors is likely to differ from the shape of the plants that die (Westoby, in White 1981).

Above a certain threshold, increases in standing crop biomass are accompanied by a reduction in plant numbers. A graph of log biomass against log population density has a slope of $-1/2$ for nearly all plants and environmental conditions investigated (White 1980, Westoby 1981). This is a consequence of Yoda's rule, above. Since, in most cases, changes in biomass (B) cause changes in numbers (N), rather than vice versa, it is appropriate to plot log N against log B. This graph has a slope of -2 (Fig. 2.2).

Thinning for plants parts (seeds, leaves, etc.) may or may not follow the $-3/2$ rule, depending on the precise relationship between plant structure and plant size (Watkinson 1980). The foliage weight of *Abies balsamea*, for example, thins at about -1 (Mohler *et al.* 1978).

(5) Plant productivity and leaf area index. The rate of net dry matter production by plants is a non-linear function of leaf area index (the ratio of leaf surface to ground area; Brown & Blaser 1968). When self-shading is important, as in well-irrigated, fertile pastures or in dense forest communities, some leaves receive so little light that their rate of respiration exceeds their rate of photosynthesis, and the graph of net productivity against leaf area index becomes n-shaped. The increase in gross photosynthesis at high leaf areas is more than offset by increased respiration.

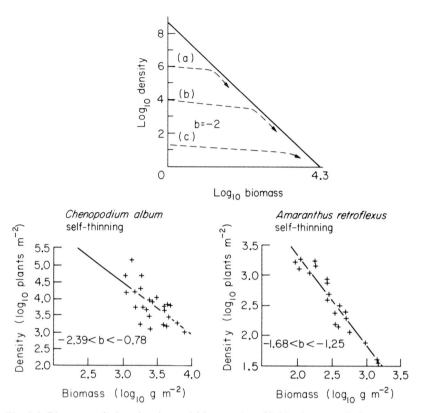

Fig. 2.2 Plant population density and biomass in self-thinning populations. When standing crop biomass is sufficiently high, increases in biomass can only occur at the expense of reduced plant numbers. Log numbers decline with log biomass with a slope of -2. (Based on data from Harper 1977.)

In less productive, nutrient-poor or arid environments, self-shading will usually be unimportant but root competition will set a limit to productivity and there will be an asymptotic relationship between net productivity and leaf area index.

(6) Mixtures and monocultures. Monocultures of the most productive of a pair of plant species will outyield mixtures of the two. This rule is more one of logic than of ecology; given that one species is more productive than the other, replacing a more productive individual by a less productive one is bound to reduce yield.

There are certain field conditions, however, where a mixture of species will outyield a monoculture. The advantages of such mixtures accrue either

Table 2.1 Seeding recommendations for British pastures. Figures in brackets are sowing rates in kg seed ha^{-1} (Anon 1981).

	Duration of pasture		
	One year	Two years	Many years
No of species	1	5	6
	Lolium multiflorum (25)	*Lolium perenne* (7)	*Lolium perenne* (12)
		L. multiflorum (7)	*Dactylis glomerata* (8)
		Trifolium pratense	*Phleum pratense* (4)
		'Broad Red' (4)	*Trifolium pratense*
		'Late Flowering' (2)	'Late Flowering' (2)
		T. repens 'White' (1)	*T. repens* 'White' (1)
			'Wild White' (1)

from risk-spreading or from mutualism. In a pasture, for instance, a mixture will be preferred when soil conditions favour different species in different parts of the grassland; when pest damage is more likely to a monoculture; when the productive season may be extended because different plants mature at different times; when nitrogen levels can be increased by planting legumes with the grasses; when mixed pasture provides a higher quality diet for the livestock (e.g. legumes are richer in minerals and proteins than are grasses). For reviews see van den Bergh (1968) and Trenbath (1974).

An interesting aspect of species mixtures is provided by the seeding recommendations for British pastures. If a grassland is to be used for just one year (a temporary ley), it is sown with only one species of grass. Two year leys contain two species of grass and three kinds of clover while permanent leys contain three grasses and three clovers (Table 2.1).

(7) Each of these generalizations can be altered by herbivore feeding; by density-dependent seed predation; size-specific defoliation; or selective defoliation of different plant species. The subsequent sections of this chapter will detail these effects.

2.1.1 Density dependence and size plasticity

A great deal of the interest in plant population dynamics centres on the kind and degree of density dependence exhibited in the rates of growth, seed set and mortality. These would be easy to categorize if plants were distributed evenly over the habitat so that every individual in the population had the same number of neighbours. This is the rule in many forestry plantations and some agricultural crops, but it is rarely observed in natural communities. Spatial heterogeneity is the rule here; high-density patches of plants are interspersed with low-density or even empty areas. Thus different individuals experience different degrees of density stress.

The patchy distribution of plants is caused by three different kinds of process; by underlying heterogeneity in establishment conditions (differences in soil structure, patchy soil disturbance, variation in seed availability) and by heterogeneity due to plant demographic processes (seeds clustering round the parent plant, lateral growth by rhizomes or tillers). Heterogeneity may also be generated by pachiness in the distribution of different growth stages of the plant (e.g. the 'stand cycle' of heather *Calluna vulgaris*; Watt 1955).

The responses of plants to high density differ from species to species. Palmblad (1968), for example, describes how different weed species respond to high seed densities. Some, like *Silene anglica*, show density-dependent suppression of germination, while others, like *Plantago major*, show high germination followed by high juvenile mortality. When germination and survival are both high, flowering may be suppressed at high densities, as in *P. lanceolata*, or the plant may show plasticity in size and seed production, as in *Capsella bursa-pastoris*. Plants of *Capsella* grown singly can produce 23 000 seeds while high-density plants produce 200 or fewer.

Detecting density dependence in field populations, however, is extremely difficult because plant size is affected by a multitude of environmental factors like soil conditions, light intensity, history of herbivore attack, and so on. Thus it is almost impossible to detect unequivocal examples of density dependence since its effects are inextricably compounded with those of background heterogeneity. It is commonplace, for example, to find two large individuals growing in contact with one another, while nearby, two tiny plants grow far apart. At five sites studied by Symonides (1978) plants of *Spergula vernalis* showed the expected strong negative relationship between mean weight and population density at three sites but a strong positive relationship in the other two! The importance of adequate unbiased sampling in highly aggregated plant populations is stressed by Law (1981); plants in local high-density patches will have quite different rates of birth and death than will a sample of larger, isolated plants.

The extent of plasticity in plant fecundity is staggering; a poppy, *Papaver rhoeas*, can produce 1 capsule containing just 4 seeds or 400 capsules containing a total of 340 000 seeds (Harper 1967) and an isolated wheat plant can produce more than 50 g of seed whereas the average plant in a crop produces only 2 g (Kirby 1968). While each individual capitulum (flower head) of ragwort, *Senecio jacobaea*, contains a relatively constant number of achenes (*c.* 70), there is great plasticity in seed production through differences in the number of capitula produced. Plants growing on poor soil at Henley in southern England produced only 68 capitula and 4760 seeds, while just a few kilometres away on rich soil at Medmenham were giant plants with 2489 capitula (174 230 seeds). The usual fecundity of ragwort plants on fertile soils is approximately 40 000 seeds and the exceptional Medmenham plants were second growth individuals that had been mown the previous year (Cameron 1935).

High population density in grasses like *Lolium perenne* leads to reduced tillering (and a consequent increase in the relative importance of the main shoot as a source of photosynthate), and also to a reduction in the proportion of assimilate directed to root growth. These changes are due to reduced light intensity in the sward, and perhaps to changes in the spectral energy distribution (Colvill & Marshall 1981).

As a result of size plasticity, a typical plant population consists of a few large plants and a large number of small ones. Sometimes, however, this is not observed and in certain coniferous trees, for instance, very high-density populations become 'locked' or 'stagnated' because of uniform competition. The failure of any plants to 'break out' has meant that self-thinning could not occur. Baker (1950) records an example of a 70-year-old stand of lodgepole pine, *Pinus contorta*, that was so dense and uniform that the plants could be pulled up like weeds; the trees were just over 1 m tall and had ground level stem diameters of under 1 cm.

The study of density dependence, like so much in ecology, is dogged by the difficulty of separating cause and effect; for instance, we cannot say whether, in general, plants are big because they suffer little herbivore attack, or they suffer little herbivore attack because they are big. Similarly, without a great deal of detailed knowledge on the plant species and the background conditions, it is impossible to say whether increases in plant density cause reductions in mean plant size, or whether reductions in mean plant size cause increases in population density by relaxing the intensity of self-thinning. It is for this reason that an experimental approach to plant population dynamics is so vital.

2.1.2 Determining *r*

We can summarize a great deal about the population dynamics of a plant by calculating its intrinsic rate of increase, *r* (per annum).

For annual plants, net multiplication rate R is simply the average fecundity of large, free-grown individuals, m, multiplied by the fraction of seeds that attains reproductive age, l; $R = lm$. The intrinsic rate of increase, defined by equation 1.1 is

$$1 = e^{-r}lm$$

$$r = \ln(lm) = \ln R.$$

For biennials, two years of exposure to the risk of death precedes reproduction and

$$1 = e^{-r}l_1 0 + e^{-2r}l_2 m$$

$$r = \tfrac{1}{2}\ln(l_2 m).$$

If, as seems realistic, the survival rate during the second year is much higher than during the first, then a biennial has approximately half the intrinsic growth rate of a similar annual plant. To outcompete the annual, other things being equal, the biennial would, therefore, have to produce the square of the seed output of the annual (not double its production).

$R_{\text{annual}} = lm$ each year while $R_{\text{biennial}} = lm$ in 2 years or $\sqrt{lm}$ per year.

Perennial plants present a much more complex picture. Many species produce seeds over an extended period, some fruiting every year, some at regular intervals and others sporadically (these are 'iteroparous' plants). Other species reproduce only once then die; these are the 'big bang' or 'semelparous' plants. For semelparous species that reproduce once at age x, we have

$$1 = e^{-rx}l_x m$$

$$r = \frac{1}{x}\ln(l_x m).$$

For species where a long period elapses before reproduction (x large) the rate of recovery from population depletion will depend critically on the age structure of the population. If all the individuals are the same age, the population will become extinct should some catastrophe befall this one class before the age of seed set. A particularly extreme example is provided by the bamboos *Sinarundinaria* and *Sannocalamus* that flower synchronously over vast areas at intervals of between 100 and 120 years. The fate of the giant pandas which inhabit the bamboo forests of China is a cause of grave concern to conservationists, since the entire habitat is likely to die and be replaced by seedlings in one sudden burst of reproduction (Seneviratne 1980).

Long-lived iteroparous plants present greater difficulties in calculating r.

If we had full survivorship and fecundity data for a species, we could simply solve equation 1.1 iteratively to obtain r. Unfortunately, fecundity and survival are poorly correlated with age for most plants so this approach is unsuitable. Also, since perennials often have lifespans much longer than ours, there is a distinct shortage of age-specific demographic data for these plants.

Despite these difficulties, it is important that we have some means of determining the intrinsic rate of increase for long-lived plants, for use in both theoretical population models and in long-term management plans for natural woodlands. Fortunately, there are several simplifying assumptions that we can make. It may be reasonable, for example, to assume that survival rate is very high for mature plants ($l_x \approx l_{x+1}$) and that a fixed crop of seed is produced regularly (every year, every fourth year, or whatever). Then we can define limits between which r will lie; the minimum is given by assuming that all reproduction after the first seed set is irrelevant (the plant is essentially semelparous), and the maximum by assuming that reproduction is continuous for ever once the threshold age at which reproduction begins, x, has been passed:

$$\frac{1}{x}\ln(l_x m) < r < l_x m.$$

Assume that a tree produces its first seed crop after 40 years and sets 10 000 seeds. Further suppose that independent study has shown survival from seed to fruiting adult to be 0.000 05. Thus we have

$$\frac{1}{40}\ln(10\,000 \times 0.000\,05) < r < 10\,000 \times 0.000\,05$$

so r lies between -0.0173 and 0.5. From these data it is clear that the trees must be left to stand for some time after their first fruiting if the population is not to decline.

The true value of r is normally much closer to the lower bound than the upper and the approximation works well when the first seed set at age x is relatively large. For species where reproduction begins with a whimper rather than a bang, it is very difficult to approximate r.

It is often said that a plant need only leave one descendant in a lifetime, and for a population in equilibrium this is statistically true, but for a genotype (and hence for an individual) it is not so much how many surviving offspring a plant leaves, but when it leaves them that is important in determining its fitness. That early reproduction was vastly more significant than later was realized 30 years ago by Cole (1954) who pointed out that lengthening the reproductive life indefinitely was equivalent to adding just one seed to the first year's effective reproduction; 101 seeds produced at age 1 serves the same purpose as producing 100 seeds per year for ever.

This raises the question as to why there should be so many long-lived plants in the world. The main advantage of the perennial habit accrues in environments where the ratio of mature to juvenile survival is high (e.g. where competition is intense), for here Cole's maxim no longer holds (Charnov & Schaffer 1973).

Straightforward age-specific survivorship and fecundity schedules are difficult to apply to many plant populations where plant size is highly variable and seed production is strongly density dependent. Also, factors like partial seed dormancy and differential survival after flowering complicate the calculations. Nevertheless, r can be determined for any life history, even if it involves laborious calculation of strings of probabilities in which the seed set from every possible history of plant development is discounted to its present value (see, for example, Hubbell & Werner 1979).

Alternatively, the determination of r may be based on size classes rather than on age classes; instead of the plants all moving up one age class in each time period, only a fraction of them grows large enough to pass from one size class to the next. If necessary, plants can be allowed to get smaller, as they might when attacked by herbivores. For details of this method see Usher (1966).

Some idea of the potential rate of increase can be obtained from cases where alien plants, freed from their enemies and competitors, have spread rampant in foreign lands. The cactus *Opuntia inermis*, for instance, spread from almost nothing to dominate over two million hectares in Australia to the exclusion of all other plants (see section 4.1). Caughley and Lawton (1981) estimate r_{max} as 2.76 on the assumption that one rootstock (weighing about 1 t) can grow to 250 t in two years. Field estimates are available for a few short-lived species in their native environments; *Phlox drummondii*, for example, has $r = 0.884$ (Leverich & Levin 1979). For eight populations of teasel, *Dipsacus sylvestris*, a biennial (but facultative short-lived perennial) with partial seed dormancy, Werner and Caswell (1977) calculated r values between 0.518 and -7.752 for age-based models and 0.957 and -1.291 for models based on stage of growth.

Even with quite detailed knowledge of seed production, as with Gardner's (1977) study of ash, *Fraxinus excelsior* (see section 2.5), we cannot estimate r because we have no independent estimate of survivorship. If we use the fecundity data (50 000 per tree spread over an area of 50 m^2) and the current population density of trees ($0.02\,\text{m}^{-2}$) to estimate survivorship

($l_x = 0.00002$) we are obviously *assuming* that $r = 0$. There is no way round this dilemma; we must have independent estimates of both fecundity *and* survival. There are no estimates of r for trees to be found in the literature.

2.2   HERBIVORES AND THE DEATH RATE OF ESTABLISHED PLANTS

Once the perils of germination and establishment are past, the risks of death decline rapidly. Mature plants succumb to extreme cold, water-logging, drought and gales and to a variety of fungal and viral diseases. However, in many populations the most important mortality factor for established plants is competition with others of the same or of different species. Deaths due to self-thinning are commonplace in dense forestry plantations but, rather oddly, in many critical ecological studies of natural plant populations as different as dune annuals and tropical rain forest trees, there is no suggestion of density dependence in the death rate due to competition (Watkinson & Harper 1978, Connell 1979). Intense competition between desert winter annuals for example, leads to reduced growth rates and fecundity, but not to increased mortality (Inouye *et al.* 1980), and over a 16-fold range of densities in laboratory populations of the grass *Festuca paradoxa*, mortality was density independent (Rabinowitz 1979).

However, density-dependent mortality, especially in the youngest age classes, has been noted in colonizing populations of the grass *Poa annua* (Law 1981), and may often be overlooked unless detailed study of the very smallest plants is undertaken.

While herbivores normally have only a slight influence on the death rate of mature plants, their potential killing power is illustrated by those cases where an alien insect has ravaged a native plant. The best known example is the virtual extinction of the Bermuda cedar, *Juniperus bermudiana*, by an accidentally introduced scale insect, *Lepidosaphes newsteadi* (Bennett & Hughes 1959).

Mass mortality of trees due to defoliation by vertebrates sometimes occurs. Rowan (1954) describes how peak populations of snowshoe hares, *Lepus americanus*, reached a density of over 100 ha^{-1}, at which level they stripped and decapitated more than one million young *Pinus banksiana* on a 10 ha plot, leaving only 40 trees undamaged. Although elephants are primarily grass-feeders, they can cause severe damage to trees when food is scarce (Petrides & Swank 1966). High-density populations of African elephant, boosted beyond their carrying capacity by immigration, can kill vast numbers of trees; in 1958, for example, only 24% of mature trees surveyed in *Terminalia glaucescens* woodland were dead, whereas by 1967 elephants had killed almost 96% (Laws *et al.* 1975).

The impact of defoliation by invertebrates on the death rate of evergreen conifers depends upon whether the insects feed on old needles or on young. Spruce sawfly, *Gilpinia hercyniae*, for example, attacks only the old needles of *Picea glauca* and *P. mariana* and the trees can withstand several years of almost complete loss of their older foliage. Spruce budworm, *Choristoneura fumiferana*, in contrast, takes both young and old needles, and several successive defoliations will almost certainly kill the tree (Reeks & Barter 1951). Late season needle regrowth in *Pinus sylvestris* allows the tree to recover from 90% defoliation by the pine sawfly, *Neodiprion sertifer*, so that even after three successive heavy defoliations there was no increase in the death rate of the trees (Wilson 1966).

Browsing in winter by fallow and roe deer and by small mammals on the seedlings of holly, *Ilex aquifolium*, can cause severe mortality; only 7.3% of the small trees survived to their 5th year under open pines and just 0.008% survived in deep holly shade (Peterken 1966).

Herbaceous perennials are unlikely to be killed by a single attack of defoliating herbivores because their perennating organs are safely underground. They only succumb if defoliation is repeated at sufficiently short intervals that their carbohydrate reserves are run down to the point where they cannot meet the needs of maintenance respiration and minimal leaf regrowth (see section 2.7.4). However, leaf-feeding by larvae of the swallowtail butterfly, *Battus philenor*, does increase the death rate of its herbaceous perennial host plant *Aristolochia reticulata* (Rausher & Feeny 1980). Also, since the death rate is greatest amongst smaller plants, any herbivore feeding that slows plant growth rate, keeps plants in the smaller size classes for a longer period, and thus indirectly increases their death rate.

Woody perennials are vulnerable to attack on their phloem and vascular cambium and on their leaf buds. The main way that vertebrates kill trees is by ring-barking. The cambial tissue and the phloem are torn away from the woody xylem and the carbohydrate supply link between the leaves and the roots is broken. Goats, squirrels, rabbits, voles and sheep are all pests of young forestry plantations because they ring-bark young trees (Shorten 1957, Keeler 1961, Adams 1975). The effect of elephant browsing on tree mortality and regeneration in East Africa is reviewed by Spinage and Guinness (1971).

Insects also kill trees by attacking the phloem and cambial layers of the stem. They may kill the tree directly by their larval feeding (as some bark beetles do) or by acting as vectors for fungi that block the phloem, as with the infamous Dutch elm disease, *Ceratostomella ulmi*, spread by scolytid beetles that killed most of the elms in southern England during the 1970s (Strobel & Lanier 1981).

Most herbivore species do not kill their food plants directly; they are essentially parasitic rather than predatory in their action. Even in the classical cases of biological weed control, the herbivores rarely killed the plants outright (see section 4.1); *Chrysolina* beetles inhibited root growth so that the *Hypericum* plants succumbed to water shortage during the dry season (Clark 1953); *Cactoblastis* caterpillars made feeding scars that allowed bacterial rots to enter and destroy the *Opuntia* cactus tissues (Dodd 1940); *Schematiza* beetles so reduced the competitive ability of *Cordia* by feeding on its leaves that the black sage was ousted by other plants (Bennett 1970).

Sometimes herbivore feeding can actually increase plant population density. The classic example of this is provided by ragwort, *Senecio jacobaea*, on the sandy soils of Breckland in eastern England. When the plants are defoliated by larvae of the cinnabar moth, *Tyria jacobaeae,* each individual produces several regrowth shoots from root buds. Population density of ragwort shoots can increase eightfold after severe defoliation (Dempster 1975).

While herbivore attack and plant competition can each act separately, it is the interaction of the two factors that creates a really important impact on plant death rate. Thus the effect of defoliation on the mortality rate is critically dependent on the size and vigour of neighbouring plants. When rabbits eat the tops from young gorse plants, *Ulex europaeus*, whose shoot bases are unshaded by other plants, the lower buds are stimulated to develop and a more bushy plant results. However, when the gorse is surrounded by dense, tall grasses, the lower buds do not develop and the plant frequently dies (Chater 1931).

Although it was a scarce and inconspicuous insect in its native Trinidad, the thrips *Liothrips urichi* when released from its natural enemies proved successful in the control of *Clidemia hirta* that had formed impenetrable thickets on the pasture lands of Fiji. Severe thrips attack caused complete die back of all young growth as well as a reduction in seed production. 'This check on the growth of the plant enabled rival weeds to outgrow it and commence to choke it, leading to further checking of the vitality of the plant... The thrips can be definitely said to have brought the weed under control over large areas not by directly killing it, but by so inhibiting its growth that it was no longer able to compete with the surrounding vegetation' (Simmonds 1933).

The death rate is frequently related to herbivore feeding rate in a complex way. Slugs, for example, will avoid the grass *Poa annua* when offered a preferred food species such as *Capsella*. When forced by hunger to attack the *Poa* they sever its shoots at soil level, eat out the basal meristem and leave the felled shoot lying on the ground. Many of the grass plants die, especially those overtopped by surviving individuals. In contrast, slug feeding on the preferred plant is responsible for a uniformly high level of defoliation but causes very few plant deaths (Dirżo & Harper 1980).

Each larva of *Oscinella frit* destroys at least one grass tiller during its development and at peak densities over 1500 tillers m^{-2} are killed each year (Henderson & Clements 1979). Since this is well under 20% of the tillers present and these ryegrass pastures suffer up to 40% yield reductions, it appears that the death of one tiller adversely affects the productivity of the survivors. Whether this is due to the frit-infested tillers being a sink for carbohydrate before their death or due to chemicals injected by the fly larvae that are translocated to the surviving parts of the plant is unknown.

2.2.1   Plant survivorship curves

Two kinds of survivorship curve have been documented for post-establishment plant populations (Fig. 2.3). In annuals like *Phlox drummondii* and *Vulpia fasciculata* the survival rate is uniformly high until flowering causes the death of most plants at about the same age (Watkinson & Harper 1978, Leverich & Levin 1979). It is hardly surprising that survivorship curves are available for few iteroparous perennials, since their longevity is of the same order as ours. What data there are, suggest that the risk of death is more or less constant and log (survivors) declines linearly with age (e.g. *Ranunculus repens*, Sarukhan & Harper 1973).

The difficulty of obtaining survivorship curves for long-lived plants can sometimes be overcome by analysing their current age structure (but see Caughley 1977). The assumption is that if recruitment is constant and age-specific mortality is neither time dependent nor density dependent, then the fraction of plants in an age class is a good estimate of survivorship to that age. The breadth of the assumptions should warn us immediately that the technique is likely to be of limited value. However, for small, mixed age class stands, where the tree species has seedlings and saplings that can survive in

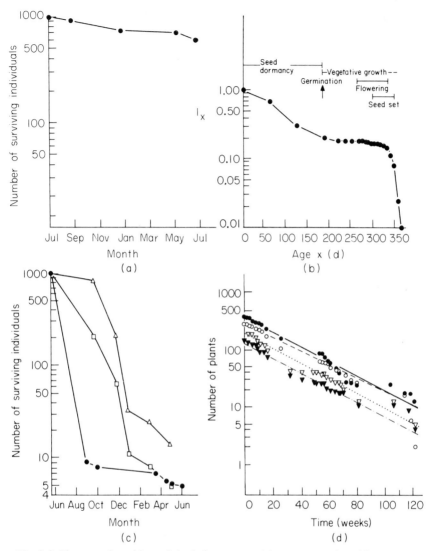

Fig. 2.3 Plant survivorship and depletion curves. Most true survivorship curves for plants (in which a cohort of seeds is followed from seed fall to seed set) relate to annuals. (a) High, age-independent survival in the dune grass *Vulpia fasciculata* (Watkinson & Harper 1978). (b) High, age-independent survival until seed set, followed rapidly by death in *Phlox drummondii* (Leverich & Levin 1979). (c) Heavy mortality of young plants followed by relatively high, constant survivorship through to seed set in (●) *Spergula vernalis*, (□) *Minuartia uniflora* and (△) *Sedum smallii* (in Harper 1977). (d) In marked perennial plants it is impossible to estimate initial cohort size or plant age distribution. In these cases depletion curves are constructed; most studies show roughly constant rates of plant mortality as in these depletion curves for four populations of *Ranunculus repens* in permanent grassland sites of 1 m^2; time in weeks after first observation (April 1969). The symbols relate to 4 different sites (Sarukhan & Harper 1973).

its own shade, the method ought to tell us something. Working with balsam fir, *Abies balsamea*, and eastern hemlock, *Tsuga canadensis*, Hett and Loucks (1976) were able to obtain satisfactory survivorship curves by fitting power functions to the age structure data (Fig. 2.4(a)). Whittaker (1975) suggests that the death rate of *Quercus alba* is roughly constant over its 300 year lifespan.

   Usually, age structure will be determined by episodes of disturbance and regeneration rather than by progressive mortality, and will reflect the history of storms, fires and herbivore outbreaks. The age structures in Fig. 2.4(b) for birch, oak and alder in northern England are typical (Tucker & Fitter 1981); they allow no estimate of age-specific mortality.

   Where the trees form pure, single-age stands, the mosaic of different aged patches reflects the timing and extent of different disturbances. Only if the disturbances were regular in time and similar in extent could the fraction of trees of a given age be taken as a guide to age-specific survivorship over the area as a whole. Also, when woody perennials regrow from stools after their stems have been felled, the rootstock will often by much older than the current trunks. This may present serious problems when attempting to age the trees in regrowth forest.

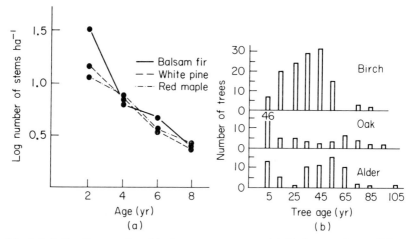

Fig. 2.4 Estimating survivorship from age structure. (a) Under certain conditions (see text) a reasonable estimate of survival rate may be obtained from the age structure of a population, as with these three American forest trees (Hett & Loucks 1976). (b) More usually, and especially in disturbed forests, the age structure tells us a certain amount about the history of disturbance and about fluctuations in recruitment, but almost nothing about survivorship; viz. these three tree species in a nature reserve in Yorkshire (Tucker & Fitter 1981).

One of the first time-specific life tables for a plant population was drawn up for gorse over a period of 10 months by Chater (1931).

	Losses	Survivors
Initial seedling population		851
Eaten by rabbits, etc.	71	780
Uprooted by birds, etc.	16	764
Choked by grass	8	756
Dried up	221	535
Rotted due to winter waterlogging	54	481
Improper rooting; surface germination	12	469
Missing for unknown reasons	85	384

Unfortunately, Chater did not report the area over which these counts were made, but seedlings thin to a density of about $200 \, \mathrm{m}^{-2}$ and mature gorse bushes are typically found at a density of approximately $2 \, \mathrm{m}^{-2}$.

2.2.2 Longevity

The longevity of many species is determined by their reproductive behaviour. In semelparous plants (annuals, biennials and 'big bang' perennials) death almost invariably follows seed set. Death can often be delayed, however, by preventing seed set. Gardeners increase the flowering period of annuals by nipping off the heads as soon as the flowers begin to fade. Groundsmen mowing putting greens at weekly intervals can turn *Poa annua* into an almost immortal perennial. *Senecio jacobaea*, normally considered a biennial, will live for three or even four years if severely defoliated by cinnabar moths (Dempster 1975).

Delay in flowering can sometimes have a dramatic effect on the fitness of a plant. In desert annuals, a delay in flowering due to herbivore feeding may mean that no seed at all is set before drought kills the plants. In other habitats, delayed flowering may alter the pollination rate; it may improve it as in alfalfa pollinated by bees in Europe, or it may reduce it as with subterranean clover in Australia (Collins & Aitken 1970). Regrowth flowers of ragwort, produced after defoliation by cinnabar moth larvae, open in autumn and are greatly damaged by frosts (Zahirul Islam 1981).

The determinants of longevity in iteroparous plants are less obvious. It is not known whether in long-lived trees such as oak or redwood (*Sequoia*) there is any relationship between longevity and reproductive output (as there is in many animals). It does seem, however, that some trees extremely stressed by wind, drought or disease, will set a large crop of seeds before dying (Matthews 1963). Whether fecundity would decline with age, as is the case with most orchard trees, if reproduction were prevented for a long period does not appear to have been tested.

Herbivore feeding on flower stalks is analogous to animal castration in its effects on increased longevity. It is much less clear how leaf feeding affects longevity. Once floral meristems have been produced, leaf feeding can reduce the supply of carbohydrate and may reduce the number of seeds filled or the weight of individual seeds (see section 2.4.1). In this case it will not affect longevity because the plant would have died anyway. If leaf feeding occurs before floral initiation, carbohydrate reserves may be reduced so far that initiation does not occur. In this case longevity may be increased as the plant delays flowering until herbivore pressure is sufficiently lax that carbohydrate levels can be restored.

Thus feeding can affect the longevity of different plants in different ways; for instance, sheep grazing increases the longevity of grasses, but reduces the longevity of woody plants like *Artemisia tripartita* in sage-brush grasslands in Idaho (West *et al.* 1979). Herbivores also affect the longevity of individual plant tissues. Sap-sucking and leaf-mining insects may cause premature leaf fall in trees (see section 2.3.5). Infestation of grass tillers by aphids can significantly reduce the longevity of individual leaves (see section 2.3.2).

The longevity of plant parts is also determined by mineral nutrition. Adding fertilizer to dune populations of sand sedge, *Carex arenaria*, greatly increased both the birth and death rates of the tillers and dramatically reduced their longevity. Tiller density was not markedly increased, and standard sampling methods would have overlooked the profound effect of fertilizer on the age structure of the tiller population (Watkinson *et al.* 1979).

Relative rates of plant and herbivore development are often critical in determining the survival of insects. In dry years, for example, *Plantago erecta* produces fewer, smaller seeds. In the following year, these smaller seeds produce small plants which senesce earlier than plants from large seeds. Mortality of the larvae of checkerspot butterfly, *Euphydryas editha*, is increased dramatically by this early senescence because they are unable to complete their development before their food supply dries up (see section 3.4.2). Many larvae die as a direct result of drought in the first year, and others indirectly as a result of reduced plant size and premature senescence in the second (Singer & Ehrlich 1979).

2.2.3   The frequency of herbivore attack

The effect of defoliating insects on the death rate of trees is known to be dependent on the age of the plants (Kulman 1971), on their position in the canopy (whether dominant or suppressed; Belyea 1952) and on their reserve levels (as affected by recent productivity, the number of previous defoliations, and so on; Embree 1967).

Different tree species respond to defoliation by the same insect in different ways; for example, a single defoliation by the gypsy moth, *Porthetria dispar*, kills 68% of dominant and co-dominant hemlocks, *Tsuga canadensis*, 28% of white pines, *Pinus strobus*, but only 5% of oaks, *Quercus* spp. Five per cent mortality is equivalent to the natural rates of loss in an unattacked but relatively crowded woodland (Stephens 1971).

The death rate increases with successive defoliations. Although just 5% of the oaks died after one defoliation, between 10 and 80% were killed by three successive heavy defoliations by gypsy moth. Most of the deaths were of small trees, since a 72% reduction in the number of oak stems, was matched by only a 37% reduction in basal area (Stephens 1971, and see Nichols 1968).

In the populations of quaking aspen, *Populus tremuloides*, studied by Churchill *et al.* (1964), defoliation by forest tent caterpillar, *Malacosoma disstria*, did not have a marked effect on death rate unless the trees were heavily attacked for three successive years. In this case, the increased death rate was suffered by the larger dominant and co-dominant trees, rising from an average of 6% to almost 30% in the 6 years since the last outbreak. The death rate amongst suppressed trees, while much higher (over 50% compared to an average 10% for the dominants), was not increased in the stands that had suffered three successive defoliations.

Mortality due to defoliation and self-thinning seem likely, therefore, to be compensatory. In forests unattacked by insects the death rate due to plant competition may be higher than in defoliated stands where competition is relaxed. The net long-term effect of moderate levels of herbivore attack on tree population density is probably very small.

2.2.4   Death rate and stress

Any factor that reduces the ability of a plant to defend itself against herbivore attack will cause an increase in the death rate; for example, *Pinus ponderosa* trees weakened by oxidant air pollution succumb to bark beetle attack. Doubly weakened, the trees then lose out in competition with *Abies concolor*, which they had previously suppressed (Miller 1973). The stem coccid *Matsucoccus pini* occurs more commonly on Scots pine, *P. sylvestris*, in areas with high levels of air pollution and spruce gall louse, *Sacchiphantes abietis* (Homoptera), is more damaging where the trees are subject to fluoride pollution (Heagle 1973).

Water stress during drought is thought to be important in affecting the susceptibility of trees to damage by defoliating insect larvae (T.C.R. White 1969, 1974). This may be due to increased levels of amino acids in the leaves improving food quality for the larvae, or to a reduction in the trees' ability to produce or mobilize defensive chemicals (see section 3.6.4).

Mortality of mature *Eucalyptus* trees due to defoliation by the sawfly *Perga affinis affinis* was almost entirely restricted to weakened trees on poor sites or to individuals which had been stressed by root damage or altered drainage following cultivation (Carne 1969).

Mechanical damage to the host tree *Pinus radiata* has an important effect on its vulnerability to attack by the wood wasp, *Sirex noctilio*. The wasp is attracted to stressed trees by the monoterpenes that vapourize from their bark, and oviposition is more successful on stressed trees because resinosis is minimal. The wasp injects mucus into the egg chamber along with spores of the fungus *Amylosterium areolatum*; the fungus thrives in the conditions of high water tension and low phloem turgor of the stressed plants and rapidly kills the tree. This creates a suitable habitat for successful feeding and development of the *Sirex* grubs (Madden 1977).

Vulnerability to herbivore attack is not always increased by stress, however. Wood (1972), for instance, found no increase in the number of bark beetles on injured, polluted or fungus-infected trees of *Pinus ponderosa*.

The death rate of plants from vertebrate herbivores may be inversely related to stress. Rabbits, for example, kill a greater proportion of large individuals of the dune annual grass *Vulpia fasciculata*; plants with several tillers and large spikes are more likely to be eaten than depauperate, single-tillered forms (Watkinson *et al.* 1979).

2.3   HERBIVORES AND PLANT GROWTH RATE

The impact of herbivore feeding on plant growth rate clearly depends upon what kind of tissue is removed and on the timing of attack relative to plant development. Leaf feeding, sap removal, tissue mining, meristem consumption, flower and fruit damage, stem feeding and root pruning are likely to have quite different effects.

Herbivores affect growth rate directly by reducing the photosynthetic area (leaf removal and leaf mining), by altering the carbohydrate balance (sap suckers and fruit feeders), by interfering with water and nutrient uptake (root feeders, xylem suckers) and by weakening the physical structure of the plant (shoot and stem borers).

The animals also exert a wide range of indirect effects by injecting chemicals into the plant, transmitting viral diseases, causing the plant to divert some of its production to wound repair or the production of defensive chemicals and so on. This section is concerned mainly with the effects of leaf removal and discussion of the other processes is deferred until section 2.7.

2.3.1   Leaf feeding

Leaf age is the most important attribute affecting its use by herbivores and also in determining the impact of its removal on plant growth. As they age, leaves change in their surface area, colour, turgor, surface toughness, thickness, and in their chemistry. Their digestibility changes and the concentrations of soluble nitrogen, tannins and different amino acids vary. The levels of chemicals acting as attractants, toxins, phagostimulants and repellants also change with age (see sections 3.6.3 and 3.7.2).

It is important to know how feeding from leaves of different ages affects plant production, survival and seed set. In the simplest case, feeding from the youngest leaves will be most damaging because both the tissue of the leaf and all its future photosynthate production are lost. The loss of senescent leaf tissue, on the other hand, may have no effect at all on plant performance. For example, removal of 75% of the foliage of mature oak trees early in the season results in a 50% reduction in wood production; removal later in the year has no noticeable effect on growth (Franklin 1970, Rafes 1970).

In other cases, especially where leaf production is continuous rather than synchronous, the loss of young leaves may be compensated by the production of new foliage (Milthorpe & Davidson 1966). Here, the timing of defoliation is doubly important. It is useful to paraphrase Fisher (1930) and define the 'photosynthetic value' of leaves of different ages; this is the discounted present value of all future photosynthetic production by a leaf, measured in terms of its effects on plant fitness. Photosynthetic value will be as difficult to measure as plant fitness (i.e. usually very difficult); it will be easier to measure for plants that have determinate leaf growth than for those where longevity is determined by defoliation.

Where compensatory growth is possible, the curve of photosynthetic value against age may show a maximum; loss of old leaves is of little consequence as they have already made most of their contribution to fitness; loss of young leaves may be relatively unimportant so long as they can be replaced in time for seed set to be completed within the growing season. When the curve does show a maximum, we would predict that leaves of maximum photosynthetic value would show the greatest investment of secondary chemicals and other defences against herbivores (McKey 1979). When there is a monotonic decline with age, young leaves should be most protected (consistent with their rapid growth and metabolic activity; see Rhoades & Cates 1976). I know of no detailed data to test these predictions.

2.3.2   Herbivores and grass growth

The grasses represent the pinnacle of grazing tolerance (Youngner & McKell 1972). Defoliation results in the production of new leaves either from the same tiller or through the development of new tillers. The regrowth leaves are formed of carbohydrate that comes either from photosynthesis of the surviving green leaves or from carbohydrate reserves stored in the roots and stem bases (see section 2.7). Grasses do suffer as a result of defoliation, of course. Their carbohydrate reserves may be depleted, tillers may die and flowering may be delayed. There is no evidence at all that the genetic fitness of an individual grass plant is ever enhanced by defoliation, when compared to an undefoliated plant nearby (cf. Owen 1980). Typical symptoms of heavy grazing on grass growth and form are listed by Peterson (1962): reduced growth rate in spring; reduced winter survival; shortened internode length; prostrate green leaves; reduced leaf width; and depleted carbohydrate reserves.

The Achilles heel of the grasses is the meristem. In most species the meristem is low down, almost at ground level, well protected amongst the basal leaf sheaths. A few species, however, produce elongated vegetative shoots in which the meristem is vulnerable to grazers. When a meristem is eaten the life of the shoot is as good as over, since shoot growth and leaf extension cease. Floral meristems of most grasses are vulnerable because they must be raised into the air to ensure pollination of the flowers.

Range grasses *Panicum virgatum* and *Andropogon scoparius* tend to decrease under heavy grazing because they produce elongated tillers, whereas species like *Poa pratensis* and *Buchloë dactyloides* are grazing tolerant since their meristems remain below the minimum grazing height.

Also, the first two species produce a high proportion of flowering tillers each year, while the latter two only flower from a few (Branson 1953).

Regrowth after meristem removal requires the initiation of axillary buds and the appearance of new shoots; even when this is commonplace, regrowth is much slower than from undamaged meristems, so plants suffering meristem predation tend to be outcompeted by those whose meristems survive.

The population dynamics of grasses are usually studied by monitoring the flux of tillers. The turnover rate of tillers varies seasonally and with the reproductive state of the plant (Kays & Harper 1974). Peak tiller production is in spring and there is a smaller peak in late summer. Peak tiller death rate occurs after ear emergence (since flowering kills the shoot; Ong *et al.* 1978). Tiller production is greatly stimulated by light intensity and is sometimes enhanced by defoliation when this increases light availability to the lower leaves or when it reduces apical dominance (Langer 1956). Close grazing of *Lolium perenne*, for example, greatly increases tillering. This partially offsets the reduced production per tiller so that heavily grazed plots return to levels of production similar to those observed on more lightly grazed areas (Grant *et al.* 1981).

Individual tillers in a 'continuously grazed' sward are actually defoliated in discrete episodes. In a *Lolium perenne* pasture grazed by sheep, for example, marked tillers were grazed every 7 to 8 days in a set-stocked sward, and every 11 to 14 days at a lower stocking rate. The interval allowed many of the tillers to flower and thereby to 'escape' from the herbivores, since flowering was associated with a marked decline in the attractiveness of the tillers to sheep (Hodgson 1966).

Westoby (1980) set up populations of *Phalaris tuberosa* of the same tiller density but differing in tiller size distribution. Populations from low seed densities had a high proportion of large tillers while those from dense sowings were more uniformly small. In *P. tuberosa* large tillers are particularly susceptible to close clipping because of their swollen bases, so under severe defoliation (clipping to 1 cm) the population derived from the lowest seed density suffered the highest rate of tiller mortality. This paradoxical result highlights the difficulty of interpreting density dependence in plant populations. Tiller death rate was not tiller density dependent since there were the same tiller densities in both treatments. The increased death rate was caused by the difference in size distribution and this, in turn, was due to the difference in sowing rate. Tiller death rate was negatively dependent upon seed density.

Defoliation also alters the population dynamics of individual grass leaves. Removal of mature leaves, for example, causes a shift in the age structure in favour of younger, regrowth tissues (often increasing average food quality for the herbivores). Feeding damage may reduce the longevity of leaves, so that litter accumulates more rapidly; for example, feeding by the aphid *Holcaphis holci* on *Holcus mollis* causes a significant reduction in leaf longevity (Packham, pers. comm.).

Alternatively, by removing leaves before they senesce, litter production may be reduced under sheep or cattle grazing (Nicholson *et al.* 1970). In any event, it is often important to take account of the flux of leaves between the sward and the litter (leaf turnover), in attempting to estimate grassland productivity. Accounting for leaf turnover in British chalk grasslands, for example, led to a doubling of the estimate of primary production, compared to estimates based on peak live standing crop biomass (Williamson 1976), and to more than a tripling of the estimate for the wetland grass *Glyceria maxima* (Mathews & Westlake 1969).

It is sobering to bear in mind when considering the impact of herbivores on grass population dynamics that grass clones can be extremely large (several hectares) and very long lived. Herberd (1961, 1962) estimates that some clones of *Festuca rubra* are over 1000 years old.

2.3.3   Defoliation and root growth in grasses

Defoliation of grasses usually stops root growth and decreases root respiration and nutrient uptake within 24 hours (Troughton 1957, Davidson 1979) because photosynthate is retained in the shoot for regrowth. The existing roots of *Dactylis* and *Bromus* decomposed within 48 hours of defoliation and radiolabelled phosphorus placed 12 cm deep in the existing root zone was not taken up and detected in the shoots until new roots grew back to this depth. This took more than 24 days in every case (Oswalt *et al.* 1959).

Under field conditions plant growth may also be limited by nutrient availability, especially in arctic and alpine environments where nitrogen and phosphorus rather than carbohydrate tend to be limiting (Chapin 1977). If nutrients do limit regrowth, then root activity should be maintained after defoliation. To test this, Chapin and Slack (1979) defoliated nutrient-limited field populations of *Eriophorum vaginatum* and *Carex aquatilis* at Atkasook in Alaska. These plants are heavily grazed by caribou and lemmings and normally possess substantial accumulations of available carbohydrate (Shaver & Billings 1975). Regrowth leaves often contain higher concentrations of nitrogen and phosphorus than the newly initiated leaves of undefoliated plants (Jameson 1963) and leaf growth rates following defoliation are as rapid as in controls, so their nutrient demands are high. With moderately intense defoliation root respiration did increase suggesting increased activity, and root nitrogen concentration remained stable while root initiation and elongation declined. At high intensities (four repeated defoliations), the mortality rate of roots increased dramatically from 35 to 95% (Chapin & Slack 1979).

2.3.4 Defoliation and net production in grasses

There is some debate over the impact of defoliation on total net production in grasses. For range plants and grasses of extensive grazings, the consensus of a large and generally confusing literature is that clipping grasses during the growing season normally produces a lower yield than a single harvest at the end of the year (Jameson 1963). However, the dryland sedge *Kyllingia nervosa* from heavily grazed parts of the Serengeti shows peak productivity when defoliated daily to a height of 4 cm (McNaughton 1979a). The productivity of intensively managed British grasslands is greatest when there are three defoliations at intervals of approximately 8 weeks (Holmes 1980).

The timing of the initial defoliation affects the total annual yield obtained from *Lolium perenne, Agropyron intermedium* and other grasses. In several species both earlier and later initial defoliation produces higher yields than intermediate timings (in early May), because plants at this time.are at a critical stage of growth, and defoliation interrupts the main phase of stem development. Close defoliation in May removes most of the apical meristems. The greatest total annual yield is obtained with the latest initial defoliation, under both close (3 cm) and lax (8 cm stubble) cutting, when harvest is delayed until mid June by which time the inflorescences are fully emerged (Binnie *et al.* 1980).

The number of defoliations for maximum yield and the intensity of plant competition are clearly related; when competition is slight as in arid, heavily grazed communities, there is little benefit to be obtained from multiple harvests (Pearson 1965). However, in fertile, well-irrigated pastures where competition for light would be severe if early cuts were not made, yield increases are to be expected. For instance, the net primary productivity of *Phalaris tuberosa/Trifolium repens* pastures in Australia is an n-shaped function of stocking density. Net production was greatest at 20 sheep ha^{-1} and least at 10 sheep ha^{-1} with 30 sheep ha^{-1} giving only slightly greater production than 10. Low productivity at low stocking rates was due to canopy closure and reduced photosynthesis because of shading. Low productivity at high stocking rates was due simply to the removal of much of the photosynthetic leaf area. At intermediate densities shading was minimal but leaf area was not substantially depleted. Also, grazing led to an increase in the photosynthetic efficiency of the sward because there was a higher proportion of young tillers at higher stocking densities (Vickery 1972).

2.3.5   Herbivores and the growth of woody plants

Leaf-feeding herbivores can reduce tree growth by depleting the plant's carbohydrate reserves, by tapping its current photosynthesis, or by reducing its photosynthetic leaf area.

Sycamore aphid *Drepanosiphum platanoidis* feeding on the unfurling leaf buds of *Acer pseudoplatanus* are drawing on stored carbohydrates. Since the number of sycamore leaves is already determined, aphid feeding causes a reduction in the size of the individual leaves. Over an 8-year period, the

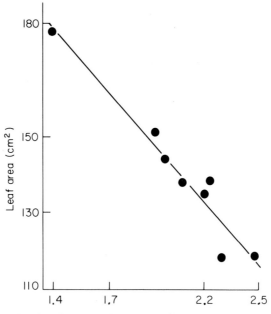

Fig. 2.5 Leaf size and aphid feeding in sycamore. Aphids feeding at bud burst deplete carbohydrate reserves leading to reduced leaf area (Dixon 1971a).

aphids caused an average reduction in leaf area of 48% on one tree and 27% on another (Fig. 2.5). This effect is not general, however. Lime trees infested by aphids have the same sized leaves as uninfested trees (Dixon 1971a, b).

Perhaps the most important herbivores in mature forests are defoliating insects (mainly moths, sawflies and beetles), which hole or skeletonize the leaves. Obviously, holes made early in the life of a leaf will expand as the leaf grows. Reichle *et al.* (1973) estimate that while herbivorous insects remove only 2.6% of net primary production of leaves in *Liriodendron tulipifera* forest, the plants suffer a 7.7% reduction in photosynthetic surface area. Similarly, while a leaf miner may reduce green leaf area very little by the mine itself, its effect on total plant production may be greatly increased if attack induces premature leaf fall (Faeth *et al.* 1981).

A great deal of information is available from studies of forest entomology on the impact of defoliating insects (Turner 1963, Vorontsov 1963) and artificial defoliation (Franklin 1970, Kulman 1971) on wood growth. Varley and Gradwell (1962), for example, plot oak growth against the density of defoliating caterpillars (Fig. 2.6). It is difficult to say from data

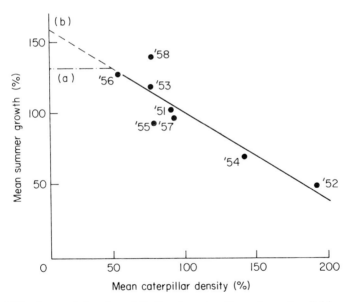

Fig. 2.6 Wood growth in oak and feeding by caterpillars. Annual radial increment of summer wood in *Quercus robur* against mean density of Lepidopteran caterpillars in Wytham Wood near Oxford (after Varley & Gradwell 1962). The curve (a) assumes plant compensation for low levels of caterpillar feeding; curve (b) assumes a linear damage function.

like these what level of wood production would be observed in the absence of herbivores. Gradwell (1974) extrapolates a linear regression of damage on density to zero insect numbers, and predicts that oak productivity would be double normal levels if insect herbivores were absent. It is equally plausible, however, to interpret his data as showing compensation for low levels of caterpillar attack. If this is the case, there would be little increase in tree growth if herbivores were excluded (see section 2.7.5).

What we require are long-term experiments where insect densities are maintained at a range of levels. A start towards this has been made by Morrow and LaMarche (1978) who worked in sub-alpine *Eucalyptus* forests in New South Wales. The trees are subject to attack by continuously high numbers of insect herbivores (mainly sawflies) which take 25 to 50% of the foliage each year. Some trees regularly suffer complete defoliation. Morrow and LaMarche sprayed parts of two species of tree with insecticide for four months, then monitored the subsequent growth of sprayed branches, unsprayed branches on sprayed trees and control trees for a further three

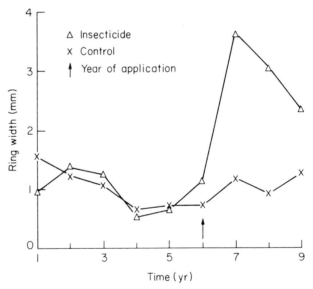

Fig. 2.7 Wood growth in *Eucalyptus stellulata* before and after exclusion of defoliating sawflies with insecticide. Insecticide added at year 6 leads to a dramatic increase in mean ring width in the following year, in treated ($\triangle$—$\triangle$), compared to control ($\times$ — $\times$) trees. Uniformly low ring widths in earlier years show the chronic effect of insect feeding on the growth of these trees (Morrow & LaMarche 1978).

years. They found that the productivities of the two tree species were strongly suppressed by phytophagous insects and that the suppression had occurred over long periods of time (Fig. 2.7).

In contrast, clipping experiments with shrubs that are regularly browsed by ungulates frequently show increases in productivity (Ellison 1960). Again, it clearly matters what kind of tissue is pruned. Removal of the lower, shaded branches of trees usually stimulates growth (up to 38% of the live crown of *Pseudotsuga menziesii* and up to 75% of the crown of *Juglans regia* can be pruned without reducing growth) but trees suffer at higher rates of pruning (Jameson 1963). Browsing by moose *Alces alces* in a pine/mountain ash forest reduced forage biomass from 181 to 109 kg ha^{-1}. Only 3.5 kg of this was due directly to moose feeding; the remaining 68.5 kg were lost due to the reduced growth rate of the damaged trees (Dinesman 1967; see also the following section).

2.3.6   Plant shape

We are surrounded by graphic examples of the influence of herbivores on plant shape; the carpet of lawn grasses, the symmetrical hedge and the bizarre shapes of topiary are horticultural extremes of a completely general process. Vertebrate herbivores often produce browse lines on trees, as do cattle on specimen trees in British parklands. Browsing by Indian elephants, *Elephas maximus*, caused crown distortion of woody plants between 2 and 5 m tall, especially in early successional woodlands (Mueller–Dombois 1972). While giraffes prefer to browse immature acacia trees, their feeding on mature trees sculptures peculiar shapes, including tall cones where the inaccessible centre of the crown grows higher than the rest (Sinclair & Norton–Griffiths 1979). Moose destroyed the leading shoots of 61 to 72% of the young pine trees in a Russian forest. This decreases the growth rate of the trees which then remains depressed over subsequent years. Since the moose attack the better trees preferentially, the weaker, malformed specimens eventually become dominant (Dinesman 1967).

Insect herbivores can cause monstrous distortions in tree shape. The pine-shoot moth, *Rhyacionia buoliana*, causes deformation of the leading shoots of young *Pinus sylvestris*, and spruce gall adelgid, *Adelges abietis*, produces pineapple-like galls on young shoots, which check growth and disfigure trees (Metcalf & Fling 1951). Sawfly feeding on young *Eucalyptus* can cause the development of multiple leading shoots and a bushy form (Carne 1969), and forest tent caterpillar, *Malacosoma disstria*, causes aspen trees to produce shorter internodes and smaller leaves, which give rise to tight clusters of foliage (Duncan & Hodson 1958).

Plant shape is determined by the nature and amount of branching. The terminal bud on a branch produces hormones that suppress branching at the lower nodes (apical dominance). When herbivores remove or kill the terminal bud, the flow of hormone stops, and one or more of the lower buds develops into an elongating shoot.

The importance of branching in natural communities depends upon the level of competition for light. In open communities like arid rangeland, clipping shoots will produce bushy and possibly more productive plants. In a forest where competition for light is intense, removal of a terminal shoot will not lead to branching lower down and will serve only to set the plant at a competitive disadvantage with its neighbours. It is extremely unlikely that attack on the terminal buds of a forest plant would increase its productivity.

2.3.7   Plant size distributions

A typical even-aged plant population is made up of a small number of large plants and a large number of small ones (Fig. 2.8). A shift occurs from the approximately normal distribution of seed weight, to the log-normal

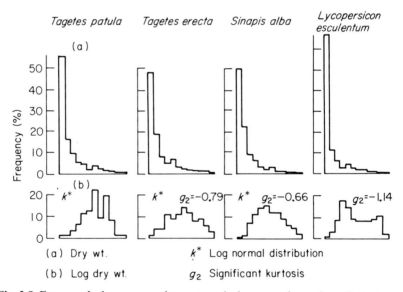

Fig. 2.8 Even-aged plant monocultures contain large numbers of small plants and small numbers of large plants. Some distributions are more or less log normal (*Tagetes*), while others are significantly bimodal (*Lycopersicon*) (Ford 1975).

distribution of the weight of seedlings grown from these seeds in an even-aged monoculture. The shift presumably reflects the emergence of a few large and dominant individuals and the suppression of larger numbers of small individuals (Rabinowitz 1979). The death rate of plants is size dependent and most of the plants that die come from the smallest size classes. In older plant

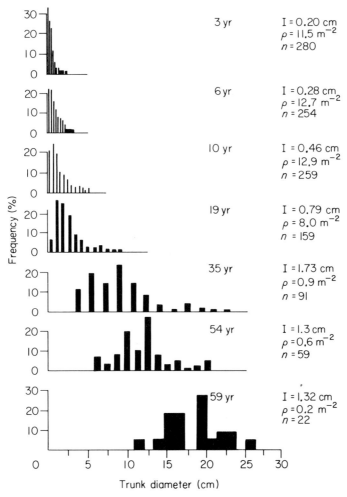

Fig. 2.9 Size distribution changes as the population ages. Frequency distribution of trunk diameter in *Abies balsamea* from 3 to 59 years as density declined from 11.5 trees m^{-2} to 0.2 trees m^{-2}. Note that death rate peaks between 19 and 35 years. I = interval of histogram bars; ρ = population density; n = sample size. (Mohler *et al.* 1978).

populations, therefore, the size distribution reflects both the relative growth rates and the relative death rates of the different classes of plants. The trend in plant size distribution (trunk diameter) is shown in Fig. 2.9 for *Abies balsamea* as the population thins from over 10 plants m^{-2} at 3 years to 0.2 plants m^{-2} at 60 years (Mohler *et al.* 1978). In mixed stands, the size distribution is complicated by the fact that old, suppressed plants appear in the same size class as young, vigorous individuals.

The fact that a particular plant size frequency distribution fits a log normal distribution cannot be taken as evidence of intraspecific competition, however. If initial seedling sizes were normally distributed (for genetic or micro-environmental reasons) and early plant growth was exponential, a log normal distribution of size would arise immediately, without any competitive supression of the smaller individuals.

Ford (1975, and see Fig. 2.8) suggests that bimodality is a feature of several plant size distributions because one class of larger plants has a positive relative growth rate, while another group of smaller plants has a relative growth rate close to zero. Rabinowitz (1979) found biomodality in the weight distribution of *Festuca paradoxa* seedlings but the bimodality was more striking at lower densities, where competition is presumably less severe. This is the opposite of the result that would be expected if bimodality were a measure of the intensity of interference between the plants. Rabinowitz concludes that the observed bimodality is due neither to bimodality in initial seed weights, nor to the timing of germinations. She suggests that at the highest densities all the plants suffer such severe resource limitation that the attainment of dominance is delayed or even prevented, so that the bimodality never develops.

The impact of herbivores on the size distribution depends upon the preferences of the herbivores and the susceptibilities of the plants. Herbivore attack on large plants will tend to even out the size distribution, giving an advantage to previously suppressed plants. Preferential attack on small plants will make them doubly disadvantaged and may increase their death rate substantially. If the small plants are not killed, herbivore feeding will tend to make the distribution even more contagious by ensuring that a larger proportion of the population appears in the smallest size class.

2.4 HERBIVORES AND PLANT FECUNDITY

Seed production is determined by the number of flower buds set, the rate of pollination, the rate of flower abortion, the number of seeds ripened per fruit and the rate of pre-dispersal fruit predation. Each of these is affected by a complex set of abiotic and physiological processes and by the feeding of different kinds of herbivorous animals.

As we have already seen, plant fecundity is strongly density dependent; plants from crowded populations are smaller and bear fewer seeds. The details of how the density dependence operates differ from species to species. In *Linum*, *Vicia* and *Agrostemma* the number of stems per plant, seeds per fruit and mean fruit weight were more or less constant over a wide range of densities, and fecundity changes were due to reductions in the number of fruits per stem. In *Helianthus*, however, the weight of individual seeds and the number of seeds per flowerhead were reduced at high density (Harper 1977).

Fecundity schedules for plants are scarce, but three basic patterns are evident (Fig. 2.10); some show a linear relationship between seed production and size (many weeds and desert annuals; Watkinson 1978, Weaver & Cavers 1980), others have a threshold size above which fecundity increases more or less linearly with size (as in certain trees), while a few seem to produce proportionately less seed per unit-shoot-weight in larger plants (e.g. *Digitalis purpurea*, Fig. 2.10).

We have very little data on the fecundity of trees over more than a small fraction of their reproductive lives, but long runs of data do exist for some crops like apples, cocoa and coconuts (Harper & White 1974). Citrus trees show a rapid rise, a long plateau and then a gradual decline in fruit (and presumably seed) production as they age (Savage 1966). Acorn crops tend to peak between 40 and 100 years and then decline slowly as the oak canopy dies back (Goodrum *et al.* 1971).

2.4.1   Defoliation and fecundity

Graphs of plant fecundity against herbivore feeding are unlikely to be intelligible unless both the timing of herbivore attack and the condition of the plant are known. In grasses, for example, a spectrum of effects has been observed. Autumn grazing of wheat, rye and oats improved subsequent seed production (Sprague 1954), while clipping *Agropyron desertorum* affected neither the number of spikes per plant nor the number of seeds filled per spike (Cook *et al.* 1958). A single defoliation by dairy cattle reduced the seed production of autumn-sown cereals by an average of 38% (Alcock 1964). Early cutting of *Dactylies glomerata* and *Phleum pratense* delayed flower production while late defoliation stopped flower production altogether (Tinker 1930).

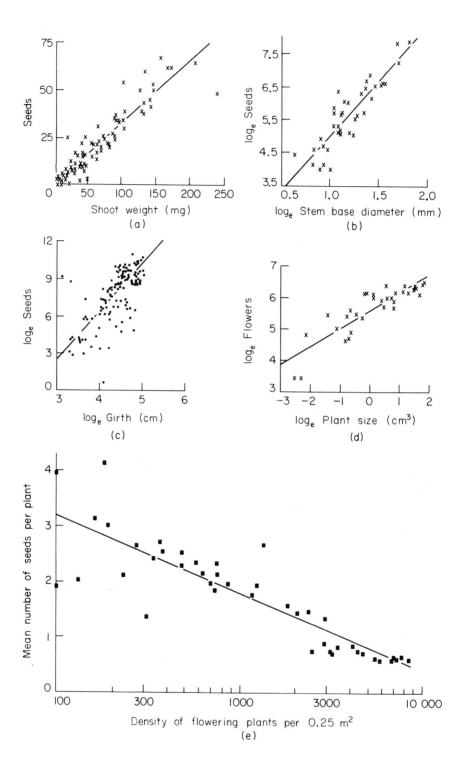

Even with a precise knowledge of the number of herbivores, it is difficult to predict the impact of feeding on seed production without a good deal of information on the density, size structure and phenology of the plants; for example, the wheat bulb fly, *Delia coarctata*, can cause complete loss of cereal yields at peak population densities of about 1.4 million ha^{-1}. Fields with 0.9 million larvae ha^{-1} give satisfactory crops and densities below 0.75 million ha^{-1} cause no detectable yield reductions (Gough 1946). In fields where the crop failed, or yield was poor, 40 to 80% of the plants and 30 to 60% of the shoots were attacked. In fields giving satisfactory yield with moderately high population densities, 6 to 45% of the plants were attacked, but only 2 to 20% of the shoots.

One field, however, produced a very good crop of seed despite a bulb fly density of 1.4 million ha^{-1}. The field had both a high initial plant density because of a heavy seeding rate, and had tillered extensively before it was attacked because of early sowing, so that both the number of plants and the number of shoots per plant were high (Gough 1946). In this case, it is the number of unattacked shoots rather than the percentage of plants attacked that is important in determining the amount of seed produced.

Reductions in fecundity under defoliation suggest strongly that seed production is limited by the availability of carbohydrate. Alternate-year fruiting is commonly observed in undefoliated plants like apple trees and woody legumes like *Cytisus scoparius* (Waloff & Richards 1977), and this is taken as indicating that fruiting makes such a heavy demand on carbohydrate stores that it takes more than one season to replenish the reserves (Davis 1957, Priestley 1970). Native trees also show some relationship between growth and fecundity; Fig. 2.11 plots Rohmeder's

Fig. 2.10 Size-specific fecundity in plants. In annuals, biennials and trees seed production appears to be a linear function of shoot weight. (a) The annual crucifer *Thlaspi arvense* (Duggan & Crawley, unpublished observations); (b) the biennial umbellifer *Daucus carota* (Holt 1972); (c) the tropical tree *Bursera sibaruba* (Hubbell 1980). In the last two cases the slope of the log/log plot is not significantly different from 3.0; fecundity $\propto$ girth^3. The data in (d) show the number of flowers in *Digitalis purpurea* against total shoot length $\times$ rosette diameter^2 (Crawley, unpublished observations). Larger plants produce proportionately fewer flowers per unit shoot weight. (e) Fecundity is density dependent in the dune grass *Vulpia fasciculata*: the smaller plants from high-density populations produce fewer seeds (Watkinson & Harper 1978).

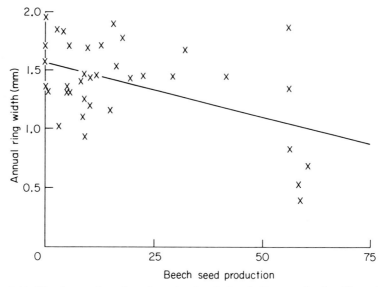

Fig. 2.11 Wood growth and seed production in beech *Fagus sylvatica*. There is some indication that high seed production is correlated with reduced wood growth ($p \leq 0.05$), but other factors clearly affect both processes (from data in Rohmeder 1967).

(1967) data as a correlation between ring width and nut production in beech, *Fagus sylvatica*. The negative correlation is significant ($p < 0.01$) but the relationship is extremely complex; out of six years when nut production was high, two years showed normal or even high wood growth. Obviously, too, there are years when wood production is low even though no nuts are produced. There is no evidence for regular periodicity in nut production in Rohmeder's data.

The relationship between percentage leaf removal and seed yield in sorghum is shown in Fig. 2.12 (Stickler & Pauli 1961); removing young, upper leaves was much more important in reducing fecundity than removing older leaves. Seed production in the field bean, *Vicia faba*, is significantly reduced by *Aphis fabae*, especially when the aphid is attended by the ant *Lasius niger*. All components of seed production are affected; the number of pods per plant is reduced from 19 to 8, the percentage of empty pods is increased from 8 to 65 and the number of seeds per pod is reduced from 3 to 1 (Banks & Macaulay 1967).

Waloff and Richards (1977) found that sucking insects (mainly the aphid *Acyrthosiphum pisum*) caused a considerable reduction in seed production by

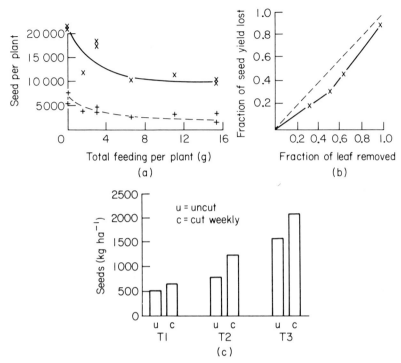

Fig. 2.12 Defoliation and seed production. Natural defoliation and defloration (a) of ragwort by cinnabar moths. Large (×) and small (+) plants at egg-laying. There is substantial compensation since all plants were stripped completely of their initial flowers (Zahirul Islam & Crawley, in preparation). Artificial defoliation of sorghum (b) led to a greater proportional loss of seed than leaf area removed (Stickler & Pauli 1961). (c) Sometimes defoliation actually increases seed yield. Subterranean clover produces more seed because seed production is only high when the pods are buried, and a high proportion of pods is buried only when the sward is regularly defoliated. Three growing seasons created by withholding irrigation 39 days after flowering began (T1), 52 days (T2) or 76 days (T3). Swards cut weekly from 53 days after sowing until 14 days after first flowering can set 30% or more seed than uncut controls (Collins 1981).

broom, *Cytisus scoparius*. The depression in fecundity is probably due to a combination of carbohydrate removal and the injection of phytotoxic chemicals in the insects' saliva (Miles 1968). There was a pronounced two-year cycle in seed production on two adjacent plots (Fig. 2.13). That the plots were exactly out of phase is good evidence that the cycle was not due to patterns of weather but to processes of plant physiology.

What few field data there are on defoliation and tree fecundity only serve

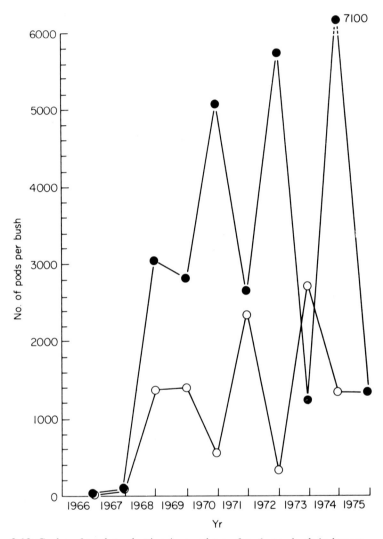

Fig. 2.13 Cycles of seed production in numbers of pods per bush in broom (1966–75). Both sprayed (●———●) and unsprayed (○———○) bushes of *Cytisus scoparius* produce large seed crops in alternate years. Sprayed bushes produce about three times as much seed as insect-infested plants (Waloff & Richards 1977).

to show that whatever does happen, it is not straightforward! English oaks *Quercus petraea* studied by Shaw (1968) set large crops of seed in some years when insect herbivores were scarce and in others when they were common. The largest seed crop he found coincided with the highest herbivore density, a fact which suggests that the same weather factor that promoted high fruiting in oak might also be conducive to high insect numbers. However, partial defoliation of young oak trees by *Operophtera* and *Tortrix* reduced their acorn production by more than 50% compared with insecticide-sprayed controls, even though an average of only 12% of leaf area was removed (Witton 1981). Similarly, controlling defoliation by *Tortrix viridana* in holm oak, *Q. ilex* woodlands in Spain, led to an increase in acorn production from 150 to $600\,\mathrm{kg\,ha^{-1}\,yr^{-1}}$ (Torrent 1955).

Many woody plants produce their leaves in a single burst and, once destroyed by herbivores, the leaves may not be replaced until the next growing season.

Coconut palms attacked by the rhinoceros beetle, *Oryctes rhinoceros*, in Western Samoa seem to follow the first pattern, and there is a linear decline in fruit yield with percentage defoliation (Zelazny 1979). The palm responds to beetle attack by producing fewer nuts of the same size as those on unattacked trees (the fruit primordia appear to 'contest' rather than to 'scramble' for the available carbohydrate).

Six species of tropical tree in Costa Rica were subjected to two artificial defoliations in the course of one year; 80% of the defoliated trees failed to fruit while 30% of the intact trees were barren in the year of defoliation. The control trees produced more fruit in 39 out of 41 pairwise comparisons (Rockwood 1973). Defoliation of the rose *Rosa nutkana* by the western tent caterpillar, *Malacosoma pluvilae*, caused a 96% reduction in hip production (Myers 1981).

It is likely that defoliation has a greater effect on seed production when plants are competing for light than when light is not limiting. Plants of *Abutilon theophrasti* grown at low density could withstand 75% defoliation without significantly reduced reproduction, whereas this level of defoliation caused a 50% reduction in seed production in high-density populations (Lee & Bazzaz 1980). The undefoliated plants at low density produced roughly twice as much seed as undefoliated, high-density plants. Thus defoliation and competition for light interact in determining fecundity (see also Table 2.2).

Defoliation can affect other aspects of reproductive performance; for instance, the chance of a plant flowering at all may be a function of its size.

Table 2.2 Defoliation and fecundity in *Hypericum perforatum*. Artificial removal of none, $\frac{1}{2}$ or $\frac{7}{8}$ of the foliage on seed production, seed viability, parent plant survival and rooting depth (Huffaker 1953).

Control	Total seed (m^{-2})	Seed production (% of controls)	Seed viability (%)	Plant survival (%)	Root depth (cm)
Control	626 844	100	80	93	74
Foliage $\frac{1}{2}$ clipped	131 932	21	65	60	61
Foliage $\frac{7}{8}$ clipped	49 144	8	71	13	34

Weaver and Cavers (1980) found that in *Rumex crispus*, the probability of its flowering was positively correlated with its rosette diameter the previous autumn. Defoliation also affects the probability of flowering in pine trees (Giertych 1970). Defoliation may influence the fitness of seeds, and removal of leaves has been found to cause a reduction in the size of individual seeds in plants as different as annual crops, herbaceous perennials and trees (Lee & Bazzaz 1980). Reductions in seed size are not general, however, and there is little clear evidence of impaired seedling performance from these smaller seeds (see section 2.6.3).

A mature tree suffering a debilitating herbivore attack may make one of two extreme responses. If the tree has occupied its site sufficiently long and grown sufficiently large to exclude plant competitors below its canopy, it would do well to put all its resources into one last lethal bout of reproduction. Its seeds fall into a gap in the canopy where they are free from established competitors, and in sufficient numbers to satiate local seed predators.

Alternatively, if the tree is too young or too small to have excluded other plants from its immediate vicinity, or it inhabits a community where other shade-tolerant species are abundant below the canopy, it would do best to respond to herbivore attack by stopping reproduction altogether and putting all its resources into regrowth so that it regains its competitive status within the canopy (see section 2.7.5).

2.4.2   Root feeding and fecundity

Although the effects of root-feeding insects are sometimes extremely obvious (as when beetle or fly larvae cause large, dead patches in grassland, or when pests such as cabbage root fly, *Delia brassicae*, devastate crucifer crops; Jones & Jones 1974), very little is known in detail about the relationship between root feeding and plant fecundity.

Root-feeding nematodes like the ectoparasitic *Tylenchorhynchus dubius* can reduce seed yields in wheat, ryegrass and pea crops (Sharma 1971) and endoparasites such as *Pratylenchius fallax* can cause severe yield reductions in cereals (Corbett 1972). The grass grubs (beetle larvae in the family Scarabaeidae) are major pests in the grasslands of Australia and New Zealand (Crosby & Pottinger 1979), especially when there is little soil organic matter to act as alternative food. The cicada *Magicicada septendecim* feeding on the root xylem of *Quercus ilicifolia* can reduce wood growth in the trunk by as much as 30% (Karban 1980), but we do not know how seed production is affected. Root pruning of the grass *Bouteloua gracilis* caused a 35% reduction in the mean rate of net photosynthesis and led to a 20% lower total biomass after 3 weeks (Detling *et al.* 1980). The proportion of ^{14}C translocated to the roots declined at first but after 3 weeks was significantly higher than in the controls. Root-feeding scarab larvae, *Phyllophaga crinita*, at a density of $45 \, m^{-2}$ reduced the cover of perennial grasses by 88% in localized areas of a short-grass range in Texas. Most patches were under 0.1 ha but the largest covered over 1 ha and almost all had recovered from scarab damage by the end of the second growing season (Ueckert 1979). Grazing by snow geese, *Anser caerulescens atlantica*, consists of the removal of roots and rhizomes of salt marsh graminoids to a depth of about 25 cm. Geese removed almost 60% of the available biomass in three different salt marsh communities and uprooted much more rhizome material, which was subsequently washed to nearby estuarine waters (Smith & Odum 1981).

Just as leaf feeding alters the distribution of photosynthate in favour of leaf growth (see section 2.7), so root pruning tends to reduce the proportion of carbohydrate channelled into shoot growth (Evans P.S. 1972, Ridsdill Smith 1977, Davidson 1979). It is to be expected, therefore, that root-feeding herbivores will depress fecundity both by reducing the rate at which water and nutrients can be gathered, and by reducing the carbohydrate supply for seed fill. As we shall see in the following section, however, trees sometimes respond to root pruning by *increased* seed production, and exceptions may well occur with other kinds of plants. Also, there is not necessarily a linear relationship between root removal and fecundity depression (see section 2.7).

2.4.3 Defloration and fecundity

There is a fundamental difference in the effects of defoliation and defloration on seed production. Defoliation reduces the supply of carbohydrate available to fill seeds while defloration and fruit predation reduce the number of sinks into which carbohydrate can be channelled.

In grazing exclosures in the Serengeti, there were between 12 and 40 flowering grass culms m^{-2} in January whereas outside there were only between 1 and 9 (McNaughton 1979b). In north Wales, only 15% of the *Ranunculus* plants that flowered in a pasture heavily grazed by cattle set any seed, while 48% set seed in lightly grazed swards (Sarukhan 1974).

Perhaps the best-known flower feeders are garden birds; bullfinches are voracious predators of flower buds on *Prunus* trees (Summers 1981), and house sparrows strip the petals from quince flowers and eat out the pistil. Most of our knowledge of the impact of herbivores on flower survival in trees, however, comes from seed orchards. Squirrels of the genus *Tamiasciurus* sever entire cone-bearing shoots from ponderosa pines and cause most damage in poor cone years; their feeding is negatively density dependent in its action (Dinus & Yates 1975). Opossums, introduced to New Zealand from Australia are important pests of Monterey pine, *Pinus radiata*, orchards. They eat large quantities of pollen shortly before anthesis and consume female strobili between pollination and fertilization, reducing seed production by up to 40%. Chaffinches can destroy 50% of developing female strobili in some orchards and the crimson rosella parrot, *Platycerus elegans*, is a pest of a similar kind in Australia. Cone borers, like the moths *Dioryctia* and *Eucosma*, seed feeders, such as the lygaeid bug *Gastrodes abietum* that kills spruce seeds inside the cone or the tortricid seedworms *Cydia* spp., and flower feeders, like shoot moths, all take their toll (Matthews 1963).

It is interesting that several of the cultural practices adopted to improve flowering in trees mimic the effects of herbivore attack. Root pruning, stem girdling (Ebell 1971), stem banding, shoot pruning, stem bending and stem shaking have all been used by foresters or horticulturalists to improve seed set (Puritch 1972, Zimmerman 1972). No one yet knows why these stresses increase seed production, but they all appear to increase carbohydrate levels and are particularly effective if timed to coincide with bud initiation.

Most unripe fruits are distasteful to herbivores, as children rapidly learn through eating green apples. Nevertheless, a variety of animals specializes in eating these energy-rich foods and another group of herbivores destroys the young fruits to get at the protein-rich seeds inside.

The palm *Astrocaryum mexicanum* is an understorey species of high evergreen forest. It suffers severe predation of its fruits both on the tree and on the ground which together amount to a 95% seed loss. A squirrel, *Sciurus aureogaster*, takes about half the fruits, and may be an important agent of dispersal (seedlings of the palm have been seen growing in the forks of large trees, 15 m off the ground, where squirrel caches may have been made; Sarukhan 1978).

Insects take a heavy toll of young fruits. The anthomyid fly *Pegohylemyia seneciella* attacks the capitulum of ragwort plants, feeding on the maturing achenes and causing a brown central spot in the flowerhead due to premature wilting of the attacked florets. In England, less than 10% of capitula are normally infested but those that are attacked suffer a 75% reduction in seed set. Further north, where its parasites may be less effective, the fly attacks about 35% of capitula in the Scottish Highlands (Cameron 1935). The microlepidopteran *Coleophora alticolella* kills about 50% of the avaliable seed crop of *Juncus squarrosus* (Jordan 1962). A weevil, *Rhinocyllus conicus*, which feeds on the capitula of nodding thistle, *Carduus nutans*, has been used successfully as an agent of biological control for this weed in Virginia, reducing the density of thistle plants by 95% (Kok & Surles 1975).

Cinnabar moth, *Tyria jacobaeae*, larvae act as both defoliators and fruit feeders on their host plant, *Senecio jacobaea*. The larvae hatching from an average-sized batch of 50 eggs will normally defoliate the plant completely before emigrating to find new hosts. Thus plants upon which egg batches are laid often produce no flowers at all at normal flowering time. Plants that escape the attentions of ovipositing adults, but are invaded by larvae emigrating from defoliated plants nearby, suffer attack on their developing flowerheads. These plants may lose between 30 and 60% of their flowerheads, depending upon the size of the plant and the number of immigrants (Zahirul Islam 1981). Once the univoltine moths have pupated, however, larger individuals of both defoliated and deflorated plants produce regrowth shoots that flower in the autumn, so that losses of seed to the herbivores are less than might at first appear (see section 2.7.5).

Feeding by bugs can reduce grass seed yield in two ways. *Leptopterna dolobrata* forms a phytotoxic salivary sheath when feeding on the culm of *Festuca rubra*. The panicle withers and turns silvery white and the few seeds that are set are sterile; this condition is known as 'silver top' (Wagner & Ehrhardt 1961). *Megaloceraea recticornis* sucks out the contents of individual seeds and reduces germination rate dramatically (Kamm 1979).

The gall wasp, *Andricus quercus-calicis*, is recently established in southern England. It passes through two generations per year on oak trees, the first in the flowers of Turkey oak, *Quercus cerris*, and the second on the fruits of English oak, *Q. robur* (Darlington 1974). English oak responds to galling of its acorn cup by shedding the whole peduncle, so that both attacked and unattacked acorns on the same peduncle are killed. On some trees acorn losses can exceed 90% and in such cases the galls can completely carpet the ground. Preliminary work has shown that parasitism of the acorn generation is extremely low (certainly less than 1%) and suggests that the gall wasp's enemies are not yet established in Britain (McGavin & Crawley, unpublished results).

It is important to know whether the loss rate of young fruits is density dependent; whether a plant producing many fruits has a greater or lesser probability of having any one fruit attacked. Feeding by the micro-lepidoptera *Coleophora albicosta* and *Laspeyresia ulicetana* and the weevil *Apion ulicis* accounted for up to 30% of the seeds of gorse *Ulex europaeus* within the pod (Fig. 2.14). These data (Chater 1931) refer to five different sites in Scotland in the same year; they are not proof of density dependence, but the correlation is highly significant ($r = 0.98$, $p < 0.01$) and extremely suggestive.

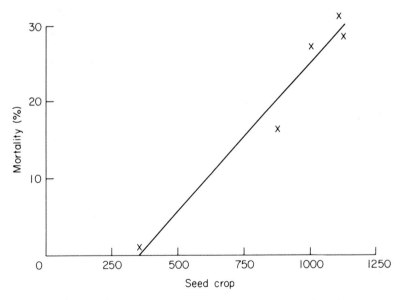

Fig. 2.14 Seed mortality and seed crop size in gorse. Insects take a higher pro-portion of the seeds of *Ulex europaeus* in areas where seed production is higher (from data in Chater 1931).

In contrast, predation on the fruits of *Astrocaryum mexicanum* by squirrels and other rodents appears to be inversely density dependent. Seed survival rate is much higher in dense palm stands (32%) than in sparse (1%, Sarukhan 1978). Again, this is not proof of density dependence and it may simply be that the same abiotic factor which makes palms scarce also makes seed survival low. However, the results do suggest that survival is higher in dense seed populations, and there is mounting evidence that predator satiation is critical for the establishment of seedling populations (see section 2.4.6).

2.4.4 Flowering time

Any factor that influences the timing of sexual maturity can have a pronounced effect on the intrinsic rate of increase. In short-lived, semelparous plants, for example, flowering time is affected by density-dependent processes such as intraspecific competition. Accelerated development at high population densities has been found in dune annuals and weeds (Palmblad 1968, Symonides 1978) and the opposite effect has been recorded in foxglove, *Digitalis purpurea*, a normally biennial plant that delays flowering at high densities (Oxley in Harper 1977).

The usual effect of herbivore attack is to delay flowering either because the flower buds themselves are damaged or because the supply of carbohydrate and protein is reduced by leaf feeding or sap sucking. Also, precocious flowering due to plant density stress may be prevented by the thinning effects of herbivore feeding.

Removing the fully expanded leaves of the clover *Trifolium subterraneum* delayed flowering by up to 30 days due to a delay in flower initiation coupled with a reduced rate of leaf appearance. By delaying flowering, severe grazing during early growth may reduce seed production to the point at which persistence of the plant in the pasture is prejudiced (Collins & Aitken 1970).

While early reproduction typically increases *r*, it can be deleterious when the plant relies on synchrony with pollinator insects; for example, the probability that a flower of *Collinsia sparsiflora* is cross-pollinated by bees was 40% for an early flower but 62% for later, mid season flowers (Rust & Clement 1977). Individuals of the shrub *Hybanthus prunifolius* in a synchronously flowering population produced an average of 658 mature fruits per individual, compared with only 62 in an asynchronous population created artificially by early irrigation. This depression in seed production was due largely to reduced pollination, but increased predation of asynchronous fruits also contributed to the lower success of the early flowering plants (Augspurger 1981).

Delayed reproduction can also lead to reduced seed production. Again, pollinators may be less common and the flowers may be exposed to damaging frosts; for example, the regrowth shoots of ragwort bearing flowers in November are very susceptible to frost damage (Cameron 1935, Zahirul Islam 1981).

Synchrony in flowering is ensured in many plant species because flowering is induced by photoperiod. Thus a grass such as *Phleum pratense*, which has tillers of different sizes, will produce inflorescences of different sizes, the smaller, younger tillers giving substantially smaller heads (Fig. 2.15). Also, a smaller proportion of young tillers produces flowers and the majority overwinter to flower the following summer (Langer 1956). Thus the size structure of the tiller population affects fecundity and, of course, the size

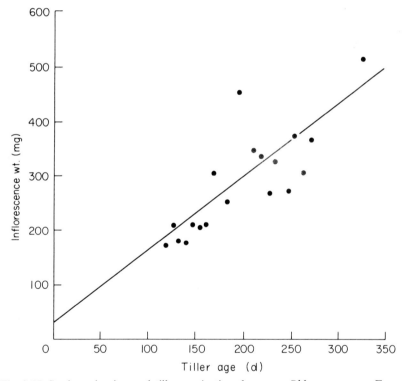

Fig. 2.15 Seed production and tiller age in timothy grass, *Phleum pratense*. Few flowers are produced by tillers younger than 150 days, but there is a linear relationship between age and fecundity thereafter (from Langer 1956).

structure is determined by the feeding preferences and the grazing intensity of the livestock.

2.4.5   Seed size

The dry weight of individual seeds is remarkably constant in many species; indeed, seeds of the carob *Ceratonia siliqua* were used as standards of weight in ancient Greece (but see Lee & Bazzaz 1980). Seed size has presumably evolved as a compromise between increased probability of survival after germination (larger seeds) and increased number of seeds, enhanced probability of surviving predators and improved dispersal (smaller seeds).

Seed size is affected in different ways by different kinds of herbivore attack. When dock plants, *Rumex crispus*, are artificially defoliated there is a substantial reduction in both seed production (from 2611.8 to 1312.9 mg) and seed size (64% of total seed weight passed through a 1.19 mm sieve for defoliated plants but only 7% of the seed of control plants). In contrast, when 75% of the flowers are removed, the average size of the surviving seeds is increased from 1.42 to 1.87 mg (Maun & Cavers 1971). Seed size in dock appears to decrease when the plant is carbohydrate limited and to increase when it is sink limited.

Hendrix (1979) found that seeds of wild parsnip, *Pastinaca sativa*, attacked by the flower-feeding moth *Depressaria pastinacella* were significantly smaller than seeds from control plants. This is because the parsnip plant compensates for seed loss by producing more fruits from its tertiary umbels (see section 2.7) and these seeds are smaller than the seeds of the primary umbel attacked by *Depressaria* (1.19 compared with 3.31 mg).

In their studies on the demography of *Rumex* species in northwest England, Bentley *et al.* (1980) have investigated the effect of defoliation by the leaf beetle *Gastrophysa viridula* on the fecundity of docks. Grazing by the beetle reduced the number of seeds produced by *R. obtusifolius* from 13 700 per plant to 3700 and reduced average seed weight from 1.96 to 1.44 mg. With *R. crispus*, however, the beetle caused no reduction in the number of seeds set, but did reduce mean seed weight. Oddly, these smaller seeds germinated just as well as the larger, and showed no signs of being competitively disadvantaged (see section 2.6.3).

2.4.6   Mast fruiting

The synchronous production of large seed crops in one year followed by a long interval when few if any seeds are set is known as mast fruiting. Mast

Chapter 2

years have been recorded sporadically in forest histories since the middle ages. From the population dynamics standpoint, we need to know whether mast fruiting is a deterministic cycle generated by demographic processes or whether it is merely due to random fluctuations in weather conditions. The weight of evidence strongly favours the latter explanation.

Harper (1977) summarizes the climatic factors that have been correlated with mast years in pines; they include mean early summer temperature three years earlier (negatively associated with cone development in *Pinus resinosa*), early summer irrigation and later summer drought two years earlier (positively correlated with strobilus production in *P. taeda*), early summer rain two years earlier (positively associated with seed production in *P. pinea* and *P. taeda*), mean temperatures two years earlier (positively correlated with seed crop in many pines) and spring droughts in the year of seed ripening (associated with young cone abortion in *P. radiata*). Drought is also important in mast cropping in Douglas fir, *Pseudotsuga menziesii* (Ebell 1967). There is no evidence that masting is caused directly by foliage-feeding herbivores.

The evolution of masting is discussed by Janzen (1976). The main advantage of synchronous seed production is predator satiation; seed predators will be unable to show a sufficiently rapid numerical response to consume the entire crop before a substantial fraction has germinated. Those predators that survive the lean period between seed crops will be swamped by food in the mast year (Silvertown 1980a). *Cassia grandis* produces mature fruits every other year, satiating its specialist bruchid seed feeders that are unable to survive the lean year in large numbers (Janzen 1971a), and the seed weevil *Pseudanthonomus hamamelidii* infests a much lower proportion of the fruits of witch-hazel, *Hamamelis virginiana*, during mast years (De Steven 1981). After fire, vast numbers of seeds are released by *Eucalyptus regnans* which completely satiate the seed-harvesting ants. In normal years there is only sporadic seed fall, of which the ants take about 60% (Ashton 1979).

Pinus edulis produces a heavy cone crop at irregular intervals that both satiates its invertebrate seed and cone predators, and allows efficient foraging by pinon jays, *Gymnorhinus cyanocephalus*, which disperse the seeds and cache them in sites conducive to germination and growth of seedlings (Ligon 1978). The scolytid beetle *Cocotrypes carpophagus* feeds on the seeds of the palm tree *Eutreps globosa* in the rain forests of Puerto Rico. The tree produces seed in synchronous bursts and lives in dense, sometimes pure, stands. The beetle passes through several generations and increases rapidly in numbers following seed fall, but suffers low survival between one seed crop and the next. Sufficient seeds germinate before the beetles build up high populations that recruitment of palms close to the parent is not prevented (Janzen 1972).

The effectiveness of seed predators in eliminating seeds produced out of phase is shown in studies with both mammals and insects. When many of the conifer species in coastal forest failed to produce a cone crop in 1962, squirrels harvested almost every cone produced by *Abies amabilis*, *Picea engelmannii* and *Pseudotsuga menziesii*. In the following year, all species except *Pinus contorta* failed again to produce cones, and the squirrels concentrated in *P. contorta* forests (Smith 1970).

If the seed feeders were very mobile (birds, for example), the synchrony in seed production would have to encompass a very large area in order to be effective. If it were only local, the herbivores would simply move from patch to patch, collecting all the seeds.

2.4.7 Fecundity and plant size variance

We have assumed implicitly that each plant in the population produces the same number of seeds per gram of shoot weight as every other; that there is a linear relationship between size and fecundity.

When the relationship between shoot weight and fecundity is non-linear, we must take account of plant size variance as well as mean plant size in

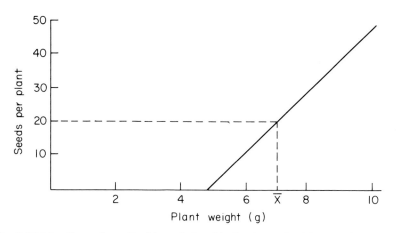

Fig. 2.16 Non-linear fecundity/size relationship in plants. In this case plants less than 5 g produce no seed. The few available data (Fig. 2.10) suggest that such non-linearities may be relatively uncommon.

predicting the seed production of the population as a whole. An example will show this most clearly. Suppose that a population of plants consists of 50 individuals weighing 10 g, and 50 weighing 4 g; average plant weight is 7 g. Now assume that there is a threshold plant size below which no seeds are set (5 g) and that fecundity is a linear function of size above this threshold (Fig. 2.16). The 50 plants weighing 4 g are too small to produce any seed but the 10 g plants each produce 50 giving a total of 2500 seeds.

If we had ignored plant size variance and attempted to predict total seed production from mean plant size, we would have anticipated the 100 7 g plants to produce $100 \times 20 = 2000$ seeds. The error of 25% is not inconsequential and, the steeper the non-linearity in fecundity, the bigger the effect of plant size variance.

When fecundity is a linear function of plant size, however, plant size variance is not important in determining seed set (though it may be extremely important in affecting the pattern of herbivore attack). If a plant produces a seeds per gram of shoot weight and the ith plant weighs W_i g, then total seed production is $\sum_i aW_i$. Under these assumptions, plant size variability makes no difference to seed production. The equation can be written $a \sum_i W_i$ which is seed production g^{-1} times total population dry weight. The effect of non-linear fecundity on population dynamics is discussed in section 4.3.9.

2.5   SEED PREDATORS AND THE DYNAMICS OF THE SEED BANK

Herbivores exert a profound influence on the dynamics of the seed populations both on and under the soil. The impact of seed losses on the dynamics of the mature plants, however, varies greatly from species to species. At one extreme, an isolated, even-aged population of semelparous plants that had no seed bank would become extinct if it were once to suffer 100% seed loss. At the other extreme, a population of long-lived, iteroparous plants could easily survive the occasional loss of all its seeds (Watt 1919, Mellanby 1968). The possession of long-lived seed banks, prolonged or repeated reproduction, the development of mixed age structure populations and the immigration of seed from other habitats, mean that the persistence of many plant populations is not critically dependent upon current seed production (Cohen 1968).

Seeds can escape specialist predators by being too small (either to handle or to allow development within one seed), by germinating too quickly to allow development to occur, by containing high levels of toxins, by being produced in such numbers that predator satiation occurs or by having dispersal agents so efficient that seeds are removed as soon as they are ripe (Janzen 1969). As we have seen, a frequent trait in trees is for the individuals to fruit synchronously with a long gap between each seed crop, so that predator survival from one crop to the next is low and predator satiation occurs in the mast year (Silvertown 1980a).

Very high seed death rates are commonly reported. Seed-feeding bugs killed 96% of the seeds of the introduced African grass *Panicum maximum* growing in the dry lowlands of Costa Rica (Janzen 1969) and the weevil *Apion ulicis* regularly eats more than 90% of the seed crop of gorse *Ulex europaeus* in New Zealand (Liller 1970). Out of 40 leguminous trees belonging to 23 species studied by Janzen (1969) more than half lost upwards of 50% of their seed crop to bruchid beetles and on a third of the trees there was over 80% seed loss.

Vertebrates also take their toll. About half the buttercup, *Ranunculus repens*, seed that Sarukhan (1974) sowed in Welsh grassland was eaten, mainly by voles and field mice to judge by the teeth marks on the empty achenes. In the arid southwestern United States, heteromyid rodents may remove more than 75% of all seeds produced (Hay & Fuller 1981). Seed-harvesting ants, too, can kill large numbers of seeds, and show pronounced preferences for some species over others (Whitford 1978).

Seed removal varies greatly from place to place within the habitat; for example, although the average rate of seed predation for the leguminous shrub *Crotalaria pallida* was about 50%, the loss rate varied from 20% up to 100% for different individual bushes (Moore 1978). Microhabitat differences in predation rate also occur. The beetle *Amblycerus cistelinus* took an average of from 12 to 42% of the seeds of the tree *Guazuma ulmifolia* on progressively drier sites. At any one site, however, between 2 and 22% at the driest site and 7 to 75% of the seeds at the wetter sites were eaten (Janzen 1975).

Gashwiler (1970) calculated the survival of naturally disseminated seed from two species in clear-cut areas of coniferous forest in Oregon (see Table 2.3).

Of the radio-tagged *Picea glauca* seeds put out in clear-cut spruce forest, 50% were lost (mainly to mice and chipmunks) when the seeds were distributed in summer, while only 20% were destroyed if the seeds were put out on the snow. The difference is probably due to burial of the winter seed during freeze-thaw and to extended availability of summer seed when dry

Table 2.3 Percentages of natural seed fall from Douglas fir, *Pseudotsuga menziesii*, and western hemlock, *Tsuga heterophylla*, lost to different causes, and percentage survival (from Gashwiler 1970).

	Douglas fir	Western hemlock
Mice and shrews (mostly deermice)	41	22
Birds and chipmunks (mostly birds)	24	3
Other agents (insects, disease, non-viability, etc.)	25	53
Survival	10	22

conditions delay germination (Radvanyi 1970). Insects took about 10% of sown Douglas fir, *Pseudotsuga menziesii*, seed, rodents ate 14% and birds removed a further 4% (Lawrence & Rediske 1962).

Providing alternative, preferred foods for seed-feeding rodents like *Peromyscus maniculatus* can lead to great reductions in the numbers of valuable seeds eaten; for example, survival of Douglas fir seed can be improved from 5% to 50% by providing sunflower seeds in clear-cut areas of forest (Sullivan 1979).

Ash trees, *Fraxinus excelsior*, in Britain are conspicuous in winter by the large bundles of fruits (keys) retained on their leafless branches. Seed fall is spread over a period of almost a year from September to August. After falling to the ground, the seed faces a period of 18 to 20 months dormancy, during which it is prey to small mammals like *Apodemus sylvaticus* and *Clethrionomys glareolus* and insects like the moth *Pseudargyrotozoa conwagana*. Survival rates for the seed over 4 years are shown in Table 2.4. Losses to *P. conwagana* are negatively density dependent and range from 15 to 75%, while losses to small mammals range from 25 to 75% and are positively density dependent. Despite a low survival rate to germination (2.5%), peak seedling numbers followed the peak seed crop with more than 30 young trees m^{-2} in 1971, consistent with the predator satiation hypothesis (Gardner 1977).

Seed predation can affect not only the abundance, but also the species composition of the seedling population. In a diverse community of winter annuals in the Sonoran Desert, seed predation considerably reduced plant

Table 2.4 Seed production and the fate of seeds of ash, *Fraxinus excelsior* in Derbyshire woodlands (Gardner 1977).
(a) The mean numbers per m² of fruit, seed and seedlings recorded in four seasons.

	1966	1967	1968	1969
Fruit formed	134	0.33	30.1	1283
Full-grown seed	133	0.33	26.9	1193
Good seed falling	101	0.08	16.2	998
Seedlings in permanent quadrats two years later	6.1	0	1.6	31.6

(b) Losses of fruit and seed and the proportion germinating in four seasons at the study site; results given as a percentage of total fruit produced.

	1966	1967	1968	1969
Undeveloped seed	1.3	0	10.6	7.0
Seed infested by moth	23.9	75.0	35.6	15.2
Loss of good seed to small mammals	70.2	25.0	48.5	75.3
Germination	4.5	0	5.3	2.5

densities, but ants and rodents had quite different effects on recruitment. Ants increased species diversity of annuals by differentially harvesting the numerically dominant species, while rodents preyed selectively on species which dominated the community in terms of biomass (Inouye *et al.* 1980).

Density-dependent seed predation can be a potent force in establishing and maintaining the spatial patterns of plants. The perennial *Mirabilis hirsuta* is a denizen of dry prairies whose seeds fall in two kinds of habitat. In undisturbed prairie establishment is negligible because of competition from mature plants but on mounds of the American badger, *Taxidea taxus*, the disturbed soil makes an ideal substrate for development (Platt 1975). *Mirabilis* seeds are large and tend to form high-density clumps close to parent plants, but because of density-dependent predation of the seeds by ants and mice, survival is usually zero in these high-density patches. Predation is much less intense at low seed densities where seedling survival is about 50% (Platt 1976). Similarly, clumped individuals of the desert annual

Ibicella parodii suffered 68% grazing mortality while plants more than 10 m from their nearest neighbour suffered a 35% death rate (Orians in Rhoades & Cates 1976).

In contrast, the prairie legume *Astragalus canadensis* suffers a lower rate of seed predation from clumps than from isolated plants. This, coupled with increased pollination success in high-density patches, tends to maintain a clumped distribution of the plants and to ensure that the rate of establishment of new clumps is low (Platt *et al.* 1974).

Wilson and Janzen (1972) used husked seeds of the forest palm *Scheelea rostrata* to test the hypothesis that seeds were more likely to escape their predators at a distance from the parent tree than when close to it. They found that piles of 50 nuts directly below the tree suffered the same rate of bruchid attack (34%) as similar piles 8 m away. For this tree, the distance hypothesis is apparently refuted. However, there was a significantly lower attack on single seeds compared to clumped ones; only 7 out of 164 isolated nuts (82 placed above and 82 below the litter) were discovered by beetles. It is also clear that the rate of attack is a non-linear function of the size of the pile. The largest pile of 308 nuts suffered only 26% attack and shows some evidence of predator satiation.

It should be borne in mind when interpreting this much-quoted experiment that the fruits had to be cleaned of their husks and pulpy exocarps by hand, otherwise the forest floor rodents would have carried them all away! If the seeds smell less attractive to the beetles in this state, the husking may have made the discovery rates of the isolated seeds artificially low. Seeds naturally husked by rodents had been found to suffer 77% mortality from bruchids in an earlier study (Janzen 1972).

There are examples of distance-responsive seed predation, however. Seeds of the solanaceous desert ephemeral *Datura discolor* are killed by nocturnal rodents like *Dipodomys merriami, Perognathus baileyi* and *Peromyscus eremicus*. Seeds immediately below the parent plant have a life expectancy of about 3 days while those only 2 m away from the canopy have an expectancy of almost 30 days. The rate of predation by rodents in areas away from cover was not increased by increasing the density of seeds.

The heavy fruits of *Datura* tend naturally to fall to the ground directly below the parent plant where they are most at risk from the rodents. The plant survives because its seeds are dispersed by harvester ants (*Pogonomyrmex californicus* and *Veromessor pergandei*) beyond the zone of rodent predation. The ants carry off the fruits to their nests where they strip off the nutritious food bodies. The seeds are not eaten by the ants but are cast out of the nest on to the midden. From here, wind or water carry the seeds away to a germination site (O'Dowd & Hay 1980).

2.5.1 Seed bank

Many annual plants maintain a dormant seed population in the soil many times greater than the peak density of vegetative individuals; for example, in just one square metre *Poa annua* had 31 300 viable seeds, *Papaver rhoeas* 22 750 and *Striga asiatica* 34 200 (Sarukhan 1974). The composition and relative abundance of species in the seed bank is often quite different from the current plant community and may even differ from the community which develops by germination following soil disturbance (Thompson & Grime 1979).

The bank consists of new seed, which will germinate as soon as conditions permit, plus seed in various kinds of dormancy. Harper (1959) recognizes three categories of seed dormancy: innate dormancy due to the presence of inhibitors, which can be overcome by a seasonal stimulus like photoperiod or thermoperiod; induced dormancy, which lasts after the stimulus that produced it has ceased to act (e.g. *Brassica alba* seeds exposed to CO_2 suffered induced dormancy which could only be broken by drying and rewetting the seeds or by removing the testa); and enforced dormancy, which is imposed by a factor like cold or darkness and lasts only as long as the factor acts on the seed.

The potential longevity of seeds is staggering. Seeds of *Lupinus arcticus* taken from the frozen silt at Miller Creek, Yukon were successfully germinated; radio-carbon dating showed the seeds to be 10 000 years old. Weed seeds of *Chenopodium album* and *Spergula arvensis* have been germinated from archaeological excavations of 1700-year-old strata (Duncalf 1976).

The existence of a large and long-lived seed bank can make for great difficulties in weed control. Attempts to control gorse, *Ulex europaeus*, with animals like the weevil *Apion ulicis* which eats the seed while still on the plant, are almost certainly doomed to failure. Seed densities as high as 10 000 m^{-2} have been recorded in the loose litter directly below the parent bushes, while less than 1 m away there were only 20 m^{-2}, and no seeds at all were found more than 2 m away from a bush (Chater 1931). Land in Cumbria which had been cleared of gorse and managed as a permanent pasture for 25 years, was ploughed up for recultivation and was immediately infested with gorse seedlings that germinated from the soil bank. However, because of the limited distance of seed dispersal by dehiscence of its pods, and the localized litter conditions in which large numbers of seed can escape predation, the seed bank of gorse is usually extremely patchy.

In arable weeds, too, the seed bank varies greatly from place to place within the field with differences in the conditions favouring seed production and preservation, and from field to field with differences in the history of cultivation and management. Weed seeds, for instance, had an average half-life of about 2 years in cultivated soil compared to about 6 years in uncultivated soil (Roberts & Feast 1973).

The seeds of light-demanding plants of woodland clearings can survive over 30 years in the soil below shade-casting trees, germinating rapidly when the crop is felled (Hill & Stevens 1981). Rather few seeds survive burial in woodland soils for more than 70 years (Marks 1974). The annual seed rain in a North American tall-grass prairie was 19 700 seeds m^{-2} compared to values of 166 and 188 seeds m^{-2} obtained in forest understoreys. In the grassland, there is a 71% reduction in seed density between the seed rain and the seed bank in the soil. Also, the seed rain was much more similar to current vegetation in the relative abundances of species than was the seed bank (Rabinowitz & Rapp 1980).

Some plants maintain an aerial rather than a soil seed bank, retaining their fruits on the branches, sometimes for may years. Lodgepole pine, *Pinus contorta*, retains its cones until fire sweeps through the forest. The heat of the fire causes the cone scales to open and the seeds fall out into a ready-made weed-free and ash-fertilized seed bed. Such pulses of recruitment after fires lead to a mosaic of single-aged stands of virtual monocultures of lodgepole (Clements 1910). *Eucalyptus regnans* maintains a huge aerial seed bank. This, too, is released by fire, when the massive seed fall satiates the ants and other seed predators (Ashton 1979).

2.5.2   Seed dispersal

Dispersal of seed reduces the likelihood of competition with the parent and with fellow seedlings and carries the seed to an area where specialist seed predators are likely to be less numerous.

Immigration of seed from other patches of the habitat has been shown to be a potent stabilizing factor in models of population dynamics (Levin 1974, Roff 1974). Over a large region containing many forest patches, for example, rare tree species might persist as fugitives, with some seeds colonizing new patches before the species temporarily disappears from the source patch. In such a case, the persistence of a species may depend critically on the sizes of the inter-patch distances as well as on the size of the patches (DeAngelis *et al.* 1979).

In many ways, dispersal is an alternative to maintaing a seed bank; some weedy species like *Spergula arvensis* or *Papaver rhoeas* persist because of their enormous seed banks while others like *Senecio vulgaris* or *Epilobium adenocaulon* survive by efficient dispersal.

The predominant modes of fruit dispersal are different in each layer of the plant community. In the ground layer where furry animals rove about, many fruits have hooks (e.g. *Mercurialis perennis, Galium aparine*); in the shrub layer where the air is still and the plants are out of reach of ground-dwelling animals, many fruits are fleshy and attractive to birds (*Crataegus monogyna, Rhamnus catharticus*); in the canopy top where air movement is more free, the wind is employed as a means of dispersal (*Pinus sylvestris, Betula pendula*). Naturally, there are a great many exceptions to the scheme; for example, many dominant trees have large fruits that cannot be dispersed by wind (*Quercus*) and many forest floor species do not rely on animals (*Geranium, Epilobium*).

The herbivores that eat fruits have two kinds of action. In most cases the fruit is eaten and the seed is either discarded, or passes through the gut without harm. In trials with four species of seed removed from the dung of baboons, *Papio anubis*, three showed significantly higher germination than fresh control seeds and one was unaffected (Lieberman *et al.* 1979). Passage of seeds through the gut of fruit-feeding birds has often been shown to increase the germination rate (Krefting & Roe 1949) but occasionally to decrease it (Livingston 1972).

The seeds of some plant species may depend absolutely on passage through the gut of a fruit-feeding animal before germination can occur. Very close mutalistic association in germination enhancement or dispersal clearly makes the plant vulnerable to the extinction of the herbivore. From a unique study of the tree *Calvaria major* on the island of Mauritius, Temple (1977) suggests that there has been no recruitment of the plant in the last 300 years because its seeds needed to be processed by the dodo *Raphus cucullatus*. The extinction of the bird had led the plant to the verge of extinction. Temple fed 17 of its seeds to domestic turkeys; seven were crushed in the bird's gizzard, but three of the remaining 10 subsequently germinated when planted in a nursery. These may well have been the first *Calvaria* seeds to germinate for more than 300 years.

In other cases, animals that normally kill the seeds are also responsible for their dispersal. The most familiar example is provided by the birds and mammals that eat acorns. Squirrels and jays are the most numerous seed feeders, and carry seeds long distances to bury them in caches (Bossema 1968, Smith 1975). Despite killing many seeds, sufficient caches remain undiscovered to ensure oak germination.

Studies of acorn dispersal have shown that a single flock of jays can move tens of thousands of seeds over distances of several kilometres in one season (Chettleburgh 1952). While such dispersal is likely to play an important role in oak dynamics, it is unlikely that oak recruitment depends absolutely on the activities of specialist dispersing animals. In the tropics, however, obligate plant–disperser relationships may be more common (e.g. Howe 1977).

Scatter-hoarding and caching are extremes of dispersal behaviour exhibited by different kinds of vertebrates. Scatter-hoarding squirrels like *Sciurus carolinensis* and *S. niger* take aggregated nuts and spread them out; they benefit because the nuts are then less vulnerable to predation by deer and to visual foragers like jays. Scattered buried nuts may also be safer from detection by other squirrels (Smith & Follmer 1972).

Seed-caching herbivores, in contrast, take relatively scattered seeds and aggregate them. About 90% of the seedlings and about 50% of mature shrubs of bitterbrush, *Purshia tridentata*, arise from undiscovered rodent caches (West 1968). The intriguing difference in survival rate between seedlings in caches and isolated individuals is unexplained. White-footed mice, *Peromyscus leucopus*, and red-backed voles, *Clethrionymus gapperi*, make caches of 20–30 pine seeds under the litter of pine needles. When seed crops are high, sufficient caches escape winter feeding by the rodents and some of the seeds within them germinate (Abbott & Quink 1970).

If seed predators are generally density responsive rather than distance responsive (see section 2.5.1), this will have consequences for the spacing of adult plants; for example young trees with small seed crops will tend to produce recruits closer to them than will large trees because specialist predators will give up searching for the sparse seeds closer to the parent plant. This effect will be further enhanced by the fact that seeds from a taller tree will tend naturally to be dispersed further.

Any mutation which tends to improve seed dispersal will reduce seed density close to the parent plant and increase it further away. If density-dependent seed predation is an important agent of natural selection, it might be predicted that seed dispersal would evolve towards one of two extremes.

Very poor seed dispersal would lead to seed densities close to the parent that may be high enough to cause predator satiation; very efficient seed dispersal would spread the seed so thinly that no specialist predator could survive on it. An intermediate level of dispersal ability would cause the plant to suffer doubly; it would not satiate the predators close to the parent and a substantial fraction of the seeds would be distributed above the predators' giving up threshold (Hubbell 1980).

2.6 HERBIVORES AND SEEDLING ESTABLISHMENT

The ecology of seed establishment has been understood since biblical times; 'some seeds fell by the wayside, and the fowls came and devoured them up: Some fell upon stony places, where they had not much earth: and forthwith they sprung up, because they had no deepness of earth: And when the sun was up they were scorched; and because they had no root they withered away. And some fell among thorns; and the thorns sprung up, and choked them: But others fell into good ground, and brought forth fruit, some an hundred fold, some sixtyfold, some thirtyfold' (Matthew 13:4). As well as cataloguing the hazards facing the seed (predation, unsuitable microsites and plant competition), the parable provides the first estimate of the intrinsic rate of increase of annual crop plants ($3.4 < r < 4.6$)!

The density of seedlings will most often be determined by the number of microsites available; in most parts of a plant's range, seeds will outnumber microsites. The availability of microsites places a ceiling on the population density of seedlings (see section 2.6.1).

Sometimes, however, seed-feeding animals will consume so much seed that potential microsites remain unfilled or are filled by competing plant species, in which case recruitment is predator limited (see section 2.6.2).

Finally, when plant growth has been slow because of drought or cold or because of intense grazing or competition, seed production may be so substantially reduced that there are insufficient seeds to fill the available microsites. Such seed-limited recruitment is only likely in those species that neither maintain a bank of persistent seeds in the soil nor are supported by immigration of seed from other habitats.

Once a plant is established beyond the stage where normal levels of herbivore feeding are likely to kill it, it may have tenure over that site for a long time. The pre-emption of sites by long-lived iteroparous plants can have a substantial stabilizing effect on local population dynamics by ensuring a more or less regular rain of seed over an extended period.

Site pre-emption also alters the number and distribution of establishment microsites; once an acorn is established, the surrounding ground is destined to be oak-shaded for several hundred years. While the physical attributes of the soil microsites may be unaltered, their light and chemical regimes are changed dramatically.

2.6.1 Germination microsites

The precise properties of a germination microsite differ from species to species, but it must provide all the requisites of germination (correct temperature, water, light and oxygen conditions), root penetration (appropriate soil particle size, compaction, aeration, pH, mineral availability and wetness) and leaf expansion (suitable temperature, light, humidity and exposure). Since the number of microsites can only be defined in relation to the requirements of a particular species, there can be no single parameter such as 'microsite density'. The maximum number of seeds germinating on a given area will be determined by the size and shape of the seeds relative to the surface roughness of the soil, and by many other factors (Harper *et al.* 1970). Even the position of the seed on the soil surface affects the probability of germination (Sheldon 1974). Thus while the number of microsites may well limit the rate of seedling recruitment, the number is probably impossible to estimate (other than by sowing an excess of seed).

Germination of seed is usually density independent, but there are examples of both enhanced and reduced germination at high seed densities (Linhart 1976). Negative density dependence through the production of germination-inhibiting chemicals is potentially regulating (Palmblad 1968) and may be of benefit to weedy plants by maintaining a bank of viable seed, and reducing the risks of damping off that might occur in high-density seedling populations. The advantages of positively density dependent germination are much less obvious, but they may include improved germination synchrony, leading to improved microclimate, or to the exclusion of competitors through site pre-emption (Linhart 1976).

The role of herbivores in determining the abundance and distribution of germination microsites is shown clearly in Cameron's work (1935) on ragwort *Senecio jacobaea* in southern English grasslands. In well-managed pastures on fertile soils, ragwort fails completely to become established even when seed is added at a rate of $1000 \, \text{m}^{-2}$. In pastures overgrazed by dairy cattle, however, the plant establishes freely with seedling densities of up to

$25 \, \text{m}^{-2}$. The greatest rates of establishment are found in the vicinity of rabbit burrows, where the bare, freshly disturbed ground makes an ideal seedbed and seedling densities of $570 \, \text{m}^{-2}$ are commonplace.

Cameron performed establishment experiments with ragwort under six different sets of conditions, sowing $1000 \, \text{m}^{-2}$ in each case; the number of young plants observed after two months were:

Treatment	Seedlings
long grass	none
long grass cut short	20
close-grazed, continuous turf	none
hard compacted soil	200
ordinary tilth, seeds uncovered	530
ordinary tilth, seeds covered	550

It is clear that ragwort requires disturbances of the soil surface and low levels of plant competition for successful establishment; both these conditions are provided by herbivores. Heavy grazing by cattle on extensive rangeland with poor soils allows the development of large areas of ragwort, which establishes in the gaps in the sward where competition is slight. Rabbit activity in otherwise well-managed grassland produces localized patches of ragwort that may do little direct harm, but which serve as a source of seed for the infestation of other areas.

In certain arid areas, the most palatable rangeland grasses are annuals. It has been a long-established range management tradition in these cases to recommend that grazing pressure is relaxed during grass flowering to ensure sufficient seed set for the following season. Experimental work on dry ranges in Israel (Noy Meir, pers. comm.) has involved excluding sheep from areas at different times in the growing season. While grazing during the flowering period did reduce seed production by about 40%, the sheep trampled many seeds into the soil. In the ungrazed plots seed production was higher, but most of the seed that fell to the ground was carried off by ants. The number of seedlings at the beginning of the following growing season was, therefore, about the same in the grazed and ungrazed areas; the soil surface disturbance by the sheep apparently compensates for the reduced seed production.

Soil surface roughness is probably the best, simple correlate of microsite

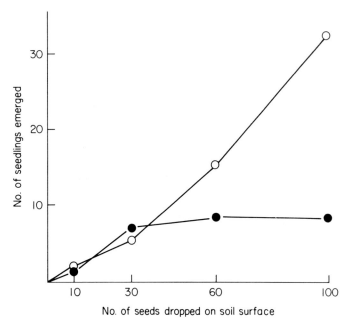

Fig. 2.17 Plant recruitment and germination microsites. On rough soils (○—○) the recruitment of *Bromus* is seed-limited, but on a hard surface (●—●) there are too few microsites (Harper 1961).

availability. The number of *Bromus* seedlings emerged as a function of sowing density is shown for two different soils (Fig. 2.17); on the dry, cracked surface seedling establishment is limited by microsites whereas on the broken surface it is limited by seed availability (Harper 1961).

The species composition of communities of competing annual plants can be greatly influenced by soil surface heterogeneity; differences in the position of seeds on microtopographic peaks and troughs can reverse the competitive advantage that is observed on homogeneous soils (Pemadasa 1976).

The size of the competitor-free space is also crucial. Although germination rate was unaffected, seedling survival was closely dependent upon the size of the gap in the canopy of the grass *Poa compressa* in which seeds of *Melilotus alba* and *M. officinalis* were planted. In 3 cm diameter holes, 60% of the seeds produced surviving seedlings after 28 days, whereas only 32% of the seeds in 0.5 cm diameter gaps survived (Caruso 1970). The gaps in the sward left by dying teasel *Dipsacus sylvestris* rosettes provide competition-free establishment sites for teasel seedlings and for other species (Werner 1977).

It may be that increased germination in fluctuating temperatures is the mechanism whereby certain species detect gaps in the plant cover that are sufficiently large to allow successful establishment (Thompson *et al.* 1977).

2.6.2   Seedling predation

If herbivores have a significant effect in regulating plant numbers via mortality it is likely that it is expressed through the deaths of seedlings rather than of mature individuals. The plant is most vulnerable to suppression or defoliation when its seed's reserves have run out and it first relies on the products of its own photosynthesis for growth and survival. Darwin (1859) recognized the importance of herbivores at this stage in plant development, pointing out that 'Seedlings.... are destroyed in vast numbers by various enemies; for instance, on a piece of ground three feet long and two wide, dug and cleared, and where there could be no choking from other plants, I marked all the seedlings of our native weeds as they came up, and out of the 357 no less than 295 were destroyed, chiefly by slugs and insects'. There is no reason to believe that a 78% death rate such as this is atypical of early successional communities.

The death rate of seedlings will often be a function of their distance from mature plants. Shaw (1974) found that the survival rate of oak *Quercus robur* seedlings was very low underneath the canopy of mature trees because they were so regularly defoliated by larval lepidoptera that fell from the foliage above. Similarly, Lemen (1981) found that elm *Ulmus parvifolia* seedlings directly under adult trees suffered 580 times the attack by leaf beetles *Pyrrhalta luteola* than more isolated seedlings. Janzen (1971b) detected greater damage by noctuid caterpillars to seedlings of the vine *Dioclea megacarpa* under parent plants than further away.

Seedlings of the giant saguaro cactus, *Cereus giganteus,* are vulnerable to predation by insects, rodents and rabbits, as well as to hazards of desiccation and uprooting. Insects killed 3 to 30% of the tiny plants and rodents took 6 to 26% in the first year of life (Steenbergh & Lowe 1969). No transplanted seedlings survived the first year outside grazing exclosures, but 12% survived within the fence. Even in the exclosures, only 5% of the plants survived 5 years (Turner *et al.* 1969).

Connell (1971) suggested that seedling predation in tropical forests might be sufficiently intense to prevent the formation of single-species stands; 'each tree species has host-specific enemies which attack it and any of its offspring which are close to the parent. The healthy parent tree supports a large population of these enemies without itself being killed, but the seedlings, whose growth is suppressed in the heavy shade, succumb to the attack of insects and other enemies which come from the parent tree itself or the soil below it'. In rain forest in Queensland he found that the death rate of small saplings was 38% when the nearest neighbour was of the same species but only 18% when it was of another. Large saplings had the same death rate irrespective of the species of their neighbours.

Subsequent experiments on density-dependent predation on seeds and seedlings, however, failed to provide backing for Connell's hypothesis; only one species out of six examined showed greater mortality near the parent tree than further away (Connell 1979; see also section 5.1.2).

2.6.3   Seed size and establishment

Changes in seed size may bring about changes in both the rate and timing of germination and in subsequent growth rate and competitive ability (Black 1957). Defoliated plants of *Rumex crispus* produced smaller seeds that germinated to produce shorter seedlings with lower dry weights of leaf, hypocotyl and root (Maun & Cavers 1971). Smaller seedlings of *Amaranthus retroflexus* were much less successful than larger ones in competition with *Lotus corniculatus* (Schreiber 1967).

The smaller seeds produced by defoliated dock plants germinated more quickly than control plants, whereas the heavier seeds from plants whence 75% of the flowers had been removed showed higher dormancy than control plants (probably because their seed coats were thicker; Maun & Cavers 1971). There is no field evidence on the fate of seedlings from large and from small dock seeds, although there must be a complex balance between the advantages of early germination (in giving small seeds a head start), and the increased competitive ability of seedlings with a larger carbohydrate reserve. It is possible that increased dormancy following defloration is advantageous as a predator-escape mechanism; large numbers of the herbivores that caused the serious defloration will starve in the next generation when few dock plants germinate.

Many trees in tropical rain forest have large seeds and rapid germination. Ng (1978) found that 75% of species in Malayan forest have seeds longer than 1 cm and 65% show rapid epigeal germination. The large seeds produce seedlings which are both more shade tolerant and more resistant to fungal attack. Out of 23 species of early successional trees from clearings and forest edges, he found all of them to have seeds smaller than 1 cm and all but one to show epigeal germination. These trees produced larger numbers of seeds,

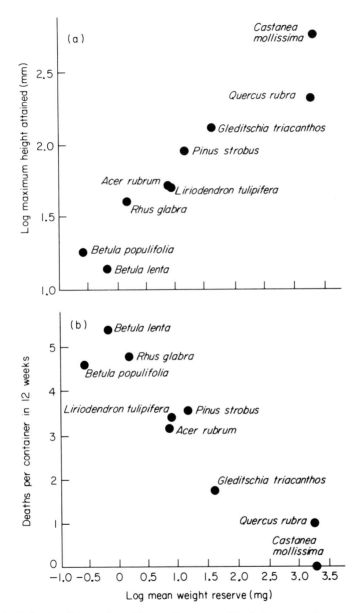

Fig. 2.18 Seed weight and seedling performance. The height growth and death rate of various tree species grown in deep shade (Grime & Jeffrey 1965).

dispersed them over greater distances and established better in compacted ground (they could obtain a root-hold in tiny cracks).

Clearly the costs of increased seed size in terms of their greater attractiveness to herbivores, are outweighed by the benefits of increased competitive ability in these rain forest trees (see also Fig. 2.18).

2.7 PLANT COMPENSATION

The relationship between the rate of animal feeding and the rate of weight loss by the plant is of critical importance in plant population dynamics. When the rate of plant weight loss is less than the rate of animal feeding, we say that the plant compensates for herbivore feeding. When the rate of weight loss by the plant exceeds the rate of animal feeding we say that the herbivore damages the plant (it interferes with the productive machinery in some way). It is possible that the rate of plant weight loss is exactly equal to the rate of herbivore feeding (see Fig. 2.19).

Plant compensation can act at the level of the individual plant or at the level of the plant population. Populations compensate when herbivore

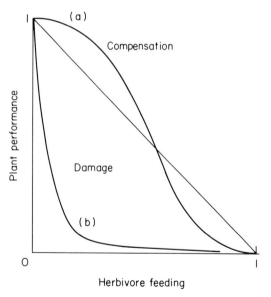

Fig. 2.19 Herbivore feeding and plant performance. Compensation occurs when low levels of feeding cause no loss in performance (a) or when regrowth after feeding replaces destroyed tissues (Fig. 2.12(a)). Damage occurs when plant performance declines faster than linearly with herbivore feeding (b).

attack on one individual allows another individual to grow more rapidly. Such relaxation of competition is commonly observed in dense agricultural crops and in forests, where total yield is depressed by less than the sum of herbivore feeding because of 'gap filling'. An individual plant compensates when, for example, herbivore feeding removes shaded leaves or causes a shift in the pattern carbohydrate distribution so that the rate of net photosynthesis is maintained (or even increased in some cases; Harris 1947a).

Ideally, one would measure the impact of herbivore on plant fitness; on the numbers of offspring per plant that survive to reproduce. Seed production may often give some idea of fitness, but the two are by no means synonymous (Salisbury 1942, 1961). Plants which have immensely high fecundities are sometimes patently 'unfit'; for instance, many of Britain's rarest wildflowers are orchids, which have huge potential seed sets (Summerhayes 1951). Again, the survival rate of plants produced in high-seed crops may be lower (when intraspecific competition is severe and debilitating, or when there is density-dependent seed predation) or higher (when predator satiation is necessary for successful establishment) than that of plants from small seed crops.

We must be wary, therefore, of attempting to predict the impact of herbivory on plant populations. Reductions in seed production may not always lead to reductions in fitness; the impact of herbivore attack on the survival rate of mature plants, or on the survival of the offspring via effects on seed size, may be just as important as any effects on fecundity.

Compensation can occur through five main processes:

(1) reduced competition with other plants (relaxed density dependence);

(2) increased unit leaf rate (net photosynthesis per unit leaf area);

(3) mobilization of stored carbohydrate or protein reserves to form regrowth tissues;

(4) altered patterns of distribution of photosynthate;

(5) a reduction in the natural rate of mortality of plant parts.

The extent to which a plant can compensate for damage may vary from one physiological process to another; for example, bean plants subject to 50% root pruning can compensate fully in their rate of transpiration, but they suffer a 50% reduction in phosphate and nitrate uptake rates (Brouwer 1963, see also Fig. 2.20). Also, plants from different habitats tend to show compensation and regrowth to different degrees. In comparing trees of nutrient-poor and nutrient-rich sites, it is typical to find that plants from nutrient-poor sites show poor regrowth but high concentrations of defensive chemicals, while on richer soils, the trees have lower concentrations of

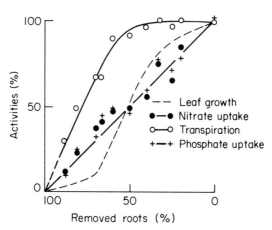

Fig. 2.20 Different physiological processes are compensated to different degrees. Root pruning of bean plants causes a linear decline in nitrate uptake, but the plant can compensate in its rate of transpiration for up to 50% root loss (Brouwer 1962).

secondary substances and show rapid regrowth from large nutrient and carbohydrate reserves (Bryant & Kuropat 1980).

2.7.1   Reduced competition

Perhaps the most widespread form of compensation occurs when herbivore feeding reduces the severity of competition between plants. It must be stressed that this kind of compensation is only likely to be important in dense monocultures or in sparse populations of high species diversity. Defoliation of one species in a dense, multi-species community like a forest, will lead to suppression of the attacked individuals by their more vigorous unattacked neighbours and to a decrease in productivity and population density of the plant compared with its competitors.

In dense monocultures, plant productivity is an n-shaped function of standing crop biomass (Fig. 2.21). At low densities, increasing the amount of leaf increases the amount of light intercepted, and productivity increases. Beyond a certain point (known as the optimal leaf area index LAI_{opt}), increasing leafiness leads to reduced productivity because every gram of new tissue added respires more carbohydrate than it synthesizes; a fraction of the leaves is below the compensation point and is a net drain on the plant's carbohydrate supplies. Black (1964) quantified these yield losses for *Trifolium subterraneum* growing in Australia. The average potential production determined by light, temperature and water availability was 27 t ha^{-1} dry matter, but an unmanaged sward produced only 8 t ha^{-1}. The lost 19 t is made up of 3 t lost because leaf area was too low early in the season and 16 t lost because of mutual shading later in the year. Much of this 16 t ha^{-1} loss should be obtainable by judicious harvesting to maintain the crop close to its optimal leaf area index.

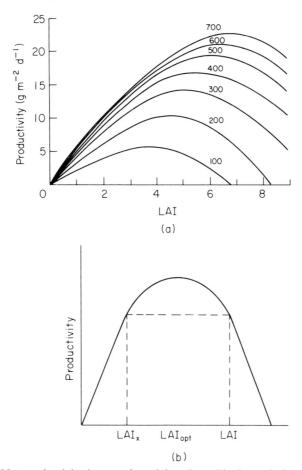

Fig. 2.21 Net productivity is an n-shaped function of leaf area index (LAI) in subterranean clover. (a). Optimum leaf area index increases with light intensity (cal cm^{-2} d^{-1}) since light penetrates deeper into the canopy and a higher fraction of the leaves remains above the compensation point (Black 1964). (b) At LAI, any reduction in leaf area that leaves a residual area greater than LAI$_x$ will cause an increase in productivity. Productivity is maximized when the leaf area is held at LAI$_{opt}$.

In any plant population above its optimal leaf area index, removal of a small amount of leaf will increase the rate of net production. The removal of leaves will continue to benefit productivity until leaf area index is reduced below LAI_x (see Fig. 2.21); productivity is greatest when the leaf area index is held at $LAI_{opt.}$ by frequent, small defoliations (Black 1964).

White clover has its optimum LAI at 3.5 whereas most grasses peak in net productivity at much higher values (e.g. *Lolium perenne* at 7.1; Brougham 1958). In subterranean clover, Davidson and Donald (1958) found that any defoliation towards the optimum leaf area of 4.5 increased the rate of leaf production, whereas defoliation of swards already below 4.5 led to drastic reduction in the subsequent rate of leaf production.

Optimum leaf area index increases with light intensity, because proportionately more of the leaves are above their compensation point in bright light; $LAI_{opt.}$ for subterranean clover, for example, is 4 at light intensities of 100 cal cm^{-2} d^{-1} and 6 at 400 cal cm^{-2} d^{-1} (Black 1964).

Almost all the examples of compensation via reduced competition come from work with agricultural pests (reviewed by Bardner & Fletcher 1974). Jones *et al.* (1955) working with sugar beet found that 'despite the delicacy of the seedling stage, sugar beet has considerable powers of recuperation from defoliation and loss of stand. Thus, the loss of half the leaf area in the 4- or 8-leaf stage caused an average decrease in root yield of only 5% and complete defoliation no more than 30%. Similarly, loss of half the initial plant population caused an average reduction in root yield of 10%, while loss of three-quarters gave a yield slightly superior to that obtained by re-sowing'. Similarly, Hussey and Parr (1963) found that the yield of cucumbers was not significantly reduced until red spider mites, *Tetranychus urticae*, had destroyed more than 30% of the leaves.

Compensation may also occur due to reduced competition between leaves on the same plant (see section 2.3.4); for example, the weevil *Phyllobius argentatus* feeds mainly on the lower, shaded leaves of beech *Fagus sylvatica* towards the centre of the canopy, and probably has little impact on net production by the tree as a whole (Nielsen & Ejlersen 1977).

The spatial pattern of herbivore attack is critical in determining the extent to which compensation can occur. If damage is spread evenly over the plant population as a whole (as it was in Jones *et al.*'s (1955) regular thinning of sugar beet), the neighbours of attacked plants will be released from competition and will rapidly grow to take up the light and root volume no longer exploited by the attacked individuals. When the damage occurs in clumps, however, the gap produced may be so large that no amount of enhanced growth by the surviving plants on the periphery can fill it up. Since most pest-insect feeding is concentrated in patches (see section 3.3.3), yield losses could be substantially more severe than predicted on the assumption of even attack. However, compensation has been reported in gappy populations of rice (Ishii *et al.* 1972), wheat (Busch & Ergun 1973) and potato (Hirst *et al.* 1973).

Reduced competition leads to relaxed density dependence in either plant death rate or in the growth rate of individual plants (size plasticity). When herbivore attack occurs before density dependence has affected the growth or death rates of the plants, there is substantial scope for competition. When a forester removes the smaller trees from a forest as thinnings, he reduces the death rate of those that remain because he is killing trees that would have died in any case (Iyer & Dosen 1974). When herbivore attack occurs after density-dependent processes have determined plant size, fecundity and survivorship, then compensation is unlikely to be important. Reductions in the size of the plants or in the density of the population will simply serve to reduce the growth rate of the population further still.

2.7.2 Timing and compensation

Synchrony between plant phenology and herbivore feeding affects both the numerical response of the herbivore and the potential for compensation by the plant; for example, defoliating caterpillars that emerge too early starve for want of opened buds; those that emerge too late find the leaves so tough and so suffused with tannins that the insects starve in the midst of apparent plenty (see section 3.7.2). A plant stripped of its mature fruits towards the end of the season may set no seed at all, whereas flowers lost early in the year may be replaced by regrowth. Conversely, a plant stripped of its old leaves may be completely unaffected by herbivore attack and a plant defoliated early may be capable of producing a new set of leaves so that its growth is little affected. However, a plant that loses its mature leaves at the stage when fruit production is about to peak may suffer substantially reduced seed production. Thus any compensatory response by individual plants will be conditioned both by how much tissue is removed and by when it is taken.

One of the most straightforward ways of reducing pest damage to crops is to contrive that the vulnerable stage of crop growth does not coincide with peak pest immigration; for example, in the United States, wheat planted too early in the autumn suffers about 40% infestation with hessian fly, *Mayetiola destructor*, whereas seed planted later emerges after the oviposition period of the fly has ended, and attack rate drops to only 4%. Thus an increase in yield of 25% is obtained simply by adjusting sowing date (Metcalf & Flint 1951).

Random dis-synchrony between natural populations of plant and herbivore due to weather conditions can be a potent factor in limiting the damage done to the plants. Grassland Homoptera, for example, only reach high densities when their reproductive period coincides with the brief spell of high nitrogen concentration in the grasses (Prestidge & McNeill 1983; see also section 3.6.5).

2.7.3   Increased unit leaf rate

Unit leaf rate is a compound parameter that subsumes many of the factors affecting photosynthetic rate (leaf age, detailed canopy structure, etc.); it is defined as the rate of production of dry matter per unit area of green leaf. Our interest in this parameter is that it can be affected by the rate, timing and position of herbivore feeding.

The ULR of fully expanded leaves typically declines as they age (Brown *et al.* 1966, Jewiss & Woledge 1967). The decline is associated with an increase in both stomatal and mesophyll resistances (Ludlow & Wilson 1971), but appears to be reversible in response to defoliation ('rejuvenation'). In a defoliation experiment with *Agropyron smithii*, for instance, Painter and Detling (1981) found that the net photosynthetic rate of heavily clipped plants increased by about 10%, whereas the net photosynthetic rate of control plants declined by about 10% over the course of 10 days. The immediate (30 min) effect of defoliation was different, in that clipped plants showed a 7% reduction in photosynthetic rate, but after 4 days, the rate had risen to 120% of preclipping levels.

Leaf rejuvenation is probably due to defoliation causing stomatal opening in the remaining leaves (Gifford & Marshall 1973). This accords with the hypothesis of Wareing *et al.* (1968) that partial defoliation reduces competition between leaves for cytokinins produced in the roots. Kinetin is known to cause rapid stomatal opening in leaves of barley, and also retards leaf senescence, probably via an effect on protein metabolism (Gifford & Marshall 1973).

The photosynthetic capacity of leaves is affected by their history as well as by current environmental conditions; for example, shaded grass leaves are thinner, and contain fewer, smaller and less closely packed cells. Their light saturation point, light compensation point and dark respiration rate all decline as the level of shading is increased (Ludlow & Wilson 1971).

The body of a plant comprises tissues that are either net producers or net consumers of carbohydrates. Mature leaves are almost always net exporters of photosynthates, and are referred to as 'sources'. Stems, stipules, petioles and green fruits also act as sources in some species. All other plant parts are net carbohydrates users or 'sinks'. The rate at which a tissue demands carbohydrate is described as the 'strength' of the sink. Rapidly filling storage organs like tubers or corms, expanding leaf buds and elongating flower stalks are all strong sinks for photosynthate. Basal metabolism of woody tissues, roots and stems are less intense, but no less important sinks for photosynthate (Wardlaw 1968).

It is now well established that plants do not produce photosynthate for which there is no demand. The rate of photosynthesis can be reduced even under ideal conditions of light, water and temperature if there is no sink for the carbohydrate. Photosynthetic rate can thus be a function of plant growth rate as well as vice versa (Neales & Incoll 1968).

This was demonstrated in a beautiful experiment by Thorne and Evans (1964) who grafted beet shoots on to sugar beet root-stocks and on to smaller, spinach beet roots. The shoots grafted to the larger sink had a higher rate of net photosynthesis. Experiments with whole plants of *Pinus radiata* by Sweet and Wareing (1966) show the effect of manipulating the sizes of sources and sinks on the rate of carbon dioxide fixation. Removing a sink was accomplished by pruning the shoot apex and all the surrounding immature needles; this depressed the photosynthetic rate of the remaining leaves at all light intensities (Fig. 2.22). Removing a source involved removing all fully expanded leaves six days before measuring photosynthesis; the photosynthetic rate of the newly unfolded leaves was 25% higher than the controls at saturating light intensities (Fig. 2.22).

One of the classic experiments on the relationship between photosynthetic rate and the balance of sources and sinks was carried out by King *et al.* (1967). They measured the photosynthetic rate of the flag leaf in wheat, and found a 40% reduction when the ear, its major sink, was removed. When an alternative sink was provided by keeping the lower leaves of the plant in darkness, the photosynthetic rate of the flag leaf recovered to levels that were not significantly different from those on intact plants. Respiration of the darkened leaves used photosynthate which the intact plants were using for seed fill (Fig. 2.23).

Removal of the major sinks does not always reduce the photosynthetic rate, however. When young flower heads of *Chrysanthemum morifolium* are removed, photosynthate is simply diverted to the root with no detectable

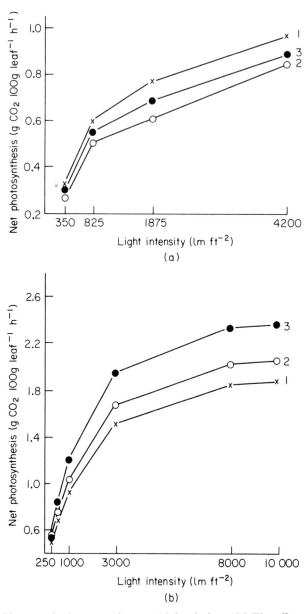

Fig. 2.22 Photosynthetic rate and source/sink relations. (a) The effect of removing sinks (plant apex) from *Pinus radiata* seedlings is to depress the photosynthetic rate; 1 initial photosynthetic rate, 2 rate after days, 3 rate 16 days after defoliation. (b) The effect of removing source leaves is to increase the photosynthetic rate of the remaining foliage; 1 controls (undefoliated), 2 one-third of fully expanded leaves removed 6 days before measurement, 3 all fully expanded leaves removed 6 days before measuring photosynthetic rate. The plants in (a) were aged 16 months and (b) 11 weeks (Sweet & Wareing 1966).

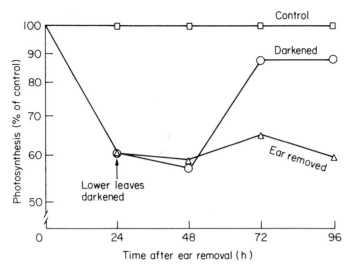

Fig. 2.23 The effect of removing the sink on photosynthesis in wheat. Removal of the ear reduced photosynthesis to 60% of control levels. When the lower leaves were shaded to create an alternative sink, the photosynthetic rate recovered to 90% of control levels (King *et al.* 1967).

drop in photosynthetic rate. Similarly, if the tuber initials of Jerusalem artichoke, *Helianthus tuberosus*, are removed, the assimilates are diverted to the roots (Milthorpe & Moorby 1974).

Different forms of herbivore feeding affect the balance of sources and sinks in different ways; leaf feeding, by reducing the source, often causes an increase in the unit leaf rate (ULR) of the surviving leaves. Even when it is the youngest, most productive leaves that are removed (as is the case with ungulates feeding on grasses), the older leaves display a rejuvenation of their photosynthetic rate (Hodgkinson 1974).

The flower-, fruit- and root-feeding herbivores reduce sink strength and therefore tend to depress the ULR (Maggs 1964, Wareing *et al.* 1968). This effect may be alleviated if new buds burst to replace the tissues consumed, or if the surviving fruits grow more rapidly (see section 2.7.7); for example, feeding by pest insects on cotton *Gossypium barbadense* provided new sinks and caused a delay in leaf senescence (Chatterjee & Chatterjee 1977).

Sap-feeding animals like aphids create sinks for carbohydrate and thereby increase the ULR of sink-limited plants. This effect may be delayed; for example, the net production rate of lime *Tilia × vulgaris* leaves in the year following aphid infestation is 1.6 times their normal level; the leaves are smaller, and darker green than on uninfested trees (Dixon 1971b).

This effect is not general, however. The rate of photosynthesis of sitka spruce, *Picea sitchensis*, needles is permanently depressed by aphid feeding so that reduced synthesis and increased respiration continue even after the aphids are removed (Kloft & Ehrhardt 1959).

It is very difficult to know whether observed changes in photosynthetic rate are due to the provision of new sinks (because photosynthesis had previously been carbohydrate limited), to reduced competition for nutrients between the surviving leaves (because the plant was previously root limited) or to adjustments in hormone balances or other such effects (see section 2.7.6).

2.7.4   Mobilizing stored carbyhydrate

Carbohydrate reserves consist of saccharides of various kinds which may be stored during periods when carbohydrate production exceeds the demands of respiration, growth and reproduction. The reserves are usually measured as total non-structural carbohydrate (TNC), determined by standard enzymatic treatment followed by acid hydrolysis (Smith 1969). Different plants store their reserves in different forms; tropical grasses, for example, typically store starches (glucose polymers) while temperate grasses store fructosans (fructose polymers).

Organic reserves are located in different tissues in different plants. In alfalfa *Medicago sativa* most carbohydrate is stored in the main tap root, though some is stored in the stubble. Within the root, the woody cortex is rich in starch and in sugars while the bark is rich only in sugars. During regrowth, half the TNC in the woody cortex was used up while only a quarter of the TNC in the crown and in the root bark was used (Ueno & Smith 1970). In a perennial grass like timothy, *Phleum pratense*, the organic reserves are stored in haplocorms at the base of young tillers and in the stubble, while the fibrous roots contain very little TNC (Sheard 1973). In trees, the reserves are stored in the cambial tissues below the bark of twigs and branches (Priestley 1962, Garrison 1972), and in the older leaves of evergreen species (Kozlowski 1971).

There is a temptation to assume that high levels of TNC are a good thing and indicate vigour in the plant. However, an increase in plant growth rate following defoliation of a plant high in TNC may simply be due to the fact that the plant was sink limited, and its high levels of TNC had actually been inhibiting photosynthesis. In this case, high TNC levels are an indication of growth limitation rather than vigour (Humphreys 1966).

The dynamics of the carbohydrate reserves are dictated by the demands of natural phenological patterns of vegetative growth and reproduction, and by the need to produce new leaves following defoliation (Coyne & Cook 1970). The starch concentration in the roots of *Andropogon scoparius* was greatest in January and lowest in March after the peak of new tiller production. Carbohydrate reserves in the roots were also drawn upon during culm elongation and when a second burst of tiller production

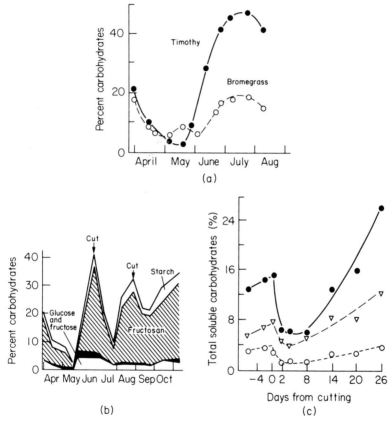

Fig. 2.24 Carbohydrate dynamics following defoliation in grasses. (a) The natural pattern of total non-structural carbohydrates shows a trough at growth initiation in spring and a peak at anthesis and seed maturation. (b) Defoliation of timothy at early heading caused a larger reduction in TNC in the stem base than cutting at inflorescence emergence on August 29 (Youngner 1972). (c) Minimal total soluble carbohydrates in roots (○), stubble (●) and leaves (▽) of *Lolium perenne* occur 4 to 8 days after cutting (Alberda 1962).

occurred in autumn (Smith & Leinweber 1971). However, water-soluble carbohydrate levels in the roots and stubble of Italian ryegrass, *Lolium multiflorum,* decreased after defoliation while starch levels remained unchanged (Kigel 1980). The carbohydrate reserves of *Lolium perenne* decline rapidly after severe defoliation or close grazing. Levels are restored after about 3 weeks of regrowth in summer time, but take much longer to recover in plants grazed or cut in autumn (Grant *et al.* 1981; see also Fig. 2.24).

The grass *Dupontia fisheri* forms a major component of the diet of arctic lemmings (see section 4.2.4) and is adapted to withstand extremely intense grazing pressure. A tiller is killed by removal of its apical meristem, but 6 simulated lemming defoliations led to a drop of only 50% in the plant's reserve of total non-structural carbohydrate. The plant also compensates for grazing by earlier initiation of active export of photosynthate from young leaves (Mattheis *et al.* 1976).

Under many conditions, carbohydrate reserves are relatively unimportant to the regrowth of grasses, especially when there is a substantial area of green leaf left after defoliation. When grazing is so severe that residual green leaf area is negligible, then carbohydrate reserves are crucial to plant survival. They serve to meet the respiratory needs of the surviving tissues and to provide the material for the construction of new leaves (Youngner 1972). However, their role in supplying structural carbohydrates for the new tissues is quickly taken over by current photosynthesis (May 1960, White 1973), for example, when two varieties of *Lolium multiflorum* were defoliated so that there was no residual green leaf in the stubble, the variety Liscate which had higher water-soluble carbohydrate levels produced higher initial leaf growth rates, and earlier development of the leaf surface, than the variety S.22 (Kigel 1980).

Regrowth tissues, of course, do not consist entirely of carbohydrate, and substantial amounts of amino acid are required for protein synthesis during leaf formation. The organic reserves must contain sufficient amide, amino acid, protein or nitrate to support tissue formation. Several studies have shown a decline in total nitrogen in grass storage organs during growth in spring and during regrowth after defoliation (Sheard 1973).

Leaf expansion in deciduous woody plants depends entirely on stored organic reserves. Many deciduous trees also flower before their leaves are expanded, which places further demands on their carbohydrate reserves. Aphids feeding on the phloem early in the season thus divert carbohydrate that would otherwise be used in leaf expansion. Since the number of leaves is determined by the number of buds that burst, aphid feeding on the limited supply of carbohydrate causes a reduction in the size of the mature leaves (Dixon 1971a; see also section 2.3.5). Defoliation of Douglas fir, *Pseudotsuga menziesii*, by tussock moths, *Orgyia pseudotsugata*, caused a reduction in the starch content of roots, twigs and surviving foliage (Webb & Karchesy 1977) and artificial defoliation of Scots pine, *Pinus sylvestris*, led to reduced starch levels in one-year old and older needles, and to reduced wood growth (Ericsson *et al.* 1980).

2.7.5   Refoliation

Many broad-leaved trees respond to defoliation by producing new leaves, but little leaf regrowth tends to occur when less than 50% of the primary leaves are removed.

Heichel & Turner (1976) studied refoliation in red oak, *Quercus rubra*, and red maple, *Acer rubrum*, after artificial defoliation; they removed 100%, 75% or 50% of the mature leaves for several successive years. Refoliation occurred from terminal buds, from lateral buds in petiole axils and from dormant buds on the previous year's twigs. The lowest rate of defoliation did not bring about any significant refoliation, presumably because the remaining leaves continued to produce the hormones that suppressed bud burst (see Table 2.5). This response also suggests that the marginal return from regrowth leaves is low compared to the productivity of the surviving first growth foliage.

Table 2.5 Refoliation of hardwood trees following artificial defoliation (data from Heichel & Turner 1976).

	Quercus rubra			*Acer rubrum*		
Intensity of defoliation (%)	50	75	100	50	75	100
Onset of refoliation (d)	30+	22–34	8–22	30+	22–27	11–22
Individual leaf size (% area of controls)	60	73	39	66	47	33
Leaves replaced (% number removed)	2	35	74	6	21	89
Final leaf area (% control)	51	44	29	53	33	28
Primary leaf area in the following spring (% controls)	85	67	56	93	76	57

Defoliation intensity determined the timing of regrowth, so that new leaves were produced most quickly when 100% of the leaves were removed, more slowly when 75% were removed, and few were replaced at all under 50% defoliation. Regrowth leaves were smaller than primary leaves and were fewer in number than the leaves removed.

The long-term impact of defoliation was felt in altered timing of bud burst and in reduced leaf areas. Defoliated trees typically leaf two or three days earlier than the controls and produce fewer leaves. Three successive years of 50% or 75% defoliation had no significant impact on regrowth ability and although three successive complete defoliations did reduce tree vigour and total leaf area, there was no increase in the death rate (Heichel & Turner 1976; cf. section 2.2.3).

The shoots of English oak, *Quercus robur*, grow in a series of very short, rapid bursts (Longman & Coutts 1974). The first, main flush of growth occurs during May and, on large trees, this is commonly the only growth of the year. Smaller trees, and larger trees in some years, have a second burst of growth in July. The second-growth shoots are distinctly different from the first; they have softer wood and are often much longer. They are frequently subject to such heavy mildew attack that the trees look grey from a distance. These are known as 'lammas shoots' although, strictly, this term should be applied to a third flush of growth that sometimes occurs in August (Longman & Coutts 1974).

Attack by herbivorous insects tends to be concentrated on the first leaves. Most of the defoliating Lepidoptera like *Tortrix* and *Operophtera* attack the leaves when they are young in May/June (Feeny 1970), while most of the gall wasps and some of the leaf miners attack them when they are mature in midsummer (Jones 1959). Since most of the defoliating insects are uni-voltine, their feeding is complete by the time the lammas shoots are produced. Late-season photosynthesis is, therefore, available to make good all or part of the losses due to defoliation by caterpillars (Gradwell 1974).

There is no evidence, however, to suggest that the sole, or even the principal function of lammas shoots is in compensation for herbivore feeding; for example, there is no correlation between the defoliation on a given shoot and the length of the lammas shoot it produces. There is, however, a significant positive correlation between mean lammas shoot length per tree and mean defoliation per tree (Fig. 2.25). Interpretation is difficult because trees from which insects were excluded with insecticides produced *longer*, rather than shorter lammas shoots (Witton 1981).

Ragwort *Senecio jacobaea* is frequently defoliated by larvae of the

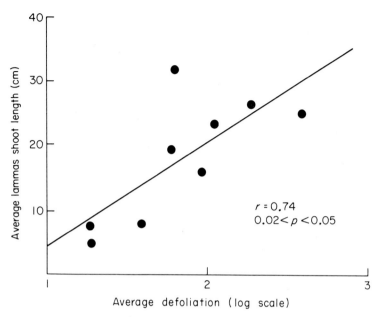

Fig. 2.25 Lammas shoot growth and defoliation in oak. There is a significant correlation between the average length of lammas shoots produced and the average level of defoliation suffered by 20-year-old *Quercus robur* (see text). There is no relationship between lammas growth and defoliation on individual shoots. While the data suggest that lammas shoots may be involved in plant compensation there are several other possibilities (e.g. trees on nutrient-rich sites put on more lammas growth and are also more attractive to herbivorous insects). Also, insecticide-sprayed trees put on *more* lammas growth; if the sole function of lammas shoots were compensation, they should have put on less (Crawley, unpublished data).

cinnabar moth, *Tyria jacobaeae*, so severely that all the plants in an area are stripped entirely of their leaves and flowerheads. There is, therefore, an apparent seed loss of 100% (Cameron 1935, Dempster 1975). On the poorest soils, and when the carbohydrate reserves of the plant are low, ragwort shows no regrowth and most of the plants die without setting any seed. In the absence of immigrant seed, these populations become extinct after two or three years of defoliation. On better soils, and when the plants were previously unstressed, ragwort can show two forms of compensation. First, it may produce new shoots during the year of defoliation which develop flowerheads and produce mature seed later in the year. Second, it may show no signs of flowering in the year of defoliation but produce new shoots from the crown of its rootstock or from root buds that develop into vigorous plants and set seed the following year.

Regrowth of ragwort from artificial and from natural defoliation by cinnabar moth was studied by Cameron (1935). Several hectares of a dense population of flowering ragwort plants were mown to within 10 cm of the ground on July 3. On October 31 the field was once again a blaze of yellow ragwort flowers on the cut area, while the uncut area was a mass of dead, brown stems; the data obtained from the two areas were:

Measurement	Uncut	Regrowth
Height	124 cm	61 cm
Number of stems	10	15
Capitula per stem	179	40
Seeds per capitulum	64	67
Seeds per plant	115 240	39 945
Seed maturation	September 7	November 25

Regrowth meant seed production was reduced by 65% instead of 100%. Regrowth plants had 50% more stems than regular plants but the regrowth stems were only half the diameter. Defoliation retarded seed maturation by 10 weeks (Cameron 1935).

In a second experiment, transplanted ragwort plants were artificially defoliated; one quarter had all their leaves removed, one quarter had all their capitula nipped off, one quarter had both their leaves and their flowers removed, and one quarter were kept as controls. The results obtained were these;

Treatment	Seed production (% of control)
All leaves removed	90
All capitula removed	40
Leaves and capitula removed	zero

The defoliated plants produced a number of small new leaves and the beheaded plants produced a new set of capitula. Probably because they had been transplanted prior to the experiment, and therefore had low reserves of carbohydrate, the plants that were defoliated and deflorated all died (Cameron 1935).

Work on the effect of cinnabar moth feeding on seed production demonstrated the importance of soil conditions on regrowth potential. On good soils, plants suffered a 60% reduction in seed output following complete defoliation, while two experiments on poor soil each resulted in a 95% loss of seed. On very poor soil, the larvae reduced the entire ragwort population to bare stems and no secondary growth occurred so that the plant population was wiped out.

In summary, ragwort can compensate for defoliation when its reserves are not depleted, when soil conditions are good and when weather conditions permit; regrowth does not occur in unfavourable weather, no matter how vigorous the plants.

Regrowth may not be immediate and defoliation may serve to increase the longevity of the plant. Those plants that grow from crown buds and flower in the year after attack may grow bigger and produce more seed than the parent plant. Nevertheless, damage to any part of the plant will reduce seed yield and delay seed maturation; this will inevitably depress the net rate of increase of ragwort (Zahirul Islam 1981).

The cost of defoliation to the plant's carbohydrate reserves is difficult to calculate. If the leaves are lost before they first balance their respiratory demands from their own photosynthesis, the costs are simply subtractive. Usually, however, defoliation does not occur until some time after the leaves have become photosynthetically independent. In this case one cannot equate the carbohydrate eaten by the herbivores with loss to the reserve, because the leaf will have repaid some, if not all, of its carbohydrate debt to the plant.

2.7.6   Distribution of photosynthate

It has long been known that the relative growth rates of the different plant organs change over the course of development as their relative demands for carbohydrate, nitrogen, water and minerals change. These changes are mediated by an alteration in the pattern of distribution of carbohydrate between the various growing points (Wareing & Patrick 1975, Trlica & Singh 1979). Root/shoot ratios often remain rather uniform throughout vegetative growth but root growth frequently slows or even ceases during flowering (Brouwer 1962). The main abiotic factors affecting root/shoot ratio are nutrient availability and soil water conditions. In nutrient-rich soils, the plant's requirements can be met in a small volume of soil and the root/shoot ratio is low. Similarly, the root/shoot ratio of drought-stressed plants will tend to be higher than that of well-watered plants. Thus water and nutrient balance interact with developmental stage to determine an equilibrium root/shoot ratio (Colvill & Marshall 1981).

Herbivore feeding typically alters the pattern of distribution of photo-synthate within the plant. When shoots are stressed by defoliation or by reduced light intensity, it is found that the root/shoot ratio is restored by channelling an increased fraction of net production to the shoot (Ryle & Powell 1975). The general rule appears to be 'stress the shoot, reduce the root', and vice versa. Defoliation of grasses often stops root growth altogether and may even lead to a reduction in root weight when roots that die naturally are not replaced by fresh growth (Ryle 1970).

Different clipping rates had no effect on the root/shoot ratio of the grass *Cynodon dactylon*, for example. It remained at roughly 0.4, except with the closest cutting regime (2.5 cm trimmed plants had ratios of 0.5; Dittmer 1973).

Infestation of pasture by leatherjackets, *Tipula* spp., and other root-feeding herbivores like nematodes increases the fraction of net production channelled into the root system, and stimulates the production of new, regrowth rootlets. These are more efficient in nutrient and water uptake than the older roots they replace, so that when the root population is pre-dominately old-aged prior to attack, root feeding may have little effect on shoot growth because the increase in average efficiency compensates for the loss in root biomass (Davidson 1979). This process is analogous to defoliation in grasses, where feeding alters the age structure of leaves in favour of younger, more productive tillers (see section 2.7.3).

The reasons for these shifts in carbohydrate distribution are plain. A defoliated plant has more root than is necessary to supply the transpiration or nutrient requirements of its small residual leaf area. It builds new leaves to redress the balance and allows respiring roots, surplus to requirements, to die without replacement. A plant suffering root pruning cannot meet the demands of its vigorous shoot system. New roots are built at the expense of replacing old leaves and the existing leaves may senesce more rapidly (Crossett *et al.* 1975).

A detailed model of plant growth under grazing requires an understand-ing of how carbohydrate is distributed between the different sinks (re-spiration, root growth, leaf production, fruits, rhizomes and so on) and how the distribution function alters with plant development and with the timing, intensity and position of herbivore attack. While we know something of the way in which photosynthate is partitioned, the mechanism whereby the redistribution is directed remains obscure. Most current models simply specify empirical root/shoot equilibria for different stages in plant development (de Wit *et al.* 1970, Evans G.C. 1972, Thornley 1976), or define a time-varying hierarchy of sinks which contest rather than scramble for photosynthate; for example, respiratory requirements are met first, then root and shoot growth, fruit fill and so on (Wang *et al.* 1977, Gutierez *et al.* 1979).

When alfalfa plants are defoliated, regrowth occurs at the expense of carbohydrates stored in the roots. Completely defoliated plants placed in the dark can restore about 17% of their previous shoot dry weight by depleting the TNC in their roots by 70%. Plants recovering from defoliation in the light return little of their early photosynthate to the roots; in the first week after defoliation, 83% of radiolabelled CO_2 remained in the shoot and even after a month, 64% of $^{14}CO_2$ was in the tops (Pearce *et al.* 1969). Root growth in *Aristolochia reticulata* is depressed by defoliation by larvae of the swallow-tail *Battus philenor* (Rausher & Feeny 1980).

Aphid feeding on the leaves affects the distribution of photosynthate in sycamore saplings. Total net production per unit area of leaf was similar on infested and uninfested plants, but 72% was incorporated into stems and roots (and was, therefore, retained by the plant) in the control plants, while infested plants lost 50% of their net production at leaf fall (Dixon 1971a).

Redistribution of photosynthate can also ameliorate the effects of patchy herbivore attack within a plant; for example, the wood growth of sawfly-infested branches of *Eucalyptus* was greater on trees where one branch had been sprayed with insecticide than on uniformly infested trees. Apparently the extra photosynthate from the undamaged branch was shared out over the whole tree (Morrow & LaMarche 1978). Similarly, when tillers of *Lolium multiflorum* are differentially defoliated, the plant responds by redirecting the distribution of carbohydrate to the defoliated tillers, restoring their position in the tiller hierarchy (Marshall & Sagar 1968). This effect is not general, however, and some grass species cut their losses when one tiller is severely defoliated, directing carbohydrate to less badly damaged parts (Ong *et al.* 1978).

In summary, 'the stress imposed by defoliation is efficiently buffered by the rapid re-organization in the carbon economy of the plant. The increase in the rate of photosynthesis by the remaining leaves, in the proportion of assimilate exported from the older leaves, and the change of distribution pattern allow the tillers to receive maximum support from the intact part of the shoot. Furthermore, the results emphasize the importance of the source-

sink balance in the physiological organization of the plant' (Gifford &
Marshall 1973).

2.7.7   Reduced death rate of plant parts

When plants are source limited there is great scope for compensation for the
removal of sinks by herbivores (Harris 1974a). In wheat crops, for instance,
the number of young shoots is many times greater than the maximum
number that can produce ears, and attack by stem-boring larvae of wheat
bulb fly, *Leptohylemyia coarctata*, kills shoots that would otherwise have
died through competition for limited carbohydrate (Bardner 1968).

   Compensation for insect-feeding on fruits may take the form of an
increase in the proportion of seeds maturing per fruit or in the proportion of
fruits reaching maturity. A good example is provided by the response of the
wild parsnip, *Pastinaca sativa*, to attack by parsnip webworm, *Depressaria
pastinacella*. Feeding by the moth larvae reduced seed production from the

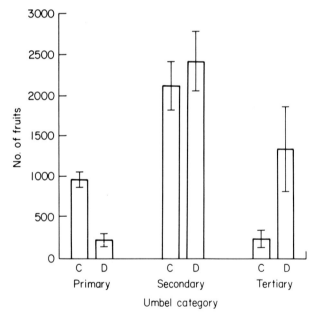

Fig. 2.26 Compensation via reduced death rate of flowers. Although most of the
flowers and fruits of primary umbels of *Pastinaca sativa* are destroyed by parsnip
webworm, damaged (D) plants produce similar numbers of fruits from their
secondary umbels and many more fruits from their tertiary umbels than do
control (C) plants (means ± s.e.; from data in Hendrix 1979).

primary umbels of large plants from 950 to 220 (Fig. 2.26), but because of compensatory reproduction due to a 6-fold increase in the number of seeds produced by tertiary umbels, the total seed production of attacked and control plants is not significantly different. Small plants were unable to compensate for damage and suffered a 50% loss in seed production when their primary umbels were attacked (Hendrix 1979).

The mechanism of compensation in this case is probably a redistribution of photosynthate following the removal of the young fruits of the primary umbel, combined with a change in sex of some of the flowers in the tertiary umbel from male to hermaphrodite following herbivore damage. In any event, the rate of abortion of tertiary flowers is greatly reduced in damaged as compared to control plants (Hendrix 1979).

Compensation for flower removal is commonly observed in legumes such as field beans, pigeonpeas, cowpeas, yellow lupins and soybeans, and several studies have shown that the plants are able compensate completely for the loss of all their flowers for periods of up to 5 weeks (Adams 1967). Compensation is due to pod set from later-formed flowers that would normally have been aborted. The amount of damage suffered after complete removal of all the early flowers is determined by how favourable the growing conditions remain, and yield will suffer if pod fill is delayed until drought or cold overtakes the plants (Sheldrake *et al.* 1979). A similar effect is observed in cotton plants, where compensation for damage or removal of bolls occurs because flower production continues for a longer period than normal. In Western Australia, insects can completely prevent the setting of a cotton crop unless it is adequately protected by insecticides. As long as each block of cotton is unprotected, no bolls are set on it, because feeding by *Heliothis* larvae causes the flower buds to be shed. Wilson *et al.* (1972) investigated the effects of delaying insecticide application for 4 and 8 weeks from its traditional application date. No cotton was set on the unsprayed plants, but they continued to grow in stature. When the insecticide was eventually applied, the later-sprayed plants produced the same number of cotton bolls, despite the fact that by this time the weather was cooler, the plants were older (and larger) and total insect damage was greater. Thus, although insecticide was essential, it was possible to minimize the period of its use.

Compensation for pod loss in soybeans occurs by an increase in the weight of individual seeds produced by the remaining pods (Smith & Bass 1972), an effect similar to that observed when 75% of the flowers of *Rumex crispus* are artificially removed (Maun & Cavers 1971).

The tephritid fly *Dacus oleae* is a serious pest of olives in Mediterranean countries because larval feeding causes the olives to fall from the trees before harvest. Recent experiments have shown that trees can compensate for 10% premature fruit fall by increasing the weight and percentage oil content of the surviving olives (Neuenschwander *et al.* 1980).

Clearly, the ability of the plant to compensate for these kinds of herbivore damage is a function of the intensity as well as of the timing of the attack. In the parsnip webworm example, the plant was able to compensate for damage to its primary umbels by producing more seed from its tertiary umbels, but at the herbivore densities studied by Hendrix (1979) the secondary umbels were not attacked and these contributed substantially to total set in both damaged and undamaged plants. If moth densities were high enough that the secondary umbels were damaged, no amount of compensation by the tertiary umbels could make good the losses.

The importance of timing of damage is shown clearly by work on soybeans (Smith & Bass 1972). When pods were removed early in the season, prior to seed fill, the plants could withstand up to 80% pod loss without a reduction in yield whereas if attack occurred late in the season, when seed fill was almost complete, pod removals of as little as 10% led to reductions in yield.

The question remains as to whether delay in flowering within a given season reduces the fitness of the seeds that are produced. It is possible, for example, that early produced seed will pre-empt germination microsites and that synchronous fruiting is more likely to satiate the seed predators. From what few data are available, it appears that any reduction in fitness of the seeds of later flowering plants will be due to losses caused by inferior pollination rates, increased risk of exposure to damaging frosts, and so on (see section 2.4.4), rather than to reduced survival of the seedlings that subsequently emerge (see section 2.6.3).

2.7.8   The prevalence and importance of plant compensation

The potential impact of compensation on plant–herbivore dynamics is described in section 4.3.7. It is clear from this analysis that plants would not be expected to compensate for continuous attack by food-limited herbivores; compensation merely makes the herbivore more abundant, without any increase in the plants fitness! Compensation might evolve in plants that have specialist herbivores, when herbivore feeding ends sufficiently long before the end of the growing season that the plant can enjoy a period of profitable, herbivore-free growth (e.g. lammas shoots on oak; see section

2.7.5). Similarly, compensation would increase the fitness of plants that were attacked by herbivores whose numbers were regulated by disease, enemies or habitat factors, rather than by food availability.

A plant species that supports a food-regulated, specialist herbivore may well increase its fitness by *not* compensating for leaf removal by producing more leaves. If it produces less leaves, its herbivore's numbers are bound to decline (since the animal is food limited).

An interesting example of how feeding and compensation alter the competitive balance between closely related species is described by Bentley and Whittaker (1979). Two species of dock, *Rumex crispus* and *R. obtusifolius*, were grown separately and in competition; *R. obtusifolius* was the superior competitor. The chrysomelid beetle *Gastrophysa viridula* eats both species but prefers *R. obtusifolius*, and it might, therefore, be predicted that it would tip the competitive balance in favour of *R. crispus* in mixed cultures. In fact the opposite happened and *R. crispus* was more disadvantaged in the competition experiment with beetle feeding than without.

This paradoxical result is due to the different ways that the two docks respond to herbivore feeding. *Rumex crispus* responds in the conventional way by altering its root/shoot ratio; it reacts to defoliation by producing proportionately more leaf. In contrast, *R. obtusifolius* responds to defoliation by putting proportionately less photosynthate into shoots; it keeps its head down. Thus although the beetle prefers *R. obtusifolius*, the abundance of foliage gradually drifts towards a preponderance of *R. crispus*. The beetles emigrate from the depleted *R. obtusifolius* and damage the *R. crispus* more and more; the more it compensates, the more they eat it up.

There is mounting evidence that defoliation may induce a general increase in plant metabolism analogous to the immune responses shown by animals, associated with wound isolation, callus production, and the synthesis, transformation and redistribution of secondary chemicals (Chew & Rodman 1979; see also section 3.7.4).

Chemicals injected into the plant phloem or into individual cells by sucking insects (Miles 1968), or saliva applied to the cut ends of leaves grazed by mammals (Reardon *et al.* 1972) may alter the unit leaf rate of the surviving leaves.

The influence of herbivore saliva on the rate of regrowth of grasses has been found in different studies to promote (Dyer & Bokhari 1976), to depress (Detling & Dyer 1981) and to have no effect upon the rate of tillering and biomass increase (Detling *et al.* 1980). These studies have been performed with insects (Capinera & Roltsch 1980) and with vertebrates (Reardon *et al.*

1974, Detling *et al.* 1980). It would only increase the fitness of individual plants to show enhanced rates of regrowth after defoliation and saliva deposition, if herbivore attack was ephemeral, either because the animals were migratory, or because they enter diapause before the plant growing season ends (see section 4.3.7).

Toxins injected into plants as saliva, in the salivary sheath or during oviposition may depress the rate of photosynthesis (Tingley & Pillemer 1977), but feeding, even by high densities of the sap-sucking insect *Strophingia ericae*, had no effect on the rates of respiration or photosynthesis of laboratory plants, nor on shoot length, dry matter accumulation, flower production or nutrient content of field populations of heather *Calluna vulgaris* (Hodkinson 1973).

The fact that herbivore feeding is active rather than passive in its effects on plant physiology means that the susceptibility of plants to herbivore attack, and their ability to compensate for damage, are affected by the age of the plant, the age of the tissues attacked, the plant's history of defoliation, its carbohydrate and amino acid reserve levels, its water status and a host of abiotic factors including physical damage from ice, fire, lightning, wind throw and atmospheric pollution.

Finally, it must be stressed that the regrowth capacity of plants is limited; even pasture grasses will be killed by early, frequent and complete defoliation, no matter how favourable the climatic and soil environments.

CHAPTER 3
ANIMAL POPULATIONS

This chapter describes how the fecundity, mortality and dispersal rates of herbivorous animals are affected by the abundance, chemical composition and spatial distribution of their food plants. The factors affecting the rate at which each individual animal feeds are considered first. Subsequently we look at how feeding success and food quality influence changes in herbivore numbers, and at how differences in food quality affect diet selection and dispersal in polyphagous animals.

3.1   FUNCTIONAL RESPONSES

The phrase functional response was coined by Solomon (1949) to describe a change in the feeding rate of individual predators brought about by a change in the abundance of their prey. It is traditional in animal ecology to describe the rate at which predators feed as a graph of prey removal rate per predator against prey density; this is a functional response curve (Fig. 3.1). In the earliest population models, the functional response was assumed to be linear (e.g. in Lotka-Volterra models; see section 4.3.1). Later, allowance was made for the fact that the feeding rate of most predators would saturate, due to limited gut capacity (Ivlev 1961) or to limited time for food handling and processing (Holling 1965).

Similar curves depicting the feeding rate of herbivores as a function of plant abundance have been widely employed more recently (see section 3.1.2). The precise form of the functional response can have an important effect on plant–herbivore dynamics, and is discussed in detail later (see sections 3.1.2 and 4.3.2).

3.1.1   The feeding rate of individuals

The most obvious factor affecting food intake is body weight, W. This sets a limit to the size of the gut and to the rate at which food can be processed (see Baile 1975 and Bines 1976, for ruminants, and Browne 1975 and Chapman 1974, for insects).

Food intake increases with $W^{0.75}$ in mammalian herbivores, reflecting the fact that the respiration rate of small animals is proportionately greater because of their higher surface area to body weight ratio (Kleiber 1961;

111

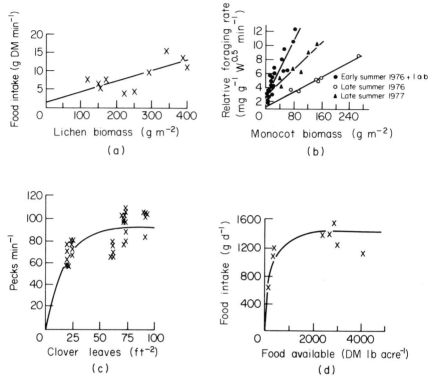

Fig. 3.1 Functional responses of herbivores to changes in plant abundance. Feeding rate increases with food availability up to a limit set by gut capacity or handling time. (a) Linear (Type 1) functional responses for reindeer in a lichen stand, and (b) brown lemmings in cotton grass/sedge communities (Batzli *et al.* 1981a, White *et al.* 1981). (c) Asymptotic (Type 2) response of wood-pigeons feeding on a clover sward (Murton *et al.* 1966). (d) Type 2 response of winter-shorn sheep on Phalaris/clover pasture (Arnold 1964). DM = dry matter; W = fresh body weight. For S-shaped (Type 3) responses see Fig. 3.26(b).

Brody 1945 uses $W^{0.7}$). The scaling factor is different in poikilotherms (von Bertalanffy 1957 gives a value of 0.67), but for insect herbivores food intake scales to about 0.8 (Reichle 1968, Schroeder & Dunlap 1970, Schmidt–Nielsen, 1979); for example, the relationship between body weight and feeding rate for the five instars of the moth *Tyria jacobaeae* feeding on ragwort *Senecio jacobaea* is shown in Fig. 3.2. Not only do the final instar larvae feed at twice the rate of fourths, but the fifth instar lasts twice as long (six days) so that over 75% of total food consumption occurs in the final instar (Zahirul Islam 1981). Similarly, Colorado potato beetles, *Leptinotarsa*

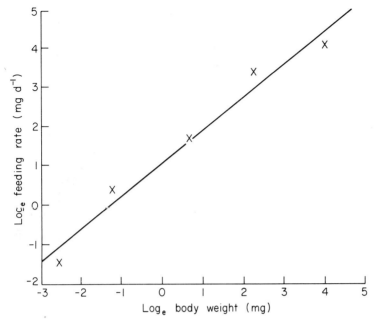

Fig. 3.2 Feeding rate and body weight in cinnabar moth caterpillars. Feeding rate increases with each instar, scaling with body weight to the power 0.88 (Zahirul Islam 1981).

decemlineata, consume 70% of the total food intake during development in the last (4th) larval instar (Chlodny 1967). In order to calculate the potential food consumption of a herbivore population, it is therefore necessary to know its size structure as well as the total population density.

Animals of the same size may differ in their food requirements. Males and females often have different activity rates and show wide variation in behaviour. Food intake of females changes dramatically with their reproductive condition. A 500 kg cow, for example, requires 3.4 kg dry matter per day, rising to 5.3 kg when pregnant and carrying a 40 kg calf. With the calf at heel and the cow producing 30 kg of milk per day, food intake rises to 16.4 kg dry matter per day (ARC 1965). A dry, non-pregnant sheep weighing 70 kg needs 0.6 kg per day rising to 1.9 kg per day when she is producing 1.8 kg of milk per day (ARC 1965).

For small mammals there are data on bank voles, *Clethrionomys glareolus* (Kaczmarski 1966), common voles, *Microtus arvalis* (Migula 1969), cotton rats *Sigmodon hsipidus* (Randolph *et al.* 1977), and *Peromyscus leucopus* (Millar 1978) showing that energy ingestion increases throughout lactation and in relation to litter size. Rats require 2.5 times their basal food intake when pregnant and over 3.5 times when they are lactating (Russell 1948). The feeding rate of both sexes may be influenced by the health of the animals; many parasitic diseases depress the appetite which results in loss of condition (Hafez 1962, Wooden *et al.* 1963), and certain dietary imbalances such as cobalt deficiency are known to cause extreme inappetite (Underwood 1956).

Weather conditions affect the feeding rate of both endo- and ectotherms. Many mammalian herbivores eat more food during cold weather to compensate for the extra energy used in keeping warm (Blaxter 1962, Ames & Brink 1977). But since the worst weather usually coincides with the lowest food quality and quantity, compensatory feeding is often impossible and the animal loses weight as its body reserves are used up. Certain ruminants like white-tailed deer, *Odocoileus virginianus*, however, reduce their feeding rate and metabolism in response to low temperatures (Moen 1978). Bad weather can have an enormous impact on the feeding of herbivorous insect populations. Low temperatures obviously mean low rates of development and low rates of feeding. For the numerous species where egg distribution is achieved by winged females, weather conditions during the flight period can determine the total amount of subsequent feeding by limiting the number of eggs laid. Thus oviposition rate and plant food availability may be quite unrelated in those seasons where good conditions for plant growth coincide with bad conditions for insect flight (e.g. dull, wet summers allow rapid plant growth but are bad for butterfly flight; Gossard & Jones 1977).

3.1.2   Feeding rate and plant density

The relationship between food availability and the rate of feeding of an individual herbivore is usually depicted as a functional response curve (Fig. 3.1). At high levels of food availability, intake is limited by the rate at which the herbivore can process the food, by the volume of its gut or by the handling time of individual food items (Holling 1965). When food is scarce, there are two possible responses; either food intake increases more or less linearly with availability, or intake increases faster than linearly. These lead, respectively, to asymptotic (Type 2) and to S-shaped (Type 3) functional response curves.

When herbivores are predator-limited or habitat-limited, the functional response will usually be unimportant because food will be present in excess. Herbivores will feed at the maximum rate determined by their size, sex and physiological condition. The functional response may be important at certain times of year and not at others, especially when there are strong seasonal patterns of food availability; for example, 'small increases in green pasture in the period of slow growth (winter) gave large increases in live-weight and wool production. Large increases in green pasture in the period of rapid growth (spring) had no effect on the animal' (Willoughby 1959).

Functional responses for mammalian herbivores have been determined in several cases. Most of the information has been gathered for sheep (Arnold & Dudzinski 1967, Allden & Whittaker 1970, Langlands & Bennett 1973) but data are available for reindeer (White *et al.* 1981) and brown lemmings (Batzli *et al.* 1981a). Reindeer *Rangifer tarandus* at different sites show a 5-fold increase in feeding rate as live plant standing crop biomass increased from 30 to 110 g dry matter m^{-2} (Skogland 1980).

As food becomes scarce, the most immediate response is to forage for a longer period; for instance, the time spent grazing by sheep on mixed *Phalaris* pastures in Australia increased from 7 h d^{-1} at 2200 kg ha^{-1} green dry matter to over 10 h d^{-1} when forage biomass was reduced to 700 kg ha^{-1} (Arnold 1964). There is a limit to such compensatory increases in searching time since, below a certain threshold in food availability, the increased energy cost of search exceeds the returns. Also, when grazing time is limiting, distractions caused by predators or irritating insects may lead to a reduction in intake (White *et al.* 1981).

Brown lemming and reindeer never reached a plateau in feeding rate at food availabilities typical of the field (Fig. 3.1). The fact that their functional response curves never saturate is strongly suggestive that their populations are food limited. If predator or habitat regulation were important, it is most unlikely that plant abundance would be depressed so far below its carrying capacity that food was never sufficient to saturate the feeding rate of the herbivores.

Rather few examples of functional response curves for invertebrate herbivores exist. Zahirul Islam (1981) determined the functional response of each instar of cinnabar moth, *Tyria jacobaeae*, fed ragwort leaves or flower heads at different rates (Fig. 3.3). In the field, adults of *Pieris rapae* searching for widely spaced cabbage plants by flight, laid only 10% of their possible daily complement of eggs because food plants were scarce (Jones *et al.* 1980).

Functional responses determined under experimental conditions often show an almost linear rise to the plateau; a Type 1 curve. This is simply because in the small experimental arenas typically employed, the animals are

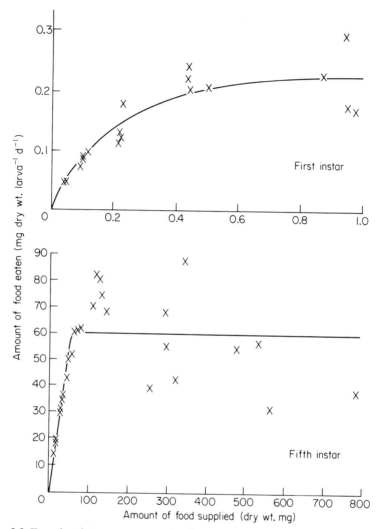

Fig. 3.3 Functional response curves for first and fifth instar cinnabar moth caterpillars fed leaf discs and unopened flowerheads respectively. Notice that the variance in feeding rate is greater at high rates of food availability than at low (Zahirul Islam 1981).

sufficiently mobile to eat up all the food provided when the supply rate is lower than their requirements. There are no refuges for the plant and almost all the food provided is edible. Under field conditions the plateau is either reached gradually (Type 2) or not reached at all (compare Fig. 3.1(a) and 3.1(b).

Functional responses that are S-shaped (Type 3) have not been reported for herbivores feeding on a single food source. They are of the highest theoretical interest, however, because at the lowest food availabilities, the attack rate on the plants is density dependent; herbivores feed at a proportionately greater rate as food availability rises. This allows for the existence of a stable equilibrium at low plant availability determined by the feeding behaviour of the herbivores (Southwood & Comins 1976, Hassell *et al.* 1977; see also section 4.3.2). An S-shaped curve could arise when an ungrazable reserve of plant tissue exists, producing what amounts to a threshold plant density below which the herbivores cannot depress the plant production (Noy Meir 1975); when the herbivores switch to less-preferred, but more abundant food sources if the availability of their preferred food plant falls below a threshold density (Murdoch & Oaten 1975, May 1977); or when herbivore searching efficiency or time spent searching declines with falling plant abundance, so that an effective ungrazed reserve of plant biomass is created (Hassell 1978).

The practical difficulties of drawing up functional response curves in the field are formidable; for example, Murton *et al.* (1966) studied wood pigeons feeding on clover in mid-winter. Feeding rate changed through the day from about 70 pecks min^{-1} in the morning to about 100 pecks min^{-1} in the afternoon. A bird watched throughout an 8-hour bout of feeding would take about 30 000 fragments of clover, plus a little grass and weed leaf, at an average rate of 74 pecks min^{-1}.

The rate of pecking (assumed to be equivalent to the rate of intake) and the rate of seaching are negatively correlated. It seems that the birds are more selective in their feeding in the morning, because they move more and peck less at this time. If, by the end of the day, the pigeon has not filled its crop, it becomes less selective, walking less and pecking at a faster rate. The birds, therefore, seem to be 'optimists' in their feeding; by feeding selectively at the beginning of the day, they allow for the possibility of taking a cropful of the highest quality food. If they fed rapidly at first then discovered a patch of high quality food late in the day, they might not be able to take full advantage of it because they were already satiated with poorer food. This is the opposite to normal selective feeding behaviour, where hungry animals are least selective (see section 3.5.2).

The extraordinary effort involved in collecting these data is the main reason why so few functional response curves for free-living herbivores have been obtained. Also, since 95% of the pigeon's diet at this time of the year consists of clover leaf, it is reasonable to equate pecks with intake of fragments of clover leaf. If the birds were more polyphagous, then observations of feeding behaviour would not, of themselves, allow estimates of the functional response. This problem had dogged many large mammal studies; it is easy to construct time budgets showing how long the animals spent feeding in different habitat patches, but it is impossible to convert these to rates of tissue removal from different plant species without laborious 'before and after' surveys of the vegetation (Walters & Evans 1979). For domestic animals on research stations, the problem can be overcome by sampling the food eaten directly from the animal's gut via an oesophageal or ruminal fistula, and these methods are now being applied to wild vertebrates (Trudell & White 1981, White *et al.* 1981).

When the turnover rate of the plants is rapid compared to the sampling period, large amounts of leaf may be produced despite low standing crop biomass. It is clearly important, therefore, that the time scales of consumption and production are equal when functional response curves are produced. Longer term effects such as plant compensation (tending to make more food available than is evident) or exploitation (tending to reduce food levels for all or some of the herbivores) must be taken into account in considering the ecology of long-lived herbivores and food plants.

3.1.3   Feeding rate and food quality

Feeding rate responds to changes in different aspects of food quality in different ways. When quality decline consists of a reduced concentration of a nutrient such as nitrogen, then feeding rate may be increased to compensate (see section 3.6). Many leaf-chewing insect herbivores behave in this way, increasing their feeding rate as the nitrogen content of their food declines so that insect growth is more or less independent of plant nitrogen (House 1972, Slansky & Feeny 1977).

Leaves of *Asclepias syriaca* fed to *Danaus plexippus* differed in nitrogen content in successive years by 19%; the larvae ate 74% more leaf in the low-nitrogen year.

When the artificial diet of larval *Celerio euphorbiae* was diluted to 85, 70 and 50% of its nutrient content, the larvae ate proportionately more, so that body weight gain was the same on each diet. When the diet was enriched with 'immoderate proportions' of certain nutrients, the larvae ate less and grew more slowly (House 1965).

Some insects increased their feeding rate to such an extent that they actually grew more quickly on the low-nitrogen diet (Schroeder 1976).

Gut retention times differ markedly with different feeding habits; for instance in the leaf-feeding howler monkey, *Alouatta palliata*, a marker took an average of about 20 hours to appear in faeces, whereas in the fruit-feeding spider monkey, *Ateles geoffreyi*, the markers appeared after only 4.5 hours. The leaf-feeder extracts the energy it requires from its food by retaining it in the gut for a longer time, while the fruit-feeder processes a large volume of food in order to obtain the protein it requires (Milton 1981).

Compensatory increases in feeding rate are also observed in sucking insects like the willow aphid, *Tuberolachnus salignus*, which feeds at a faster rate when nitrogen levels in the phloem sap are low (Mittler 1958). Even on the most dilute diets (e.g. xylem sap), increases in amino acid content lead to reductions in feeding rate (Horsfield 1977).

The response is not universal, however, and the feeding rates of some insects fall with declining quality (Schoonhoven 1972). Larvae of the beetle *Paropsis atomaria* feeding on the young foliage of 13 species of *Eucalyptus* showed no correlation between feeding rate and leaf nitrogen content. Although they fed at more or less the same rate, the animals on the higher nitrogen diet grew more quickly (Fox & Macauley 1977).

When reduced quality is due to reduced digestibility of the food, the potential for compensation may be reduced. Large vertebrate herbivores show two types of response to reduced digestibility. In non-ruminants like horses and zebras, the feeding rate is increased to compensate for reduced quality; because the rate of throughput of food is increased, the efficiency of removing the nitrogen from each gram of tissue is reduced, and animals typically show lower assimilation efficiencies on low-quality food (McNeill & Southwood 1978).

With ruminants, the reverse response is observed, and food intake increases more or less linearly with digestibility (over the range 50 to 75% for domestic livestock; Holmes 1980). The feeding rate of ruminants is often regulated by the distension of the gut, so when food quality is low and the rate of passage of food through the gut is reduced, intake is reduced (Blaxter *et al.* 1961). There are two main causes of this; when the food is of low digestibility (low nitrogen and high fibre) it stays bulky and remains longer in the rumen. Alternatively, when the decline in quality is due to increased secondary chemicals, the functioning of the gut microflora might be impaired; for example, resins reduce microbe numbers and thus slow the rate of cellulose fermentation and increase rumen retention time. Resins also inhibit caecal functions in browsing birds like ptarmigan, limiting protein assimilation and reducing the rate of feeding (Bryant & Kuropat 1980).

As we have seen, the time spent grazing is affected by food availability, so that sheep, for example, feed for a longer period when plant biomass is low (Young & Corbett 1972). When food quality declines, sheep must spend a greater time in rumination (White *et al.* 1981), so that if quantity and quality decline simultaneously, the increased time necessary for rumination may reduce the time available for foraging so much that intake is reduced. Thus ruminants are particularly sensitive to reduction in food quality because they are often unable to compensate by feeding at a faster rate.

Certain specialist insect herbivores that are keyed to particular feeding stimulants increase their feeding rate with increases in the concentration of the stimulant. Experiments with cabbage aphid *Brevicoryne brassicae* have shown that feeding rate is positively correlated with the concentration of allylisothiocyanate in the phloem sap (van Emden 1978). Plant strains that differ in the concentration of the substance in their sap are thus differentially affected by cabbage aphid attack. Aphid resistance in the brussels sprout variety 'Early Half Tall' compared with 'Winter Harvest' is thought to be almost entirely due to its lower allylisothiocyanate content; at low levels of the secondary chemical the aphids grow less rapidly and are less fecund.

When reduced food quality is due to an increase in the concentration of toxins, then, clearly, continued feeding would be debilitating. As we might expect, the ability of animals to perceive differences in toxin levels between different foods is well developed. Small mammals can discriminate between plants of different alkaloid contents, probably by their bitter taste. Meadow voles, offered 14 different clones of reed canary grass *Phalaris arundinacea*, fed most from those with the lowest alkaloid contents and least from those with the highest (Kendall & Sherwood 1975).

For insects that share a similar feeding habit, feeding rate is closely correlated with development rate. In arctic Lepidoptera, feeding on *Betula* leaves, for example, *Oporinia* larvae feed at 70% of body weight per day while *Dineura* take 21% (food dry weight to fresh body weight). *Oporinia* may take less than one month to develop, and puts on the final 90% of its body weight in about 10 days while *Dinuera* takes over 43 days for the same equivalent gain in weight (Haukioja 1981). This difference is accounted for by the fact that *Oporinia* feeds on the young leaves of birch and these have a much higher ratio of nitrogen to total phenolics (Haukioja 1981). Other arctic Lepidoptera feed even more slowly than *Dineura*; for example, *Byrdia groenlandica* eats *Dryas* and *Saxifraga* leaves, taking only 13% of its body weight per day and developing over a period of about 10 years (Ryan in Haukioja 1981).

3.1.4   Availability

The term 'food availability' is used in the literature as if its meaning was obvious. However, there are major practical difficulties in scaling the x axis of functional response curves determined in the field. Measurement of the above-ground dry weight (standing crop biomass) of a plant species will frequently lead to an overestimate of food availability. Plant tissues may be unusable for a vast number of reasons: because of fouling by dung or urine; because they are physically inaccessible to the animals; because they contain toxic or repellant chemicals; because they are hard to find amongst other plants because of their absolute or relative sizes; or because the animals show extreme selectivity of parts within the plant. Conversely, if the turnover rate is rapid, the standing crop biomass may underestimate plant availability (as in closely grazed mesic grassland; see section 3.1.2).

Almost every aspect of the natural history of the herbivore influences the likelihood of its encounter with a particular plant, and hence the availability of that plant as food. First, most herbivores exhibit habitat selection, avoiding areas where they are exposed to extremes of weather or are especially vulnerable to predators (see section 3.5.1). Plants in these areas are not available. Ovipositing insects avoid certain habitats that contain apparently suitable host plants (Shapiro & Carde 1970, Benson 1978).

Precisely the same distribution of food plants will be attacked in quite different ways by herbivores that differ in mobility, sensory acuity and in their responses to the habitat cues of light, moisture, and the like.

In Australian sheep country, for example, two patches of equally palatable grass differ in their availability depending on their distances from drinking water (Goodall 1967). Different clumps of grass of exactly the same age and nutritional quality differ in their availability to Aldabran giant tortoises if they are different distances from shade (Coe *et al.* 1979).

Physiographic factors also affect the pattern of feeding. In cool places, the sunny side of a hill may be grazed in winter no matter how much good forage is available on nearby shady slopes (Hercus 1961). Ridge tops and sunny slopes may be more heavily browsed by deer because the snow is shallower than elsewhere (Horton 1964).

In sheep pastures, the presence of certain disliked species can strongly reduce the utilization of liked species (Arnold 1964). Changes in the stature of the plant community can also affect availability. If Australian *Phalaris* pastures are allowed to grow tall by maintaining a low grazing pressure, the green tissues that would normally be preferred, are enmeshed in dry grass and, although there may be 500 kg ha^{-1} green dry matter, virtually none is eaten (Arnold 1962). Similarly, deer will not select green forage from amongst the standing litter of *Agropyron spicatum* when alternative food sources are available (Willms *et al.* 1981).

Frequently, the size of an individual plant affects its availability. In patches of wild parsnip, *Pastinaca sativa*, the plants on the edges grew larger, and were more severely attacked by parsnip webworm, *Depressaria pastinacella*, than the plants in the centre. Attack rates were higher on the edge plants because the unopened umbels necessary for successful oviposition were available here for a longer period (Thompson 1978).

Relative size can sometimes be more important than the absolute size of a plant. Working on the swallow-tail *Papilio machaon* at Wicken fen, Dempster and Hall (1980) found that large plants of *Peucedanum palustre* were unavailable to oviposition butterflies when they were overtopped by tall sedges, yet tiny plants growing on pathsides, where the surrounding vegetation was low, were found by the insects and eggs were laid on them.

Neighbouring plants of other species may reduce the availability of a particular individual by hiding it from view, by protecting it amongst spiny branches, by masking the chemical cues whereby the herbivore finds the plant, or by deterring the herbivore directly by its own smell (see section 5.4 on 'associational resistance'). Similarly, a plant may be more vulnerable in a group of its own kind than when it grows in an isolated position (see section 5.4 on 'resource concentration').

Several studies have shown that herbivores are much more selective in their feeding than they were thought to be. Scientists clipping the young tips from heather shoots in an attempt to mimic grouse feeding, consistently failed to gather diets as high in protein quality as those obtained by the birds (estimated from their gizzard contents). An estimate of grouse food availability based on the weight of clippings per hectare would therefore seriously overestimate the true heather shoot availability (Moss 1972).

An obvious problem is that without destroying the plant and subjecting it to extensive chemical and physical analyses, it is impossible to know what levels of nitrogen, minerals and carbohydrates it yields to the animal, and what concentrations of toxins, deterrents and fibres it contains. Thus two apparently similar plant communities with the same standing crop of plant material may differ widely in availability if one contains, say, higher nitrogen levels than the other, because of varietal differences between the plants, soil conditions or different histories of fertilization.

The availability of forage is also a function of the ability of the animal to process it. Old red deer on the island of Rhum were the first to starve when the food quality declined, because their cheek teeth were too worn down to chew the tougher grasses (Lowe 1969).

Thus an intimate knowledge of the natural history and behaviour of the animal must be gained before an accurate estimate of food availability can be obtained for use in constructing functional response curves and in studies of feeding preferences (see section 3.5). Only in the simplest agricultural systems is total plant biomass likely to be a reasonable estimate of food availability. Even when food items are discrete (seeds, fruits) and uniform in quality, their spatial distribution is likely to make them differentially available to herbivores (see section 3.3.2). A great deal of thought must be given to the way that food availability is measured in the field.

3.1.5   Feeding rate and herbivore density

Both density-dependent enhancement and density-dependent depression of feeding rate can occur. Usually, the more herbivores there are, the less well each individual performs because of competition for food or mutual interference. In a few cases, herbivores feed more successfully in groups; this process is known as 'social facilitation'.

The classic example of facilitation is provided by bark beetles. These animals feed on the relatively nutritious wood immediately beneath the bark of many forest trees. If one adult bores into a healthy tree, its offspring stand little chance of survival; the tree may isolate the phloem and meristematic cells to localize the damage and it may clog the beetles' feeding tunnels with resin. However, the defensive resources of the tree are limited. If sufficient bark beetles attack the tree simultaneously, they will swamp its defences and survive to reproduce. To ensure that many beetles attack the same tree, the first colonists release pheromones which attract large numbers of other adult beetles. If population density is sufficiently high, enough beetles will arrive to ensure a successful mass attack, and the tree will almost certainly be killed. Thus the reproductive rate of these beetles is positively density dependent; the more of them there are, the more successful is their reproduction (Berryman 1976, Wood 1980).

Another example is provided by a fascinating experiment performed by Way and Cammell (1970). They set up first instar cabbage aphids on the underside of leaves. Half of the cages had an aggregation of aphids on the upper surface and half were without an aggregate. The isolated individuals on the underside opposite an aggregate of aphids grew larger and produced heavier progeny than those aphids reared on the same leaf without an aggregate above them. Group feeding clearly improved the food availability to the solitary aphid, presumably because of a non-linear relationship between food quality and salivary injection or because of enhanced carbohydrate flow rates.

Another kind of social facilitation in insects is described by Ghent (1960). Apparently, only a small proportion of the first instar larvae of the sawfly *Neodiprion pratti banksianae* is capable of breaking through the tough epidermis of the pine needle, therefore when cultured singly most of the larvae starve. In aggregations, however, larvae take advantage of the holes in the epidermis made by the few 'first biters' and continue to feed successfully.

The pine needle miner, *Exoteleia pinifoliella*, uses empty mines of the previous generation as oviposition sites; thus trees infested in the previous generation tend to become more heavily attacked while other trees escape entirely (Finnegan 1965).

For vertebrate herbivores, the main benefit of group feeding may be that proportionately less time need be spent on the look-out for predators (Sinclair 1977). In some cases, however, there are clear indications of interspecific facilitation. The rate of grazing of Thomson's gazelles, for example, was more than three times as great in areas that had been closely grazed by wildebeest than in ungrazed, rank vegetation (McNaughton 1976). Here, the feeding rate of the gazelles increases with the population density of wildebeest.

The negative effects of herbivore density on feeding rate can be due to exploitation, mutual interference or 'pseudo-interference'. When the available plant food is less than the amount required for growth, maintenance and reproduction of the herbivore population, then intraspecific competition is inevitable. At one extreme, every individual herbivore feeds at about the same rate until all the food is eaten up. Depending upon the stage of development they have reached at this point, they may all starve, or they may all suffer similarly reduced fecundity. This is 'scramble' competition.

At the other extreme, there is a rigid, hierarchical exploitation of the food. One animal eats its fill, then another and so on until the food runs out. Obviously animals that obtain no food starve to death, but those that do feed, grow to full size and produce a full complement of offspring. This is 'contest' competition (Hassell 1975). The two forms of exploitation have different effects on population dynamics (Fig. 3.4); contest is strongly stabilizing and scramble is relatively destabilizing (May 1973, Hassell 1978).

Herbivore feeding rate may also be depressed because of intraspecific

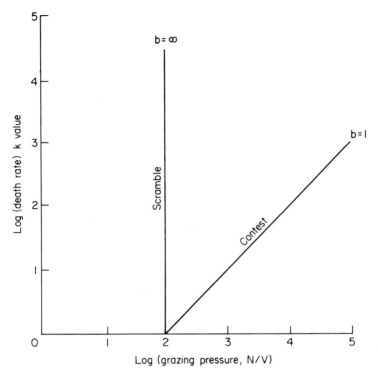

Fig. 3.4 Scramble and contest competition. As the number of herbivores (N) increases relative to the amount of vegetation available (V) there comes a point (2 in this example) when there is insufficient food to satisfy demand completely. In scramble, each herbivore takes 1/N of the food; no herbivore obtains enough to eat, and all die (the death rate is infinite). At the other extreme, the animals feed in turn, contesting the resources, so that each eats its fill before another feeds. There is no wastage of food and the *k* value increases linearly with log (N/V) with slope b = 1 (after Hassell 1978).

contacts, even when food availability is high; for example, Gruys (1970) found that nocturnal contact between the larvae of *Bupalus piniarius* led to reduced growth and subsequently reduced fecundity. Mutual interference between the loopers operated at densities well below the level at which competition for food was likely to begin. Density stress in caged populations of rodents (Krebs & Myers 1974) and rabbits (Myers *et al.* 1971) is the result of an extreme form of interference.

Any tendency of insect larvae to occur in aggregates (because of the oviposition behaviour of the adults (see section 3.5.3) or other factors) may mean that the caterpillars suffer competition due to food depletion even when the total amount of foliage in the habitat appears unlimited (Rafes 1970). When aggregation in patches of high food density leads to their being so intensely exploited that competition between the herbivores results, then the process known as 'pseudo-interference' has occurred (Free *et al.* 1977). The clumped distribution of the food and the aggregative behaviour of the herbivores means that the average feeding success per herbivore over the area as a whole declines with increasing herbivore numbers, even though there is no behavioural interference at all; the effect is due entirely to differential exploitation of patches (Hassell 1978).

3.2   NUMERICAL RESPONSES

A numerical response occurs when a change in plant availability causes a change in herbivore population density.

Most of the unequivocal examples of numerical responses come from populations where food availability is relatively easy to measure; for example, the reproductive success of desert rodents is related to the abundance of winter annuals. Beatley (1969, 1976) found that after adequate autumn rain, there were more than 100 plants m^{-2} and rodent populations subsequently rose to high levels. Dry autumns led to plant populations of only $30\,m^{-2}$ and to disproportionately low rodent numbers (only 14% of

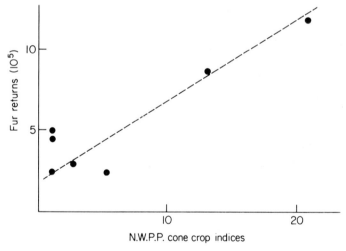

Fig. 3.5 Numerical response of red squirrel to the abundance of cones of *Picea glauca*. Alberta fur returns as a function of the North Western Pulp and Power cone crop indices for the years 1961–76 ($r = 0.911$; $p < 0.05$; Kemp & Keith 1970).

their peak density). In Alberta, there is a significant correlation between the size of the cone crop of white spruce, *Picea glauca*, and red squirrel, *Tamiasciurus hudsonicus*, populations as indexed by annual fur harvests (Fig. 3.5.). Evidently summer rainfall the preceding year reduces cone production which in turn reduces squirrel numbers. The fact that weather patterns in Alberta are widespread explains the observed synchrony in population fluctuations over large areas (Kemp & Keith 1970). The Siberian nutcracker *Nucifraga caryocatactes macrorhynchos* emigrates into Europe in large numbers when its most important food plant, *Pinus cembra sibirica*, fails to produce a cone crop (Formosov 1933). The crossbill *Loxia curvirostra* fluctuates violently in abundance in Finland, increasing dramatically in years when the cone crop of spruce is abundant, and emigrating, often in spectacular numbers, in times of food shortage (Newton 1970). Bank voles, *Clethrionomys glareolus*, in continental Europe show numerical responses to mast years of beech and acorns (Turcek 1960).

The interaction between different factors in determining numerical responses is demonstrated in Newsome's (1970) attempt to produce a mouse plague. In South Australia, breeding conditions are favourable in summer because food is abundant, but burrowing is difficult because the soil is hard and dry. In winter, when burrowing conditions are ideal, there is too little food to allow breeding. Newsome was able to create a plague by feeding an enclosed population through the winter; the control population showed the normal winter decline while numbers in the enclosure with added food rose to $1100\,ha^{-1}$. Newsome did not attempt to prolong the outbreak by providing good summer burrowing conditions, so the population crashed in the summer when the soil became hard.

Food supplementation in *Microtus townsendii* increased population density twofold by increasing reproductive rate, shortening the winter non-breeding period, increasing immigration and reducing home range. A viable surplus of voles seems to exist in natural communities, and population density is probably limited by both behaviour and food (Taitt & Krebs 1981).

Food supplementation of a field population of *Peromyscus maniculatus* improved the production of young and increased slightly the survival rate of adults. Juvenile survival was unaffected and, although total numbers increased, the number of adult males remained the same on control and treated areas (Fordham 1971). This lends support to Sadleir's (1965) suggestion that the male portion of the breeding population is regulated by aggressive behaviour.

Higher levels of protein, calcium, phosphorus and sodium in the diets of *Microtus ochrogaster* living in an alfalfa community rather than in grassland or prairie, led to greater peak densities but did not prevent population declines. The voles in alfalfa showed higher individual growth rates, bred earlier and produced more young than grass-fed animals (Cole & Batzli 1979).

Food supplementation led to increased densities in other studies of small mammals (Watts 1970a, Hansson 1971, Flowerdew 1972) but not in all (Krebs & DeLong 1965).

Food supplementation (with frozen leaf spinach!) brought about improved survivorship in populations of the pond snail, *Lymnaea elodes*. Snails at high density were smaller and laid fewer eggs per egg mass, suggesting that intraspecific competition for food was important in population dynamics (Eisenberg 1970). Since each snail only received a few milligrams dry weight of spinach, it seems likely that some unknown nutrient like a vitamin or a mineral rather than the extra calories was responsible for the increase in survivorship.

Estimates of the maximum rates of increase of herbivore populations are available for a few large vertebrates. Most of the data come from attempts to establish new deer or elk hunting reserves in North America. Two expanding populations of white-tailed deer, *Odocoileus virginianus*, gave r values of 0.547 and 0.444 per year, while three elk *Cervus canadensis* populations grew more slowly, but just as consistently (r values of 0.292, 0.178 and 0.172). These values are calculated from data summarized by Gross (1969). In the arctic, caribou have $r = 0.27$, while musk oxen, which reach sexual maturity later, have $r = 0.18$ (White *et al.* 1981).

For forest-defoliating pest Lepidoptera such as *Zeiraphera* spp., *Bupalus* and *Choristoneura* r values typically lie between 1.0 and 1.5 (Anderson & May 1981). For smaller herbivores the rates can become astronomical; *Aphis fabae*, for example, can increase 4-fold in two weeks, which is equivalent to an annual r value of over 36 (Banks & Macauley 1967)! Hassell *et al.* (1976) summarize the rates of increase of 24 insect species which range between $r = 0.26$ and $r = 4.32$.

3.2.1 Animal growth and food

The growth of all animals is determined principally by the quantity and quality of food they obtain. Larval growth of winter moth *Operophtera brumata* is affected by the quality of the oak leaves it eats. Individuals that hatch late in the spring grow less quickly because their diet is richer in tannins (see section 3.7.2). Larvae fed on artificial diets with casein as the protein source suffered a reduction in growth after the addition of as little as 1% fresh weight of oak leaf tannins. These larvae pupated at a lower weight than normal and experienced reduced survival and lowered fecundity as adults (Feeny 1969).

A rather more subtle timing effect has been noted in the moth *Oporinia autumnata* feeding on birch *Betula pubescens* in Scandinavia. Caterpillar mortality is high on late-flushing trees because of a shortage of feeding sites. A tree that was defoliated the previous year will flush up to 7 days later than normal so more larvae might be expected to starve after a population peak than in a normal year. However, larval growth rate is density dependent, so the pupae produced after defoliation are smaller than normal. The smaller pupae take longer to develop and therefore the adults emerge about 5 days later than normal. Thus the density-dependent reduction in larval growth rate acts to increase the synchrony between foliage availability and larval food demand in the following season (Haukioja 1980).

Density dependence in growth rate is probably widespread in herbivorous insects. Gradwell (1974) reports that pupal weight of winter moths declines as larval density decreases; Carne (1969) describes a decline in the weight of Eucalyptus sawfly *Perga affinis* pupae with increasing defoliation rating of the host tree (Fig. 3.6). Pupal weight is critical in Lepidoptera and sawflies because eventual egg complement is determined at this stage; smaller pupae produce smaller adults which lay fewer eggs (Hinton 1981).

Autumn breeding of sycamore aphid is high when spring populations were low, and vice versa. This is not due to a reduction in the food quality of the sycamore leaves, but is probably a result of intraspecific competition. High early aphid populations produce smaller, less fecund individuals when they emerge from diapause in autumn (Dixon 1975).

One of the paradoxical features of microtine rodent cycles is the fact that the animals in peak populations are heavier than at other times in the cycle. One explanation is that this is simply the result of the smaller, subordinate animals having emigrated, but Krebs and Myers (1974) consider that the weight increase is real and not merely a statistical artefact. Also, if the increase in average weight were due to emigration of the smaller animals, then variance in weight would be lower in peak populations; this is not observed.

Krebs *et al.* (1973) have estimated growth rates of voles *Microtus pennsylvanicus* in increasing and decreasing populations. Growth rate declines with size, but is consistently higher in increasing populations than in

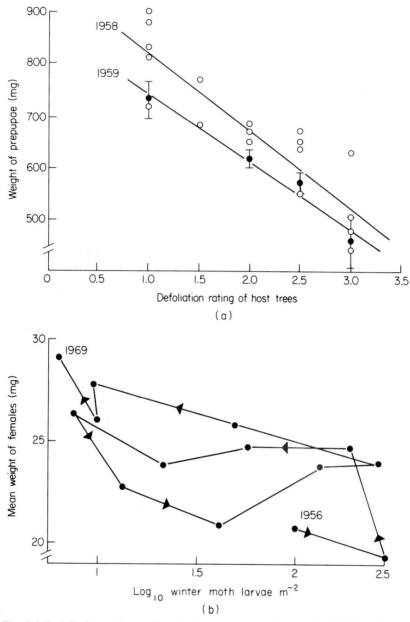

Fig. 3.6 Defoliation and growth in herbivorous insect larvae. (a) Weight of pre-pupae of *Perga* sawflies as a function of the defoliation rating of their *Eucalyptus* host trees (roughly proportional to larval density; Carne 1969). (b) Adult female weight as a function of larval density in winter moth over the period 1956–69 in Wytham Wood, Oxford; Gradwell 1974). Both sets of data highlight the significance of intraspecific competition for food in reducing larval growth rates (and hence pupal weights and finally adult fecundities). It is in-teresting to note that weights are greater in decreasing populations (b) at a given density than in increasing populations (cf. Fig. 3.7).

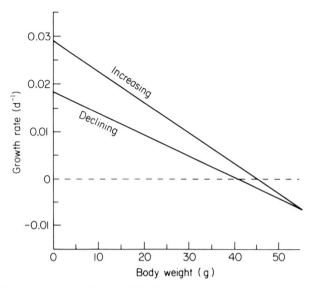

Fig. 3.7 Hysteresis in growth rates. The instantaneous relative growth rates of male *Microtus pennsylvanicus* of different body weights in increasing and declining populations. Males of a given size grow more quickly in an expanding than in a contracting population. The two slopes are significantly different ($p < 0.01$; Krebs *et al.* 1973).

declining populations (Fig. 3.7). These data provide strong evidence for the importance of competition in vole cycles. Large voles are most competitive at all densities. In the increase phase, growth rates are higher at any particular population density because they reflect the integration of low-density effects (more food, less stress and so on). At the same density in a declining population, growth rates are lower because they reflect the integration of high-density effects (less food, more stress, etc.). On the grounds of this simple time-lag effect we would expect the animals to be largest in peak populations.

Density dependence in the growth rate of domestic livestock is less well documented. We should expect that as stocking rate is increased, the yield per animal will decrease. Evidence for such a decline is difficult to find, however, because improved swards, higher fertilizer input and better management often accompany increased stocking rate, and the onset and impact of competition are lessened (Holmes 1980).

A light-hearted, but, nonetheless, interesting example of the importance of weather conditions and herbivore growth is given by Guinness and Albon (unpublished data) from their work on the red deer of Rhum. Earliness of

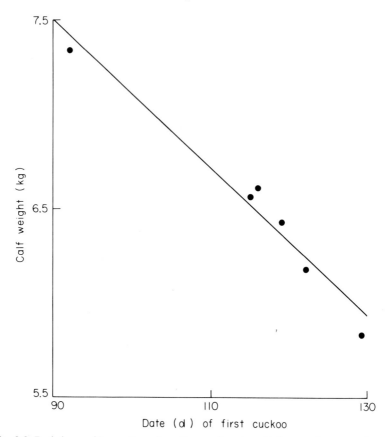

Fig. 3.8 Red deer calf growth and earliness of spring. Calves perform much better in years when spring (as measured by the first record of cuckoo song) comes early (Guinness & Albon, unpublished data).

spring was known to be important in affecting calf growth so, inspired by the Times correspondence columns, they correlated calf weight with the date of arrival of the first cuckoo (Fig. 3.8).

3.2.2   Development rate and food

Reductions in the rate of development have profound effects on the value of *r*, both directly by delaying sexual maturity and indirectly by introducing dis-synchronies between the herbivore and its food plants.

The grasshopper *Poecilocerus pictus* suffered an extension of its hopper period from 75 days to 113 days and passed through 7 rather than 6 instars when its rations of *Calotropis gigantea* were reduced to 25% normal levels (Muthukrishnan & Delvi 1974). Larvae of the moth *Pseudaletia unipuncta* took 3 days to pass through the 5th instar when fully fed but 10 days when fed half rations of corn foliage (Mukerji & Guppy 1970). The age of first reproduction in red deer increases with population density presumably as a result of a reduced plane of nutrition (Gibson & Guinness 1980).

The length of the breeding season in penned populations of wild rabbits in Australia was affected by food supply; when the animals were provided only oaten hay in excess, they bred for 77 days while on unsupplemented pasture the season lasted 106 days. Pastured rabbits whose diet was supplemented with fresh green feed in excess bred for a period of 141 days (Stodart & Myers 1966). Onset of sexual maturity in arctic lemmings occurs later in declining than in increasing populations (Krebs & Myers 1974; see also Fig. 3.9).

The number of generations through which an insect passes each year (its voltinism) is determined in part by the length of time for which food is

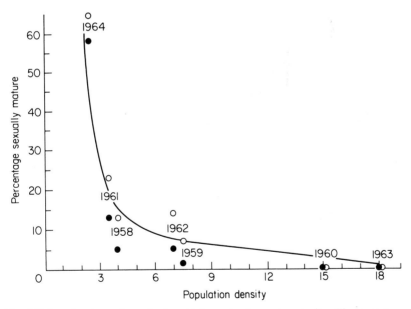

Fig. 3.9 Density dependence in the attainment of sexual maturity. The percentage of *Clethrionomys rutilus* becoming sexually mature in their first summer drops virtually to zero at population densities greater than 9 voles snap-trapped per 100 trap-nights in May (males (○) and females (●); after Koshkina, in Krebs & Myers 1974).

available. At high temperatures, the length of each generation is reduced and in less seasonal environments, the food plant is available longer. Thus frit fly, *Oscinella frit*, has only one generation per year on grasses in the northern USSR, two per year in Britain (the first on young tillers, the second on developing panicles), three per year in central Europe and four per year in the southern USSR (Tischler 1965).

Weather and food quality often interact in affecting insect development. The resin canals in the bud scales of Scots pine *P. sylvestris*, develop in response to the short days at the end of summer (they are triggered by photoperiod). The larvae of the pine-shoot moth, *Rhyacionia buoliana*, cannot attack resin-rich buds so larval survival is high only in those years when summer temperatures are high enough to permit rapid larval development so that the buds are attacked before they become resinous. This process may account for the fact that pine-shoot moth is a pest in North America where summers are warm but not in Western Europe where they are cooler (Harris 1960).

Reduced development rate also increases the risk of predation because the animals are exposed to their enemies for a longer period (and additionally may be less able to defend themselves or escape). Damage done to soybean by the Mexican bean beetle was greater by 11% on a variety that was high in tannin because of compensatory feeding by the beetle (see section 3.5.2). When a predatory pentatomid *Podisus maculiventris* was introduced to prey on the bean beetle the outcome was reversed. The predators ate more of the slow-growing larvae on the tannin-rich variety which consequently suffered only one third as much damage as the control (Price *et al.* 1980).

The contrasting costs and benefits of early moulting to the adult stage were investigated for milkweed bugs, *Oncopeltus* spp. by Blakley (1981). Animals that moult early suffer lower larval mortality but produce adults which are smaller and have lower fecundities and (presumably) lower energy reserves for dispersal. This is particularly important for females that emerge on seedless milkweed plants, since they must migrate to a fruiting plant before they can reproduce (Dingle 1972).

3.2.3   Fecundity and food

There are clear differences in fecundity in species that are found in two or more habitats which are due, at least in part, to differences in food quantity and quality. The mule deer, *Odocoileus hemionus*, for example, has an average of 1.65 fawns per doe in shrubland managed to produce herbaceous species, but only 0.77 fawns per doe in chaparral, a poor range with low food availability (Taber & Dasmann 1957). In Britain, sheep on poor-quality hill pastures average 0.8 lambs per ewe while on productive, lowland pastures they average 1.7 (Eadie 1970).

Food shortage affects the reproduction of invertebrates mainly by the production of smaller, less fecund adults (see section 3.2.1). Poor feeding conditions for the immature stages are often reflected in density dependence in fecundity (Fig. 3.10).

Fecundity is also density dependent in many vertebrate populations; ovulation rates per doe elk vary between 1.9 ova per adult at low density and

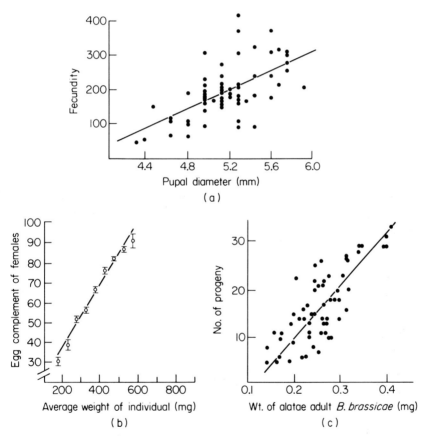

Fig. 3.10 Density-dependent fecundity in insects. Intraspecific competition for food leads to smaller adults, and smaller adults lay fewer eggs. Data for (a) cinnabar moth, (Dempster 1971); (b) Eucalypt sawfly (Carne 1969); (c) cabbage aphid (Way 1968).

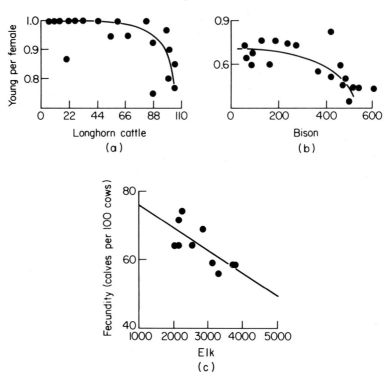

Fig. 3.11 Density-dependent fecundity in vertebrates. Both longhorn cattle (a) and bison (b) show distinctly non-linear density-dependence (Fowler 1981), whereas (c) the fecundity of elk (measured as calves per 100 cows) declines more or less linearly with population density (Boyd, in Gross 1969).

1.35 ova per adult at high density, and the number of fawns produced per 100 deer of both sexes also declines with density (Fig. 3.11). Fecundity depression in vertebrates tends to be affected by feeding success during adulthood rather than by mature body weight alone. Other effects include the resorbtion of embryos; in rabbits, for example, up to 60% of the litters may be lost in massive pre-natal mortality (Brambell 1944).

The major effect of food quality on fecundity is through changes in the level of available nitrogen. Fecundity has been correlated with plant nitrogen for deer, sheep, cattle, voles, grasshoppers, aphids, plant hoppers, weevils, moths, butterflies and many others. It seems that wherever this relationship has been looked for, it has been found. The weight of evidence suggests that aphid fecundity is directly related to the soluble nitrogen content of the phloem sap. Many aphids show microhabitat preference in

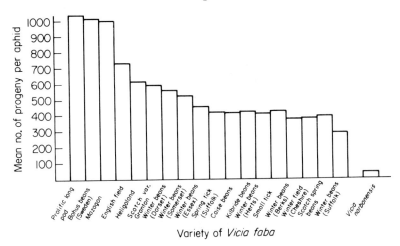

Variety of *Vicia faba*

Fig. 3.12 Fecundity and food quality in insects. Mean number of progeny per aphid for parthenogenetic bean aphid on different varieties of *Vicia faba* and on *V. narbonensis* (from data in Davidson 1922).

favour of young or senescing leaves where the rates of nitrogen translocation are highest (Kennedy 1958). Fig. 3.12 shows the variability in fecundity of *Aphis fabae* on different varieties of its host plant *Vicia faba* and on a different species, *V. narbonensis*. Polyphagous froghoppers in fertilized grasslands responded to the increased amino acids in the phloem with increased fecundity (Prestidge & McNeill 1983).

Enhanced levels of foliar nitrogen, caused by atmospheric pollution with oxides of nitrogen from engine exhausts, may have been responsible for outbreaks of herbivorous insects on trees on motorway reservations in Britain (Port & Thompson 1980).

Laboratory rats are sensitive to the protein content of their diet; the pregnancy rate, the number of litters, the size of the litters and the survival rate to weaning all show depression at low and at very high concentrations of protein. From data in Russell (1948) it is possible to calculate r values for rats on different diets. Fig. 3.13 shows maximum rate of increase at 14% protein and, although more recent research has shown that the maximum occurs at about 18% protein (Sadleir 1969), the existence of the maximum is not disputed.

Plant secondary chemicals also affect reproduction. Infertility in sheep as a result of feeding on pastures rich in subterranean clover is well known to Australian farmers (Bickoff 1968). Laboratory experiments with the vole *Microtus montanus* have shown that cinnamic acids and vinylphenols

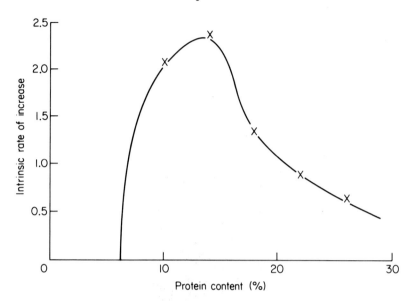

Fig. 3.13 Intrinsic rate of increase and food quality. Rat populations grow most quickly ($r = 2.5 \, \text{yr}^{-1}$) at dietary protein levels of about 14% (from data in Russell 1948).

extracted from winter wheat and added to a standard diet can depress the number of litters per female, the number of young per female and the fraction of females breeding after 100 days. The chemical additives were not toxic to the voles, since control and treated animals maintained the same body weight and apparent condition throughout the experiment. The likely mode of action of these plant substances is as inhibitors of uterine and follicular development. In the field, reproduction by the voles ceases in autumn as their food plants die back. This evident reduction in quality is correlated with increases in the levels of cinnamic acids within the plants (Berger *et al.* 1977).

3.2.4   Mortality and food

Mass mortality is generally associated with climatic extremes, such as exceptionally severe winters that kill vast numbers of birds in Britain or prolonged droughts that kill the ungulates of African savanna. Mass mortality due to demographic causes is less noticeable although the consequences of epidemic disease are often gruesomely apparent (e.g. the spectacle of hoards of debilitated rabbits suffering from myxomatosis).

Deaths due to depletion of food plants are witnessed much less frequently, mainly because the immediate response to declining food supplies is emigration (see section 3.3.1). Also, the signs of death, in the form of corpses or skeletons, are rarely left for us to find, and the band of nocturnal predators and scavengers is extremely efficient.

In ungulates, it is well established that the death rate of juveniles increases with population density and with reduction in food availability, as in wild sheep (Geist 1971) and numerous species of deer (Guinness *et al.* 1978). Poor maternal conditions as a result of low food availability leads to high rates of mortality amongst offspring in deer, sheep and other livestock (Sadleir 1969); for instance, juvenile red deer *Cervus elaphus* on the island of Rhum suffer an average winter mortality of 11%, following losses of 18% between birth and autumn. During the first 6 months of life, calves were more likely to die if their mother was young or old (mortality was lowest among those with mothers 7 to 10 years old), and if the calf was lighter or born later in the season. Winter mortality of juveniles was higher when adult female density was higher (Guinness *et al.* 1978; and see Fig. 3.14).

One of the classic examples of density dependence in vertebrates comes from experiments in shooting wood pigeons, *Columba palumbus*. Shooting did not increase winter mortality above the level experienced in the absence of shooting, nor did cessation of shooting lead to an increase in bird numbers. Pigeon survival was determined by food availability, in the form of grain stubble or clover leys, and immigrant birds moved in to take advantage of any unexploited food (Murton *et al.* 1974).

The few examples of mass starvation due to over-exploitation of plants that do exist have come from studies of insects. In a 5-year study of cinnabar moth Dempster (1971) estimated no mortality due to starvation in 3 years, but 20% and 50% in the other two. In the year of peak insect numbers the ragwort plants were defoliated, but, because the plants had been large, only 20% of the caterpillars starved. In the following year the plants were small, regrowth forms and the high moth population that had survived from the previous year rapidly defoliated them. This 2-year outbreak pattern was also observed for cinnabar moth by Cameron (1935).

In his study of the butterfly *Euphydryas editha* R. R. White (1974) obtained 25 estimates of prediapause larval starvation rate over a period of 4 years at 14 sites. The larvae fed on either *Penstemon, Plantago, Castilleja, Collinsia* or *Pedicularis*; one plant species formed the predominant food at each site. Out of the 25 cases, larval starvation rates exceeded 50% in 10. Sometimes the high rate of larval starvation was associated with significant

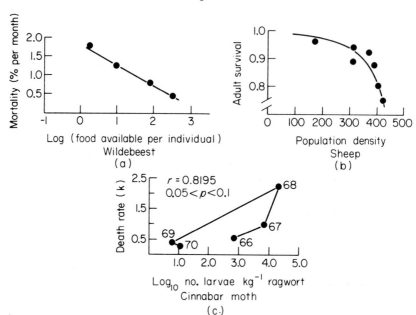

Fig. 3.14 Density-dependent mortality. Increased death rate at high population density has been described for vertebrate and invertebrate herbivores.
(a) Reducing the food available per individual increases the death rate of wildebeest from less than 0.5% to over 1.5% per month (Sinclair 1979). (b) Death rate increases non-linearly with density in adult Soay sheep (Grubb, in Jewell *et al.* 1974). (c) Death rate in cinnabar moth larvae is proportional to the number of larvae per kg of food plant (Dempster 1971).

defoliation and reduced seed production by the plants (in 4 cases more than 50% of the food plant crop was consumed, 2 cases were on *Collinsia* and 2 on *Pedicularis*). More often, larval starvation was associated with low rates of defoliation, and was caused by low rainfall leading to early senescence of the food plants (at Jasper Ridge, starvation rates exceeded 95% every year for four years, but the *Plantago* never suffered more than 3% consumption; Singer 1972, R. R. White 1974).

This butterfly shows an unusual adaptation to food shortage in that some larvae exhibit what is known as 'repeated diapause' (Singer & Ehrlich 1979) so that 'a single larva can, with two weeks feeding each year, outlast the tenure of the graduate student attempting to write a dissertation on its habits'. Populations can, therefore, survive through a year when no host plants are available to prediapause larvae.

The grasshopper *Poecilocerus pictus* suffered 11% mortality in the course of development on full rations, but 42% of the insects died when fed only 25% rations of *Calotropis gigantea* (Muthukrishnan & Delvi 1974). Mortality of the moth *Pseudaletia unipuncta* in the 5th instar only rose appreciably when the caterpillars were fed every fourth day with a fixed ration of corn foliage; whereas less than 10% died if fed more frequently, 80% died if fed every four days (Mukerji & Guppy 1970).

Survival may be affected by the feeding success of the parent, as well as by current food availability. Morris (1967) reared larvae of the fall webworm *Hyphantria cunea* on different aged leaves from the same apple tree. Survival and fecundity were both higher in animals reared on young foliage. The progeny of these insects were then reared; half on a nutritious food (leaves of speckled alder, *Alnus rugosa*) and half on a deficient, synthetic diet. The larvae whose parents had been well nourished survived equally well on the two diets, but the larvae whose parents had been reared on low-quality, late-season foliage, suffered greatly on the deficient diet. Egg viability and early larval establishment were so low from the late-fed batch that no larvae survived at all.

Egg viability is also affected by herbivore density. Waloff (1968) noted only 50% viability in the eggs of the twig-mining moth *Leucoptera spartifoliella* on broom in the year following a peak of abundance, whereas viability had been 95% of the previous year. Klomp (1966), however, in a 15-year study of the pine looper, *Bupalus piniarius*, in which viability varied between 99 and 70% detected no hint of density dependence. Wellington's (1960) work on western tent caterpillar, *Malacosoma pluviale*, showed a good correlation between egg density and egg viability, with emergence falling from 93% in 1956 when the population peaked, to 74% in 1958 during the decline.

3.2.5   Phenology and numerical responses

Breeding success in many herbivore species depends critically on the synchrony between reproduction and the availability of high-quality food.

Aphids hatching at bud burst and feeding in the unfurling leaves can achieve twice the adult weight and produce more rapidly maturing progeny than aphids that emerge later. In sycamore *Acer pseudoplatanus*, however, egg hatch of *Drepanosiphum platanoidis* and bud burst do not occur at the same time relative to one another each year, and aphid mortality can be high on trees where hatch occurs before bud burst (Dixon 1976). The advantage of early feeding to winter moth larvae is shown by the great cost that the species pays to ensure synchrony with young foliage; in most years up to 90% of the larvae hatch too soon to enter the oak buds and consequently perish. If feeding on the more mature leaves did not significantly reduce fitness, there would obviously be strong selection pressure for later hatch (Feeny 1970).

The intrinsic rate of increase of many grass-feeding insects may be extremely low under most conditions. When increases in the nitrogen content of grass phloem (e.g. at the flowering time) coincide with a high proportion of the insects being in breeding condition, then very large numbers of progeny may be produced. Several hoppers (Auchenorrhyncha) show a highly mobile lifestyle, moving from patch to patch within the grassland as food quality improves temporarily (Prestidge & McNeill 1983). Other sucking insects show the same kind of opportunism in selecting ephemeral, high quality feeding sites. The birch aphid, *Euceraphis punctipennis*, is mobile and aggregates in spring and autumn on the leaves of *Betula pubescens* that are temporarily of high food quality (Wratten 1974).

Timing of nesting in greylag geese, *Anser anser*, is related to the peak growth rate of the grasses that form the food of adults and young. The earlier the population as a whole bred, the greater the proportion of clutches that hatched and the greater the number of fledged young per nesting pair (Newton & Kerbes 1974).

Reproduction is not always timed to coincide with peak nitrogen, however; for instance, peak growth and reproduction in the vole *Microtus agrestis* occurred outside the narrow timespan when grass protein levels were at their peak (Evans 1973).

In many seasonal environments, the onset of plant growth is extremely unpredictable on the basis of daylength. It would be highly advantageous if reproduction could be triggered by a precise predictor of forthcoming plant production and curtailed by a predictor of the end of the growing season. Such a system has been detected in the vole *Microtus montanus*. Reproduction ceases in the autumn when the concentration of cinnamic acids in the food plants rises (Berger *et al.* 1977) and is triggered in the spring by the appearance of 6-methoxybenzoxazolinone (Sanders *et al.* 1981). These two chemical cues allow reproduction to be finely tuned to the production of food.

3.2.6   Age structure and numerical responses

The rate at which a herbivore population can respond to an increase in food availability depends upon the fraction of the population that is of reproductive age. Growth will be slow when most of the population is made up of juveniles or of elderly adults, and extremely rapid when most of the animals are of peak reproductive age.

Extreme effects of age structure are to be found when all the members of a population belong to the same age class. The classic example of this phenomenon is provided by the periodic cicadas *Magicicada* spp. In the northern part of the eastern United States these insects take 17 years to develop, and in the southern part their life cycle lasts 13 years. In one area up to three coexisting species of *Magicicada* will be perfectly synchronized, and adults will be seen only once every 17 years (Lloyd & Dybas 1966). Less spectacular examples are provided by the cockchafer *Melolontha* which takes four years to develop in Europe, and the oak eggar moth *Lasiocampa quercus* which shows 2-year periodicity in adult emergence (Bulmer 1977).

It is difficult to understand how such uneven age structures developed (May 1979). Once established, however, it is rather more straightforward to understand how they might be maintained. In the case of the periodic cicadas, it is likely that the periodicity is due to predators; adult cicadas released into areas where the resident cicadas are nymphal are all taken by predators (Marlatt in Lloyd & Dybas 1966). Above-ground predators are satiated in the year when the adults emerge and below-ground predators like moles cannot build up to high population densities by feeding on cicadas alone because of the periods of extreme food shortage that occur when the nymphs simultaneously leave the soil. It would be difficult for a mole population to maintain its density through the period between adult emergence and the time when the cicada's offspring attain moderately large size. Hoppensteadt and Keller (1976) and May (1979) define the kind of predation regime that would create periodicity in simple, age-structured population models.

It is possible that periodic emergence in other long-lived insects is caused by competition between the age classes where the depressive effects on the smaller individuals of between-age-class competition are more severe than those of within-age-class competition (Bulmer 1977). It has also been suggested that cannibalism of small larvae by the older individuals may contribute to the maintenance of periodicity in cockchafer populations (Fidler 1936).

Examples of such bizarre age structures are few, however. Only 3 of the 1500 or so species of cicada are known to show periodic emergence, for instance (May 1979). There are no known cases of periodic populations in vertebrate herbivores.

3.3 IMMIGRATION, EMIGRATION AND FOOD

The spatial distribution of the plants and the spatial distribution of herbivore attack per plant are critical determinants of plant–herbivore dynamics. Different herbivores show different kinds of spacing behaviour depending on the distribution of their main plant foods.

3.3.1 Emigration

Emigration is the most immediate, and often the most important, numerical response to declining food availability; for example, crossbills move locally each year, leaving forest patches where the cone crop is relatively poor and aggregating in regions where seed production is higher. Mass emigrations of Scandinavian populations occur when large numbers of birds (due to previous good seed crops) coincide with a poor spruce cone crop, and on these flights, birds may be found up to 4000 km from their home range (Newton 1970). The emigration rate of sub-adult and juvenile fox squirrels, *Sciurus niger*, is low in years with a favourable mast crop and high when little seed is set (Nixon & McClain 1969).

At clover densities much below 500 leaves m^{-2}, wood pigeons leave the fields in search of alternative foods like brussels sprouts. Also, in snowy weather when the low-growing clover is inaccessible, they switch to feeding on the less nutritious but more conspicuous brassicas (Murton *et al.* 1966). Dark-bellied Brent geese, *Branta bernicla bernicla*, spread out further from one another and forage longer once the cover of their food plant *Zostera maritima* falls below 15% (Charman 1979). Pallas's sandgrouse, *Syrrhaptes paradoxus*, emigrates from the Asiatic steppes in vast numbers when its food is scarce, occasionally reaching western Europe. Presumably the factors affecting sandgrouse food and human food production are similar, because the nomadic tribesmen have a saying that 'when the sandgrouse fly by, wives will be cheap' (Thompson 1926).

Direct evidence of immigration and emigration is hard to come by, especially in studies of small animals. Dempster (1971) obtained an ingenious estimate of dispersal in adult cinnabar moth by comparing the predicted number of eggs (based on a knowledge of pupal density and a regression of adult fecundity on pupal size) with the observed egg density. Egg density was lower than expected following a peak year when emigration was suspected to have been important, and higher than expected following a low-density year when immigrant moths were seen.

Post-defoliation dispersal of larvae is widespread amongst Lepidoptera. The butterfly *Melitaea harrisii* feeds on *Aster umbellatus* in fields and grassy clearings in Maine. The adult lays its eggs in clusters of 20 to 400 but there is no relation between the size of the plant and the number of eggs laid on it. The plants are too small to allow full development of larvae, even from the smallest egg clusters, so larval emigration in search of new food plants is inevitable. The death rate during emigration is very high, and Dethier (1959b) never observed larvae to find new host plants more than 1.2 m away from the defoliated plant they were abandoning. Thus a population of *Aster* plants at a spacing greater than 1.2 m cannot support a butterfly population unless numbers are replenished each year by immigration.

It is hard to see why a mutant *Melitaea* has not arisen that lays its eggs in sufficiently small clusters that development of the larvae could be completed on a single *Aster* plant. In Dethier's 1956 survey of one field, the adult butterfly population was 24 and these produced 57 egg masses averaging 200 eggs each. So while there was a potential larval population of 11 400 only 57 plants were attacked out of a total plant population of 9462.

If a ragwort plant has more than about 60 eggs of cinnabar moth laid on it, even large plants will be stripped completely of their leaves and flowers, and all surviving larvae will have to emigrate in search of new host plants. Large plants that receive small egg batches may not lose all their larvae. Plants that escape attack by ovipositing adults are therefore vulnerable to secondary attack by dispersing caterpillars. The probability of one of these plants becoming infested is a function of the distance to the nearest plant on which eggs were laid that could act as a source of emigrants (Fig. 3.15). In this case the spatial pattern of the plants as well as their density affect the impact of feeding on ragwort seed set (Zahirul Islam 1981).

When small plant size is correlated with high plant density, larval survival during emigration may be relatively high. Wiklund and Ahrberg (1978) found that orange tip, *Anthocharis cardamines*, butterfly larvae were more likely to emigrate from small crucifer species like *Arabidopsis thaliana* but that the risk of failing to find a new host plant was small because the plants always grew in large, dense stands.

Dispersal is critically important in microtine cycles. When Krebs *et al.* (1969) enclosed populations of *Microtus pennsylvanicus* and *M. ochrogaster* to prevent emigration, density rose to abnormally high levels and the vegetation was severely overgrazed. Since overgrazing of this kind is uncommon in unfenced populations, they argued that emigration must usually occur before the vegetation is damaged.

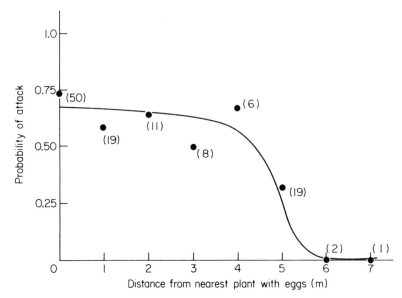

Fig. 3.15 Probability of a ragwort plant being attacked by dispersing larvae as a function of distance from the nearest plant on which cinnabar moth eggs were laid. Ragwort plants more than about 5 m from the nearest egg batch are unlikely to be attacked by dispersing larvae. Figures are the number of plants from which the proportion was estimated (Zahirul Islam & Crawley, in press).

If the grassland is in equilibrium, of course, there will be just as many voles trying to immigrate into one patch as are trying to emigrate from another. Dispersal can only regulate a population when there is a net loss of animals to the habitat as a whole. This usually comes about because animals are forced into marginal or hostile environments where their survival is very low. The dispersal behaviour evolved, presumably, because the fitness of the animals would have been even lower had they stayed. The kinds of animals which disperse are typically the young of the year; they are smaller and less aggressive than the resident adults. In deteriorating conditions they have no chance of leaving descendants if they stay where they were born and, despite the risks, their fitness is increased if they emigrate.

If the risks of emigrating are high, the costs of staying behind may be even greater. Dempster (1968) studied the psyllid *Arytaina spartii* on broom *Cytisus scoparius* and found that when population density on the bushes was low, few animals emigrated and the fecundity of the residents was high (267 offspring per female). Only one year before this, when the density on the bushes was nearly 35 times higher, there was mass emigration. The females that stayed behind, however, produced, on average, only 6.4 progeny each.

Dispersal from the central (and best) area of the habitat by ground squirrels, *Spermophilus armatus*, was a principal factor in population regulation; emigration may have provided the only opportunity for reproduction at peak densities (Slade & Balph 1974).

Density-dependent emigration not only reduces density but also changes the composition of the population; for example, in enclosed populations of *Microtus townsendii* in British Columbia, adults and sub-adults dispersed in spring, leading to a seasonal decline in numbers. The animals that dispersed, however, tended to be smaller than the over-wintered, heavyweight voles that remained. It is likely that the cause of emigraion was aggression in this case too, and the outcome was an increase in the mean size of the remaining vole population (Beacham 1979).

Emigration also occurs because of declining food quality. Pine bud moth larvae are known to be deterred by resin produced by the buds of *Pinus sylvestris*. Voute and Walenkamp (1946) cut certain branches to prevent resin flow and left others intact. All the larvae on the cut branches established themselves in buds, whereas all the larvae emigrated from the resinous branches.

The abandonment of deteriorating host plants is accomplished in aphids by switching from a wingless (apterous) to a winged (alate) adult form. Alate production has been correlated with aphid population density, reduced soluble nitrogen in the sap, senescence of plant tissues, increased carbohydrate content of the leaves, water stress, photoperiod and other factors (van Emden *et al.* 1969). Emigration is not a straightforward density-dependent process, because alate production continues after population density has declined below the level at which wing production was initiated. Wing production and emigration appear to be irreversible consequences of a deterioration of food plant quality (Mittler & Sutherland 1969). Much of the interest in aphid wing formation centres on the fact that the aphids seem to be able to predict the deterioration in food quality; winged forms are produced while the plant is still suitable for growth and reproduction. The great flexibility of the aphid life cycle derives from the fact that although the winged forms are determined, whether or not they emigrate from the plant is facultative, and the alatae can stay to produce young if conditions suddenly improve (Johnson 1969).

Many herbivores follow a vagrant lifestyle, exploiting patches of high-quality food in an opportunistic way. The seed-feeding bird *Quelea quelea* moves over considerable distances in Africa (Ward 1971); the wildebeest of the Serengeti plains follow the rains from one patch of green to another in a seasonal cycle that takes them over thousands of kilometres (Maddock 1979). The infamous African armyworm *Spodoptera exempta* follows a similar pattern. It is a pest of graminaceous crops like maize, sorghum, millet and rice that moves with the rains, flying downwind distances of up to 100 km in a single night; because of this flight behaviour, the moth tends to accumulate in rainy areas and the hatching of its larvae coincides with the flush of young crop growth that follows the rains (Brown *et al.* 1969). Barren-ground caribou make long-range seasonal migrations between their summer feeding grounds on the tundra heaths and their wintering range in the forests of the taiiga (Bergerud 1971).

Plant density affects the progress of outbreaks of many insect herbivores. The psyllid *Cardiaspina albitextura* attacks red gum *Eucalyptus blakelyi* in eastern Australia and can kill the tree if repeated defoliations occur at intervals of less than three years. The tree responds to heavy psyllid attack by shedding all its leaves and then, after a time-lag, producing a new crop of regrowth foliage.

When the trees are closely spaced the insects can move readily from crown to crown, taking advantage of the regrowth on other trees. Numbers therefore remain at high levels and the vigour of the trees is gradually impaired, leading to die back and eventually to death. When the trees are more spaced out, the psyllids are unable to emigrate when high densities have made the foliage unsuitable for oviposition or adult feeding, so great mortality is suffered when the *Eucalyptus* casts its leaves. The regrowth leaves are only lightly infested and the psyllids do not build up to damaging levels before the carbohydrate reserves of the tree are replenished (Clark 1964).

The tendency to disperse may be governed genetically and populations may well be polymorphic for this trait (Southwood 1977). Dispersal in the giant tortoise *Geochleone gigantea* on Aldabra appears to have approximately balanced costs and benefits (Swingland & Lessells 1979). Animals that migrate to the coastal areas of the island to take advantage of the flush of grass growth that occurs there following the rains, run the risk of increased mortality because of the lack of shade trees near the coast. Animals that stay inland have a lower feeding success, but a higher survival rate (98% versus 84%).

Emigration may be caused by deterioration in other aspects of habitat quality such as increased predator numbers (Grant 1972), fouling by faeces

(Freeland 1980), increased human disturbance (Owen 1972) or nuisance from ectoparasites (White *et al.* 1981).

3.3.2 Plant patterns and herbivore dispersal

The way in which herbivores respond to the spatial distribution of their food plants is of the greatest significance to the dynamics of both their populations. Where the mobility, sensory acuity and host-plant-finding ability of the herbivore populations are high, then all plants are likely to be equally 'findable', irrespective of their spatial pattern. In such a case the distribution of herbivore feeding may match plant distribution perfectly; when this ideal of distribution behaviour occurs, the herbivore is said to treat the plant population in a 'fine-grained' way (MacArthur & Wilson 1967), and there is a perfect positive correlation between plant abundance and herbivore use (Fig. 3.16b). Real populations of herbivores are unlikely to be perfectly fine grained in the way they treat their food plants, and will show varying degrees of ill correlation. In some cases, the herbivore will aggregate

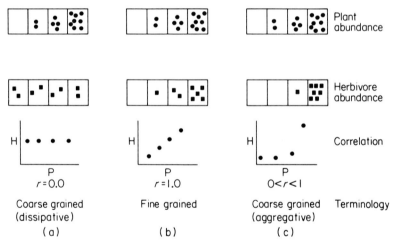

Fig. 3.16 The concept of grain. When herbivore use is directly proportional to plant abundance the population is said to treat the plants in a fine-grained way. There is a perfect positive correlation between herbivore density (H) and plant density (P). Coarse-grained attack may be due to aggregation (c) of the herbivores in the highest-density patches, or to dissipation of herbivore feeding (a) which, due to factors such as territoriality, predator avoidance, etc., leads to good patches being under-exploited and animals spending time in patches of low food density (figure by Soberon & Crawley).

in patches of high plant density (see section 5.4); this will result in positive correlations with low values of the correlation coefficient when herbivore aggregation is very pronounced (Fig. 3.16(c)). Alternatively, the distribution of herbivores may be so sensitive to the distribution of predators, the need for shade or water, the need to avoid human disturbance, or to more subtle effects of neighbouring plant species (see section 5.4) that herbivore feeding and plant abundance are almost completely uncorrelated (Fig. 3.16(a)).

If herbivore numbers are food limited, density will be greatest for fine-grained herbivore species because they exploit the greatest proportion of the food available. Herbivores that aggregate in high-density plant patches 'waste' the food available as isolated plants (they may also compete more intensely with one another within the high-density patch; see section 3.1.5). Herbivores that do not distribute themselves in relation to food availability 'waste' large amounts of food where the plants are aggregated, because of the mismatch between high food availability and high herbivore numbers. In such cases the herbivore may suffer intense competition for food in the occupied patches, despite an apparent excess of food elsewhere. While equilibrium herbivore abundance is positively correlated with the degree of fine-grainedness with which they exploit their resources, population stability may often be negatively correlated. By providing a refuge from over-exploitation, the 'wasted' plants in coarse-grained systems act as a powerful stabilizing force (Hassell & May 1974; see also section 4.3.7).

Grain is not the same as plant spatial pattern (cf. Pielou 1969). It is possible for a herbivore to treat a very patchy plant population in a fine-grained way, as long as the animal is mobile and adept at host plant finding. Similarly, if the herbivore is restricted to part of a uniform crop (as certain pests are concentrated in the edges of fields), then it treats a regular distribution of plants in a coarse-grained way.

When the average distance between one plant and another is long compared with the dispersive abilities of the animals, then plant pattern can have a profound impact on herbivore survival. Consider, for example, cinnabar moth larvae emigrating from defoliated ragwort plants (Fig. 3.15). At high densities, all the plants in a regularly spaced population will be within range of caterpillars emigrating from neighbouring plants, so plant attack rate will be high and insect survival will be good. In an aggregated plant population, a certain number of isolated plants is almost bound to escape, and insect survival will be slightly lower. In low-density, regularly spaced plant populations, most of the plants will be separated from one another by more than the threshold distance of caterpillar dispersal. Thus only a small proportion of the plants will experience secondary infestation and the insects will suffer extremely high mortality. In an aggregated low-density plant population, the fraction of insects that survives will depend on the spatial pattern of oviposition. If eggs are laid on plants at random, or laid preferentially on plants in high-density patches, survival rate will be relatively high because a larger proportion of insect eggs is laid within range of a host plant. If the adult insects oviposit preferentially on isolated plants, the death rate of dispersing larvae will be even higher than when plant pattern is regular. Thus, over a series of generations, both plant and insect numbers are likely to fluctuate much more widely when the plant population is regularly spaced than when it is more patchy.

3.3.3 Spatial patterns of herbivore feeding

It is an integral part of predator–prey theory (Hassell 1978) and optimal foraging theory (Pyke *et al.* 1977) that predators should aggregate in regions of high prey density.

For herbivores, this process operates primarily through habitat selection (see section 3.5.1). Within a habitat the relationship between plant density and herbivore numbers is much less clear; for example, the same pattern of plant density is exploited in quite different ways by different species of herbivore (see Cromartie's (1975) study of the pests of *Brassica* crops; see section 5.4).

Two important aspects of host plant density affect the pattern of feeding by invertebrate herbivores. First, it appears from studies in pest control that the number of herbivores per plant is lower in dense crops than in sparse (Way & Heathcote 1966); this may simply be because the plants are smaller, or because a fixed number of immigrant pests per unit area is distributed over a larger number of plants. Second, the density, size and species make-up of neighbouring, non-host plants can influence the feeding or oviposition behaviour of the herbivore. Host plants may simply be hidden from view, or their chemical cues might be masked by the smells of the other species (see 'resource concentration', section 5.4).

Several insect species prefer to lay their eggs on isolated plants rather than on plants in high-density patches (see section 3.5.3). Usually, however, this is a preference for larger individual plants, rather than a response to plant density as such; for example, Thompson (1978) found that for plants of the *same* size, the attack rate was higher in a patch than for an isolated individual, despite the fact that, per plant, an isolated individual was more prone to attack.

Rausher and Feeny (1980) established high- and low-density field populations of *Aristolochia reticulata* in order to compare attack rates by the swallow-tail butterfly, *Battus philenor*. They found no difference in either the number of leaves per plant that was eaten, or in plant growth or death rates at the two densities.

In vertebrates, the within-habitat distribution is often affected by the previous feeding activities of other species (see section 3.5.1), by social factors, or by the species' own history of exploitation of the food plants; for example, in pastures lightly stocked with sheep, selective grazing is concentrated initially in the better parts of the pasture. The vegetation in the poorer areas thus becomes taller, older and less attractive to the sheep, so that feeding becomes more and more aggregated into the better areas as time goes on (Arnold 1964).

In general, then, herbivore feeding is patchy even in uniform plant populations. The map in Fig. 3.17 shows the spatial pattern of *Aphis fabae* infestation in a field of *Vicia faba* beans in southern England; the insects are aggregated at the edges of the crop, especially in the south-west corner, while the centre of the field is almost pest-free (Way & Cammell 1973). The edge effect is due to the prevailing wind (south-westerly), turbulence round the hedgerow and settlement behaviour of the aphids (Lewis 1965).

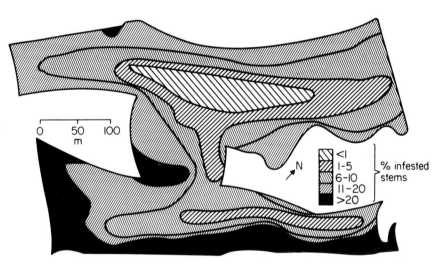

Fig. 3.17 Patchy distribution of pests in uniform crops. The numbers of *Aphis fabae* on field beans on June 1st (see text for details; Way & Cammell 1973).

3.4   MONOPHAGY AND POLYPHAGY

Many herbivorous animals eat a limited number of plant species (oligophages). Others, mainly invertebrates, are restricted to one food plant species (monophages). Some species eat a wide variety of plants (polyphages; see Fig. 3.18). We should distinguish between absolute monophages, which feed on only one plant species over their entire geographical range, and functional monophages which concentrate on one species in a particular habitat, but take different plant foods in different places. Also, many invertebrate herbivores are restricted in their movements during the feeding stages, and may be confined to a single plant for their entire larval development. Thus, while the adult may lay her eggs on several plant species, each individual feeds and develops on just one; the species is polyphagous but the individuals are monophagous.

All herbivores show some degree of selectivity in their feeding, choosing between plant species and between different tissues within the plants. Even archetypal polyphages like plague locusts and goats show strong preferences

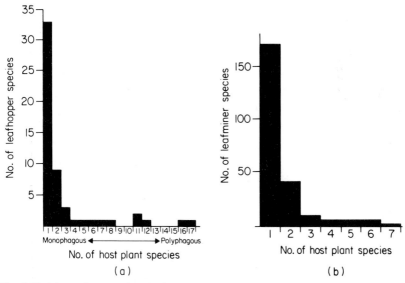

Fig. 3.18 Monophagy and polyphagy in insect herbivores. For mesophyll-feeding leaf-hoppers and leaf-miners on British trees most species are monophagous and there are few broad polyphages (Claridge & Wilson 1981). Vertebrates and larger, leaf-chewing insects tend to show much less pronounced monophagy.

and will avoid certain species entirely (Mulkern 1967, Spatz & Mueller-Dombois 1973).

It is because the class of oligophagous herbivores is so large, and covers such a range of feeding behaviours, that I use the word *specialist* to describe species that feed from a small number of plants, perhaps from only one genus, or from one well-defined, characteristic family of plants (e.g. Cruciferae). I use *generalist* to mean broadly oligophagous (so that generalist and polyphage are virtually synonymous).

The range of food plant specialization is illustrated by the British butterflies. A few species like the black hairstreak, *Strymonidia pruni*, feeding on blackthorn, *Prunus spinosa*, and white admiral, *Limenitis camilla*, feeding on honeysuckle, *Lonicera periclymenum*, are strictly monophagous. Some, like the swallow-tail, *Papilio machaon*, which feeds in Britain only on *Peucedanum palustre*, are functionally monophagous. Others, while they eat several species, are confined to one taxonomic group of plants, as the green-veined white, *Pieris napi*, is restricted to crucifers. A few butterflies can develop on numerous kinds of plants that apparently have little in common. The green hairstreak, *Callophrys rubi*, for example, feeds on rockrose (Cistaceae), purging buckthorn (Rhamnaceae), gorse, broom, *Genista* (Leguminosae), bramble (Rosaceae), dogwood (Cornaceae), *Vaccinium* and *Calluna* (Ericaceae). Thus most British butterflies are oligophagous, a few are monophagous and a few are broadly polyphagous (Ford 1945).

One of the difficulties in discussing the dietary habits of herbivores is that botanists and zoologists understand monophagy and polyphagy to mean different things. Thus to an animal ecologist, koala bears are monophagous because they only eat *Eucalyptus* foliage, while to a plant ecologist, studying the population dynamics of one species of *Eucalyptus*, they are polyphagous because their numbers are affected by the abundance of more than one species of *Eucalyptus*. Similarly, an entomologist would classify the dipteran leaf miner *Phytomyza ranunculi* as monophagous since it only attacks *Ranunculus*, but since it feeds on *R. acris*, *R. repens*, *R. bulbosus*, *R. ficaria* and *R. flammula*, and all these species might occur together in the same moist meadow, a plant ecologist studying the dynamics of one of these species would be bound to consider the fly to be polyphagous.

It is difficult (and not particularly informative) to compare the degree of polyphagy exhibited by different herbivore species; for example, the number of plant species eaten by a herbivore is often affected as much by the composition of the local flora, as by the preferences of the herbivore.

Animals specializing on well-represented plant groups are bound to appear more polyphagous than those which specialize on monospecific genera or on uncommon plant families (see section 5.5.2).

3.4.1 Advantages of polyphagy

The advantages of polyphagy are more obvious than those of monophagy. Probably most important is the fact that food for a polyphage is easy to find, since almost every plant encountered is edible and thus the costs of search for food are low. Similarly, an individual is unlikely to starve because of fluctuations in the abundance of one food plant species, and a polyphagous herbivore is able to switch to plant foods that become temporarily abundant. Food quality varies so greatly within a growing season and from place to place within a habitat, that a polyphagous herbivore may well be able to obtain high-quality food for a greater period of the year, by selecting the most nutritious plant tissues in sequence. A polyphagous species may obtain all these benefits by exercising seasonal or opportunistic habitat preference, so that by dispersal, it explores a temporal succession of food species.

Polyphagy will also be favoured when extended search for food plants is associated with a high risk of predation; small mammals like rabbits and voles would suffer greatly if they had to move long distances between one food plant and another, or move a long way from cover in search of food. A broad diet will also be advantageous when individual plant foods fail to provide all the necessary nutrients, amino acids, vitamins and trace elements required for development (Westoby 1974). Also, when many of the food plants are mildly toxic, taking small amounts from many species may be preferable to taking large amounts from one species. In this way the body's detoxification systems will not be overwhelmed (Freeland & Janzen 1974; see also section 3.7.2).

In vertebrate herbivores there is a broad correlation between body size and polyphagy; polyphages tend to be large and monophages to be small. There is little evidence of such a trend amongst the invertebrates, and within any insect group, monophagous and polyphagous species are roughly the same size (cf. Wasserman & Mitter 1978).

The body size of vertebrate herbivore species that feed on low-quality diets tends to be greater than that of similar animals feeding on more nutritious foods; for example, amongst the African ungulates, there is a negative correlation between body size and dietary protein content (Bell

1970). Low nitrogen is associated with high fibre, low moisture and high secondary chemical concentrations in the diet (see section 3.6.2). Large animals are relatively better able to cope with these problems because of their lower metabolic rate, their ability to accommodate larger and hence more complex digestive tracts, and the greater physical strength of their mouthparts (Mattson 1980). In ruminants, the smaller species have higher relative consumption rates, shorter life cycles and tend to feed more selectively, taking nutrient-rich, more digestible tissues (Jarman & Sinclair 1979).

3.4.2   Advantages of monophagy

So long as a plant food remains abundant, accessible and predictable, selection will favour finer and finer specialization towards monophagy.

The main advantage of monophagy is that by the development of specialized feeding anatomy or behaviour, the herbivores can exploit resources not available to other animals and, in so doing, free themselves from competition.

Monophagy is also advantageous when food plants occur in large, monospecific patches. Here the costs of search are negligible (indeed it would be costly to be polyphagous in such a habitat, because it would require dispersal from one patch to another). A fine example of specialization in virtual monoculture is provided by the crossbills whose peculiar beaks allow them to extract seeds efficiently from the cones of the dominant conifer trees (Newton 1970).

There is little evidence to suggest that monophagous herbivores are any more efficient at detoxifying plant chemical defences (Auerbach & Strong 1981; see also section 3.7.10); for example, forb-feeding Lepidoptera larvae grow faster (0.2–$0.3\,\mathrm{mg}^{-1}\,\mathrm{mg}^{-1}\,\mathrm{d}^{-1}$) than larvae feeding on tree leaves (0.05–$0.2\,\mathrm{mg}^{-1}\,\mathrm{mg}^{-1}\,\mathrm{d}^{-1}$). Larvae of generalized feeding habits were no less efficient than species with more restricted feeding habits, so that any costs that might be involved in generalists having a higher detoxifying capacity, have at most a trivial effect on larval growth rates (Scriber & Feeny 1979).

Monophagous herbivores that have overcome the plant's chemical defences can use its specific chemicals to aid host plant selection (reducing the costs of search), can sequester them for use as pheromones (reducing the costs of low herbivore density), or sequester them for use as defences against predators (reducing the cost of being predictably associated with one plant species; see section 3.7.2).

3.5   SELECTION AND PREFERENCE

One of the most difficult aspects of plant–herbivore dynamics lies in attempting to predict the impact of a polyphagous herbivore on the abundance of its various food-plant species. The plant species differ in their abundances, their accessibilities, their spatial distributions and in their nutritional qualities. On top of this, the herbivore has behaviourally determined preferences for the different species which may be fixed or variable. A knowledge of both availability and preference is vital if the diet of a polyphagous herbivore is to be predicted and its effects of plant population dynamics and herbivore performance evaluated.

Preference can operate at the level of the habitat (see section 3.5.1), the plant species (see sections 3.5.2 and 3.5.3) or the within-plant micro-habitat (see section 3.5.4). The impact of herbivore feeding on the plant population is determined by the interaction of processes at these three levels.

3.5.1   Habitat selection

The spatial distribution of herbivores over a mosaic of habitat patches reflects the outcome of several opposing forces. In an ideal world, where animals had a perfect knowledge of the distribution of food in their habitat, and food was the only factor important in determining their spatial distribution, the animals would congregate in the best patch of the habitat (the patch where their rate of energy intake was maximized). As more and more animals gathered in the patch, however, there would come a point when the individuals would suffer reduced feeding rate, either because of interference or because of exploitation of the food. Animals would then do equally well to occupy the next best patch of the habitat, where feeding conditions are inferior but competition less intense. Thus as herbivore population density increases, the habitat patches should be occupied in a sequence that reflects the ranking of their value as feeding sites, and if numbers decline, the worst areas should be the first to be emptied. This pattern of dispersal behaviour has been christened the 'ideal free distribution' (Fretwell & Lucas 1969).

While few herbivore species can have a perfect knowledge of food distribution, and even fewer are distributed solely in relation to food availability, the germ of the idea is sound. Hunter (1964), for example, found that hill sheep in southern Scotland distributed themselves over six sward types in the pattern predicted by the ideal free distribution. The best patches

(*Agrostis/Festuca* grasslands on well-drained, base-rich soils) were first to be occupied and carried the highest densities of animals. The difficulty with all rankings such as these is that there is a temptation to *define* the best area as the one with the highest herbivore density. In such a case, the ideal distribution is entirely tautological.

A corollary of the ideal free distribution is that reductions in herbivore numbers will lead to the poorest areas of the habitat being abandoned first. There will thus be an increase in the average performance of the population (improved individual growth rate, fecundity and so on), even when numbers are already low. There is little evidence to back up this prediction, even from closely monitored stocks of domestic livestock (Holmes 1980).

While some herbivores are attracted to suitable habitat from long range by sight or smell (by taxis), most habitat selection operates by kinesis; the animals move more slowly, and turn more frequently, in good habitat than in bad.

In a study of horses and ponies, Archer (1973) timed the visits of the animals to 30 different grass/clover plots. The animals showed a distinct preference for patches of most strains of *Lolium perenne* but the well known variety S23 was well down the list.

Wood-pigeons become less choosy in their habitat selection as food stocks decline. In January there is a good correlation between clover density and pigeon density, but by February the food stocks have been depleted and the pigeons are seen even on the poorest fields (Murton *et al.* 1966; see also Fig. 3.19).

Seasonal patterns of habitat selection are exhibited by several African ungulate species which move from one region to another as soil water conditions change, following the flush of growth from the higher ridges immediately after the rains, to the more rank, but moister low ground during the dry season (Bell 1970). In places where the rainfall is higher and differences in plant availability less pronounced, movement between habitats is much less evident (Field & Laws 1970). The order in which the ungulates move into each habitat is determined by their feeding preferences; zebra are the first to enter the tall-grass communities, feeding on stems and leaves. Wildebeest arrive next, feeding on the grass leaves which are now more accessible. Finally, Thomson's gazelle arrive to exploit the short growth grasses and herbs which develop after wildebeest grazing (Vesey-Fitzgerald 1960, Gwynne & Bell 1968, McNaughton 1976).

Habitat choice by geese *Anser anser albifrons* in winter is based upon freedom from snow and isolation from disturbance. Within the habitat,

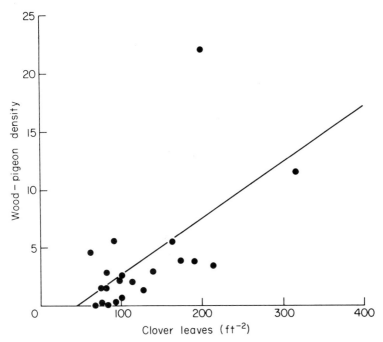

Fig. 3.19 In January there is a good correlation between clover density and wood-pigeon numbers. When clover density is depleted in February the correlation breaks down (Murton *et al.* 1966).

geese prefer certain vegetation zones within a salting pasture. Within a vegetation type, the birds prefer *Puccinellia maritima* to other grasses like *Agrostis stolonifera* and *Festuca rubra*. It appears, however, that selection of food species is of less importance than maintaining a high rate of food intake (Owen 1971, 1972).

Autumn grazing by cattle affects the spatial distribution of deer in spring on the *Agropyron spicatum* ranges of British Columbia. Deer displayed a preference for grazed over ungrazed grasslands, once the green shoots had exceeded the height of the stubble (Willms *et al.* 1981).

Having chosen the habitat in which to feed, there is little evidence that herbivores aggregate in areas of high food plant density (see section 3.3.3); for example, several studies with rodents searching for seeds have shown that scattered food is found as well or even better than clumped food (Reichman & Oberstein 1977, Taylor 1977). Tephritid flies *Urophora* spp. do not aggregate in areas where resources (flowerheads of knapweeds *Centaurea* spp.) are more abundant, but their attack (galls per flowerhead) is distributed contagiously over individual flowerheads (Myers & Harris 1980).

Skylarks, *Alauda arvensis*, foraging in arable farmland in winter preferred areas of high food availability, but they did not exploit areas in sequence of declining profitability. The birds used apparently suboptimal areas to a considerable extent, perhaps to sample food availability, or to make good certain nutrient deficiencies (Green 1978).

Even within a virtual monoculture, food quality can affect the spatial distribution of herbivore feeding; for example, patches of heather *Calluna vulgaris* rich in phosphorus were most heavily grazed by mountain hares *Lepus timidus*, while patches richest in nitrogen were favoured by red deer *Cervus elaphus* (Moss *et al.* 1981).

Habitat choice may be conditioned by predator avoidance (Thomson's gazelle in the Serengeti, Kruuk 1972; snowshoe hares in Newfoundland, Grant 1972), by the need to avoid annoying insects (caribou avoid areas where biting flies are abundant, White *et al.* 1981), or by the need for shelter from severe winter winds (see section 3.1.4).

3.5.2   Detecting feeding preferences

An animal exhibits a preference for an item when the proportion of that item in the diet is greater than its proportion in the environment. A tissue is avoided if it makes up a lower percentage of the diet than it represents in the environment.

Preference has been detected in all groups of herbivores; in insects where the adults search for food plants by flight (Dethier 1959a; Kogan 1977), or by differential re-take-off (Kennedy *et al.* 1959); where nymphs and adults move through a matrix of food and non-food plants (e.g. grasshoppers, Mulkern 1967; locusts, Bernays & Chapman 1978); where larvae move from one host plant to another (Thorsteinson 1960, Soo Hoo & Fraenkel 1966); in vertebrates where ruminants discriminate between plants by their smell and taste (Arnold *et al.* 1980); and where primates select between tissues of different toxicities (Goodall 1963, Montgomery 1978, Oates *et al.* 1980).

Selective feeding is most often detected by plotting the proportion of food intake made up by a particular species against the proportion that the food makes up of the available plant tissues (Fig. 3.20). Vertebrate ecologists use the phrase 'preference ratio' which is the proportion of the food in the diet divided by the proportion of the food in the habitat: values greater than 1 show preference, values less than 1 show avoidance. Values close to 1 show that the herbivore takes the food in the proportion in which it occurs in the

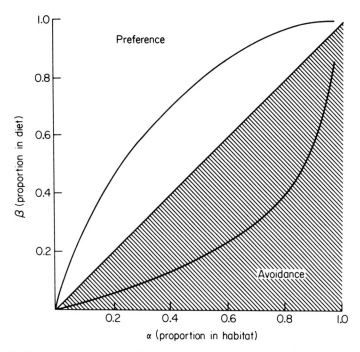

Fig. 3.20 Detecting preference. The proportion of a food item in the diet
(β) as a function of the proportion in the environment (α), i.e. the amount in
the environment divided by the total amount of all *food* species. Neutral feeding
occurs when foods are eaten in the same proportion as they occur. Preference is
shown for species that appear in the diet with a greater frequency than in the
environment. Species appearing less frequently in the diet than in the set of
available foods are said to be avoided.

environment. For ungulates, preference ratios tend to lie between 10, for the
most preferred plants, and 0.1, for avoided species (Heady 1975).

Inadvertent inclusion of non-food plant species in the set of available
foods can alter the assessment of the preference status of a food species,
especially when the non-food species is abundant; for example, suppose that
the percentages of three species in the environment and the diet are:

species	A	B	C
percentage habitat	1	9	90
percentage diet	5	95	trace

When species C is included as available food, then A and B are both strongly
preferred (preference ratios of 5.0 and 10.6). However, when C is omitted

from the analysis, the habitat contains 10% species A and 90% B; B is still preferred (slightly), but A is quite strongly *avoided* (ratios of 1.1 and 0.5, respectively).

Despite this shortcoming, this is probably the best measure of preference when exploitation is insignificant, but more complex measures are necessary when feeding substantially reduces the availability of certain foods (Cock 1978).

The fractions of different foods in the diet can be determined with reasonable precision from the faeces of large herbivores, once allowance has been made for different rates of mastication, different hardnesses and different rates of digestion of the various plant tissues. When oesophageal or ruminal fistulae are employed, very accurate determinations of the fractional make-up of the diet can be made (Golley & Buechner 1968).

The difficulty lies in estimating the points on the x-axis of Fig. 3.20. An accurate determination of the fraction of a food item in the environment demands not only an accurate estimation of the availability of the species in question, but also a good estimate of the availability of all the other potential foods. This will often be extremely difficult in natural habitats (see section 3.1.4).

Practical differences aside, it is clear that all generalist herbivores do show rankings in the rate at which they eat different food plants when given a free choice in experimental tests.

True feeding preferences can only be determined under the very strictest controlled environmental conditions, when all differences in availability between the different foods are eliminated. Thus ground-up piles of precisely the same weight of food (or the same volume, but preferably both) are presented to the animal in free choice areas (so called 'cafeteria trials'). Different feeding rates then reflect different preferences (Westoby 1974).

There are extremely difficult problems of interpretation with experiments like this. Not least, of course, is the fact that real herbivores do not eat ground-up plants. As soon as intact leaves or fruits are presented to the animals, it is all but impossible to ensure that they are equally available (helpings of the same weight will almost certainly differ in surface area, and so on). Similarly, in experiments that run for a long time, the animal becomes progressively more choosy as it becomes satiated. Also, the order in which the foods are sampled by the animal is important. A food encountered after feeding from a palatable species may be treated with disdain, whereas the same food, encountered after noxious material, may be ingested freely.

There are two extreme forms of selective behaviour in trials like this. In the first, the animal moves over the experimental habitat and judges what species are present. It then proceeds to eat its most preferred species until either it is satiated or the food is depleted. If food runs out, the animal moves to its next most preferred species and so on. Under this scheme of hierarchical preference, intake of a plant food is determined not by its own abundance, but by the abundance of the other, *more preferred* foods (Pyke *et al.* 1977).

In the second extreme, the animal samples all the foods; since they are equally available, the proportion of the diet taken from one species is an accurate, quantitative measure of preference. Even here, however, it would be difficult to predict what would happen if the experiment were repeated with one of the more preferred foods omitted. The precise mechanisms that cause the animal to stop feeding from one food and move on to another are extremely difficult to fathom.

In any field experiments, the effects of availability and preference will be inextricably intertwined. At intermediate levels of availability, with alternative foods of widely different preferences, it will be practically impossible to say whether a plant failed to appear in the diet because of low preference or because of low availability. There is a danger that the difficulty in separating preference and availability effects leads us to attribute all discrepancies to whichever factor is measured with least precision. It is all too easy to assume, because one group of plants is avidly attacked by herbivores while a nearby patch escapes attack altogether, that the plants differed in subtle ways, unmeasured by us, but detectable to the herbivores. Hypotheses of this kind should be tested by transfer experiments (see section 3.3.2). Interpretation of field results is further hampered by the inherent patchiness of much herbivore damage; perfectly suitable host plants may escape attack simply because of the non-random feeding behaviour of their herbivores (see sections 3.5.1 and 3.5.4).

Much of optimal foraging theory has been based on the assumption that natural selection will favour individuals that maximize their net rate of energy intake (Schoener 1971, Pyke *et al.* 1977, Krebs *et al.* 1981). Typical predictions of this body of theory are that large foods should be preferred to small; near foods to distant foods; patchy foods to spaced-out foods; calorie-rich foods to calorie-poor foods; digestible foods to indigestible foods; and foods that are easy to ingest to foods that are hard or time-consuming to ingest.

All this is eminently rational and many of these predictions will be true for herbivores, but the ideas were developed largely for predator–prey systems where variations in food quality (other than size and energy richness) were assumed to be negligible. Food quality varies so much for herbivores that maximizing the rate of caloric intake is unlikely to be the principal concern in most cases (Pulliam 1975, Belovsky 1978).

Optimal foraging predicts that food species should be strictly ranked, and new plants should be included in the diet only when more preferred kinds are depleted. The diets of generalist vertebrate herbivores are not structured like this, but contain complex mixtures of species (Westoby 1974). When the plants providing the bulk of the energy needs do not contain certain essential nutrients, an optimal herbivore may show partial preferences (Pulliam 1975). Alternatively, diets may be varied simply by the animal's need to sample its food environment continuously in order to remain aware of the changing phenology (and hence food quality) of the plants in its habitat (Westoby 1978).

Invertebrates may also show mixed diets even when the opportunity to specialize is apparently open to them; for example, the plate limpet, *Acmaea scutum*, preferentially consumes mixtures of foods even when each of its foods is readily available. In theory, this ought to drive the rarer of the two foods to extinction, and is clearly a destabilizing mode of feeding if grazing reduces plant abundance. However, detailed observations show that limpet feeding does not significantly deplete the abundance of the two main species of encrusting microalgae, *Petrocelis middendorffii* and *Hildenbrandia occidentalis*, despite the fact that the limpets take about 60% *Petrocelis* and 40% *Hildenbrandia* over a wide range of relative densities (Kitting 1980).

The fundamentals of selective feeding are summarized in Ivlev's (1961) masterly and far-sighted account; satiated animals are more choosy than hungry animals; animals are more selective when food density is high than when it is low; animals may show switching after exploitation has reduced the availability of a preferred and previously abundant food source; interference between animals at high densities can lead to their becoming less selective even when the amount of food per animal is maintained; animals are more selective when foods are patchily distributed than when they are regularly spaced-out; changes in the spatial distribution of foods may bring about alterations in preference and lead to switching (see section 3.5.5).

3.5.3   Food selection by vertebrate herbivores

Different species select food on different criteria. Sheep and cattle select leaf in preference to stem, and green material in preference to dry or old (Arnold

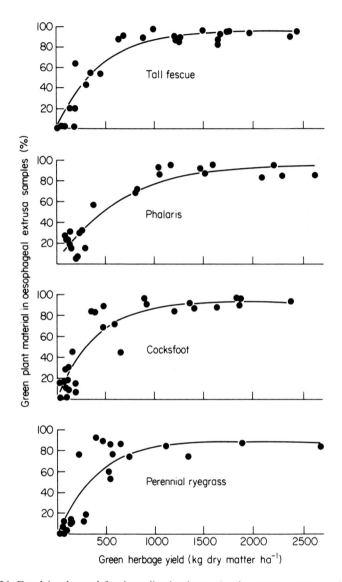

Fig. 3.21 Food intake and food quality in sheep. As the green matter available in the sward falls from 2500 to 500 kg ha^{-1} the sheep are able to keep their diets high in green matter by feeding more selectively. Below 500 kg ha^{-1} they are unable to maintain diet quality by selective feeding and the percentage of green matter in their diets falls rapidly (Hamilton *et al.* 1973).

1964). Compared to the forage as a whole, the selected material is usually higher in nitrogen, phosphorus, sugars and gross energy, and lower in fibres (Arnold 1964). Sheep also discriminate between one species of grass and another. Soay sheep, for example, preferred *Festuca rubra* and *Poa* spp. (preference ratios of 2 to 5), but avoided *Agrostis* and *Holcus* spp. (both 0.5; Jewell *et al.* 1974).

It is difficult, however, to generalize about the feeding behaviour even of particular species; while sheep are considered to be highly selective grazers in Australian grasslands (Arnold 1964), they are thought to be less selective than cattle in American short-grass prairies (Schwartz & Ellis 1981).

The influence of pasture quality of food intake in sheep is shown in Fig. 3.21. When there is more than about 550 kg ha^{-1} dry matter of green herbage, the sheep select a diet significantly more digestible than the pasture as a whole. Below this level of availability, the sheep obtain a diet of lower digestibility than the average green matter in the pasture because the sward is so diluted by dead grass that they are unable to select between the different types of green material (Hamilton *et al.* 1973). In fact, both ungulates (M.R.S. Price 1978) and primates (Clutton-Brock 1977) typically select diets that are higher in protein than the average levels in the foliage available to them.

Recent studies with other groups of herbivores, however, suggest that they base their feeding preferences not on maximizing positive nutritive value, but on minimizing the concentration of toxins, repellants and digestibility-reducing substances in their diets. Bryant and Kuropat (1980), for example, studied the winter browsing habits of a variety of arctic animals; three species of ptarmigan, three grouse, capercaillie, two kinds of hare, and the moose. In each case their conclusion was the same; all these animals ranked their foods on neither energy nor nutrient content. Instead, their preference was strongly and negatively correlated with the ether-extractable resin content of the tissue (terpenes and phenolic resins). The animals preferred buds, catkins and twigs of willows to those of birch or alder, despite the fact that willows were the 'least nutritious' of all the species in terms of calorific value and nutrient content.

The costs of obtaining protein-rich foods are sometimes greater; for example, the nuts of shagback hickory are richer in protein and more digestible than acorns but are harder to open. Thus, squirrels take acorns in winter when they are energy limited, and hickory nuts in the breeding season when their protein requirements are highest (Smith & Follmer 1972). Red grouse avoid feeding on both the youngest (most 'nutritious') and the oldest heather plants, but these preferences are related to the heather's height and the cover it provides, as well as to its nutritional quality (Savory 1978).

Desert rodents select seeds on the basis of their size and ease of handling (Rosenzweig & Sterner 1970). Larger mice tend to take larger seeds, but collection of seeds of different sizes may well be an indirect consequence of microhabitat selection rather than a direct feeding preference (M. V. Price 1978, Wondolleck 1978, M'Closkey 1980, Hay & Fuller 1981).

The detection of feeding preferences in field populations is hampered, as we have seen, by the fact that the amount of food taken from one species is affected by the abundance and attractiveness of alternative foods (see section 3.5.2). This effect is illustrated by experiments with voles *Microtus agrestis* feeding on cyanogenic and acyanogenic forms of *Lotus corniculatus*. The voles ate both forms when fed *Lotus* alone but ignored the cyanogenic morphs when more palatable alternative food was provided in the form of oats and carrot (Crawford-Sidebotham 1972).

The rate of feeding from a particular plant will often depend upon the abundance of more preferred food species, rather than upon the abundance of the plant in question. When more preferred foods are abundant it may appear that a particular plant does not form part of the diet, and only when alternative foods are quite absent will it be possible to interpret changes in the feeding rate as pure responses to changing availability.

In one of the few field examples where the availability of a range of plant species was known to be approximately equal, a clear ranking by preferences was evident. The experiment happened by accident; a forest plot had been laid out so that each of four tree species grew within a few metres of any point, and all the trees were 5 years old. Deer broke through the fence and browsed the trees. They preferred *Pinus banksiana* to *P. strobus*, took little food from *P. resinosa* and ignored *Picea glauca* completely. Even here, though, *P. banksiana* had grown considerably taller than the other species and may well have been more available to the deer (Horton 1964).

The abundance of more preferred natural foods is often important in pest control; for instance, the damage caused by bud-feeding bullfinches, *Pyrrhula pyrrhula*, in orchards is negatively correlated with the abundance of ash seeds, *Fraxinus excelsior*, during January (Summers 1981).

So long as the preferred and non-preferred food plant species are known in advance, reasonably precise predictions can be made about diet composition. For moose *Alces alces* on Isle Royale, it was found that the nutrient content of the plants, the size of the food items and their relative abundances allowed a close prediction of diet composition (Belovsky 1981). Non-preferred species that were predicted to form a larger part of the diet on the basis of their nutrient content and item size, were avoided presumably because they were distasteful or toxic to the animals (see section 3.5.2).

Preferences will normally be conditioned by foraging costs; herbivores will be more selective in the foods they gather from greater distances. The beaver, *Caster canadensis*, is more selective in the sizes of trees it fells with increasing distance from its pond. Close to the pond it takes a wide range of sizes, but prefers smaller species like witch-hazel, *Hamamelis virginiana*, the further from the pond it moves (37% of the trees cut beyond 70 m were witch-hazel but these comprised only 16% of the trees available; Jenkins 1980).

It is of interest to know whether herbivores make 'sensible' dietary decisions and what criteria they adopt in choosing foods. Natural selection will presumably have weeded out those who made poor decisions. While there is a good deal of experimental evidence to show that preference is strongly correlated with performance, there is no reason to expect that preference and performance will be correlated when animals are presented with foods they do not encounter in the field (see section 3.7.6). For herbivores in alien or man-made environments, it will be impossible to judge the suitability of novel foods by visual, olfactory or tactile clues. The only way to find out how good (or bad) they are is to try some and see. Food sampling behaviour has been extensively investigated in rats (Rozin 1977) and has been analysed in the context of large vertebrate herbivores (Westoby 1974, Belovsky 1978, Maiorana 1978).

Contrary to optimal foraging theory, the diets of large vertebrate herbivores are typically composed of many plant species. Apparently 'suboptimal species' may be added to the diet for a number of reasons;

(1) the abundance of preferred foods is too low to allow the animals to restrict their feeding to the most preferred species alone;

(2) it is not worth being more choosy; the costs of discriminating (increased searching and handling time) are greater than the rewards;

(3) the species that yield the maximum rate of energy return do not supply all the dietary requirements and extra species are eaten to provide vitamins, minerals or specific amino acids (see section 3.6.6);

(4) while the abundance of food may change slowly through the season, its quality may alter very rapidly; the animals need to sample a wide range of plants so that when preferred foods become unpalatable or are used up, the herbivores can switch to those of the remaining plants that are of highest quality;

(5) some moderately toxic species might only be edible in small quantities (see section 3.7.2);

(6) the foods may not differ sufficiently that the animals need (or can) discriminate between them.

3.5.4 Host plant selection by invertebrate herbivores

The large literature on host plant selection by insects is accessible through reviews of feeding attractants (de Wilde & Schoonhoven 1969) and feeding inhibitors (Chapman 1974). Volatile chemicals may attract the ovipositing female from a long distance (Saxena & Goyal 1978) and other chemicals may stimulate or inhibit egg laying once she has alighted (Riddiford & Williams 1967, Lundgren 1975). In other groups like aphids, the female selects host plants by a process of random alighting followed by differential re-take-off.

Adult aphids dispersing by flight have rather limited ability to choose where they land, although many species are attracted to yellow objects like senescing leaves, water traps or washing hung out to dry. There is clear evidence, however, that *Myzus persicae* alights on host and non-host plants with equal frequency, and that the final distribution of aphids over the plant population is brought about by differential re-take-off and low-level flight (van Emden *et al.* 1969). Re-take-off or arrestment follows a period of probing by the aphid, during which the suitability of the plant is tested. Since it takes at least 15 minutes to reach the phloem, and most bouts of probing last only 1 to 5 minutes, it is evident that host plant sampling is superficial. The choice of staying or leaving is based on olfactory, tactile and taste cues of the epidermis rather than on a sample of the food itself. Aphids frequently leave plants of the right species that appear to be in ideal physiological condition for attack; for example, Muller (1958) found that 99% of *Aphis fabae* leave the resistant bean variety *Vicia faba* 'Rastatt' within a few minutes but 90% of them also left the susceptible variety 'Schlanstedt'.

Insects that show host plant alternation in different generations have adults that show different acceptance criteria. One British butterfly, the holly blue *Celastrina argiolus* exhibits this curious habit; the insect is bivoltine and the larvae of the first generation feed on holly in June and July while larvae from the second brood of eggs develop almost entirely on ivy (Ford 1945).

The adult insect determines not only which species of plant are attacked, but also the distribution of eggs per plant (and hence the impact of herbivore feeding on plant performance). There is no evidence to support the once widely held belief that adult ovipositional behaviour is conditioned by the kind of food plant on which it developed as a larva (Wiklund 1975, Claridge & Wilson 1978).

Oviposition patterns vary within a group. Of the British butterflies, for example, the majority lay their eggs singly, spaced out on different plants. A few (5 species) lay their eggs in small batches of 5 to 15, while 9 species lay very large egg clusters of 100 eggs or more. Two species whose larvae feed on abundant grasses broadcast their eggs while flying low over the pasture (Ford 1945). Cluster laying in different Lepidoptera is correlated with unpalatability and aposematic colouring of the larvae (Fisher 1930), with reduced parasitism (Dowden 1961) and with obligate ant-association (Kitching 1981).

Many butterflies avoid laying eggs on plants where other eggs have already been laid. This tends to produce a regular distribution of eggs per plant as shown by orange tip, *Anthocharis cardamines*, and its principal food plant *Cardamine pratensis* (Fig. 3.22). This butterfly further enhances the regularity of its larval distribution by being aggressively cannibalistic; it is rare for more than one larva to survive per plant.

Other insects are deterred from oviposition by later signs of herbivore damage; for example, the buttercup leaf miner, *Phytomyza ranunculi*, avoids leaves containing mature larval mines but cannot discern recent egg scars (Sugimoto 1980).

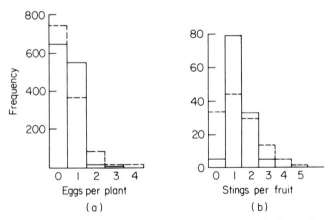

Fig. 3.22 Regular patterns of attack by insects. (a) Orange tip butterfly eggs on plants of *Cardamine pratensis* (Wiklund & Ahrberg 1978). (b) *Dacus tryoni* oviposition stings on loquat fruits (Monro 1967). (Observed, ——; expected Poisson distribution,-----.) These patterns arise by avoidance of plant parts already attacked.

Oviposition pattern is often affected by the spatial pattern of host plant distribution; for instance, the orange tip butterfly prefers isolated plants and plants on the edges of clumps (Wiklund & Ahrberg 1978). Parsnip webworm, *Depressaria pastinacella*, is a specialist seed-spinning moth that feeds on the umbellifer *Pastinaca sativa*. Thompson and Price (1977) found that isolated plants supported 50% more larvae per gram of umbel, suffered an 80% higher attack rate per plant, and were subject to a 40% higher attack rate per umbel, than plants in patches of 30 to 40 m^{-2}. This preference for isolated plants is due to the fact that the adult moth oviposits on to unopened umbels, and these are available in greater numbers and for a longer period on isolated plants.

Several butterflies display two rather odd, apparently maladaptive traits. They seem to make 'ovipositional mistakes' involving errors of both commission and omission. First, they lay some eggs on plants that are toxic to the larvae; *Papilio machaon* lays on the umbellifers *Bifora radicans* and *Levisticum officinale* (Wiklund 1975), and *Pieris napi macdunnoughii* lays on the crucifer *Thlaspi arvense* (Chew 1977) on which there is 100% larval mortality. The simplest explanation of this behaviour is that, since the plants are recently introduced aliens of the right plant family, they exhibit the correct ovipositional stimuli, but selection has not operated strongly enough or long enough to produce discriminating adult behaviour. Also, many mistakes may be due simply to scarcity of the preferred host plants. Most insects will oviposit on anything rather than hold back their eggs.

Second, as Fig. 3.23 shows clearly, adults do not oviposit on many species that would apparently make suitable hosts for larval development (Wiklund 1975). The possible explanations for this are numerous. In the first place, larvae can be fed experimentally on plants that come from places that the adult insects never go (because of light or humidity conditions, altitude, exposure, vegetation structure and so on); since eggs can never be laid naturally on these plants there will not have been selection for the adults to recognize them, or for the plants to be defended against larval feeding. Further, the fact that a larva can survive on a food is no evidence of the strength of selection acting in the field, nor of how fecund such insects would be as adults (larval survival may be a poor index of genetic fitness).

Many kinds of *Eucalyptus* have mature and juvenile growth-form leaves that differ in morphology and chemical composition. Although adults of the chrysomelid beetle *Paropsis atomaria* do not oviposit on the young mature-form leaves of *Eucalyptus bicostata* in the field, these leaves are as good as, or superior to, the leaves of its normal host *E. blakelyi* as larval food. However,

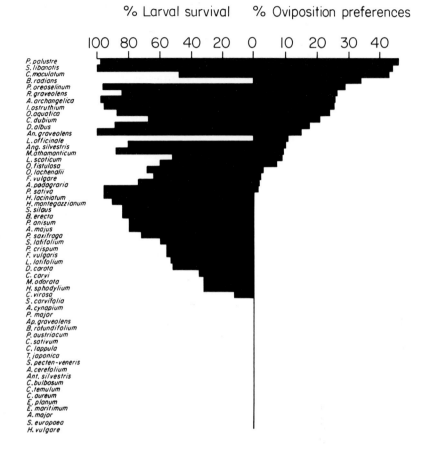

Fig. 3.23 Preference and performance in a butterfly. There is a broad correlation between oviposition preference and larval performance, but swallow-tail butterflies lay eggs on umbellifers like *Bifora radians* and *Levisticum officinale* where larval survival is very poor. They fail to lay eggs on plants like *Pastinaca sativa* and *Heracleum laciniatum* which allow high larval survival. These 'ovipositional mistakes' are discussed in the text (Wiklund 1975).

the larvae grow very slowly and suffer high mortality when fed the juvenile-form leaves of *E. bicostata* (Carne 1966). The butterfly *Battus philenor* changes its ovipositional behaviour as the season advances; as the foliage of its preferred species *Aristolochia reticulata* declines in quality as larval food as it ages, the butterfly lays progressively more of its eggs on *A. serpentaria*. The reduced quality of older leaves is due to their low nitrogen content and high toughness rather than to digestibility-reducing plant secondary compounds (see section 3.7.2). For *A. serpentaria*, mature and young leaves are equally acceptable to the butterfly larvae (Rausher 1981a).

Many species display great variation from one individual to another in their oviposition behaviour. Wiklund (1981), for example, found that females of *Papilio machaon* showed a strict hierarchy of oviposition preferences and that the ranks were correlated with the suitability of the

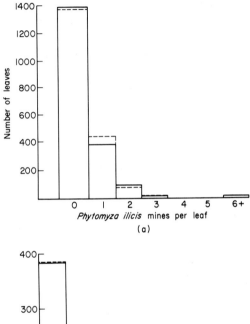

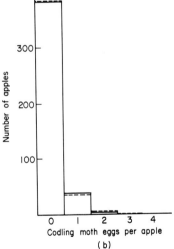

Fig. 3.24 Random attack distributions. (a) Holly leaf-miner on the leaves of holly (Crawley, unpublished observations). (b) Codling moth eggs per apple (Geier 1963). (Observed, ———; Poisson expectation, ------.)

plants as food for the larvae. Individual females, however, differed in the thresholds of availability at which new plant species were accepted. Some females concentrated almost entirely on the most preferred species, while, at the same relative plant densities, others laid eggs on many species, including some on which the larvae could not develop.

For discrete tissues like fruits and leaves it is possible to investigate the statistical distribution of herbivore attack. If the probability of attack was unaffected by previous oviposition, we should expect the number of eggs per fruit to follow a Poisson distribution; this is the distribution observed for codling moth, *Cydia pomonella*, in apples and *Phytomyza ilicis* mines in holly leaves (Fig. 3.24). While the holly leaf miners do not appear to compete, the codling moth is cannibalistic, and only the largest larva will survive in a fruit (Geier 1963).

Eggs of the Australian fruit fly *Dacus tryoni* are aggregated in hard-skinned fruits because the adult fly finds it easier to lay into previously punctured skins (Pritchard 1969). The same fly, however, appears to avoid fruits of loquat *Eriobotrya japonica* which have already been stung, thus giving rise to a regular distribution (see Fig. 3.22; Monro 1967).

It is difficult to interpret these attack distributions critically unless one can be reasonably certain that the plant tissues are uniform in quality. It is possible, for instance, that unattacked tissues possess a subtly different chemistry or experience different microclimatic conditions from those that suffer greater impact. If this is true, random attack of the susceptible plants will appear as statistical aggregation (random attack with added zeros; see Pielou 1969). The main experimental problem is that herbivore attack itself can alter plant chemistry, so post-hoc analyses tell us rather little about plant chemistry at the time the adult insects made the decisions about which plants to attack and which to avoid. Again, unless we know exactly what constitutes an available plant, we shall be able to say little about adult choice (see section 3.1.4). Comparison of the performance of transplanted larvae on a set of plants that were avoided, and a set that were selected, might produce evidence of food quality differences which were perceptible to the adults.

The pattern of larval distribution does not always conform to the pattern of oviposition. In cinnabar moth, for instance, the eggs are laid in clumps of up to 100 on the underside of the rosette leaves. The first instar larvae are cryptic in their colouration and feed on the leaves, resting and moulting in dense aggregations. In later instars their feeding behaviour and their spacing behaviour both change. They move up the plant to feed on flowerheads and buds, they develop striking orange-and-black aposematic colouration and they are much less aggregative. Some larvae disperse in search of other plants long before population density rises, or food quality declines, to levels that stimulate general emigration (van der Meijden 1976). Similarly, Singer (1971) found that early larval stages of the checkerspot butterfly *Euphydryas editha* are immobile and show no host plant selection while older, post-diapause individuals are mobile and capable of host plant discrimination. Here, adult oviposition behaviour determines the distribution of young larvae, while larval selection and host plant density determine the distribution of the larger forms.

We know almost nothing about the preferences of most grass-feeding insects (but see Mulkern 1967, Bernays & Chapman 1978). This is partly due to a traditional reluctance on the part of entomologists to identify these plants, and also to the difficulty of collecting insects from individual grasses; 'swept from low herbage' is as close to botanical precision as labels from some entomological collections come! Some insects like *Holcaphis holci* appear to be taxonomic specialists, while others, like the moth *Noctua pronuba* are almost certainly broad generalists. Some grass-feeding invertebrates are specialists not on taxonomic species but on foods of different qualities. The mobile frog-hoppers and leaf-hoppers (Auchenorrhyncha) are opportunistic on plant sap of different nitrogen concentrations, moving from species to species and from tissue to tissue as plant nitrogen levels change (Prestidge & McNeill 1983). The frit fly, *Oscinella frit*, selects grass tillers by their age, rather than by the species to which they belong (van Emden & Way 1973).

Preference is correlated with performance in invertebrate herbivores; for example, measuring larval performance by weight gain, the range caterpillar *Hemileuca olivae* grew best when fed sorghum, sudan grass, blue gama and corn, and worst when fed Japanese millet, oats, barley, Russian wild-rye and Kentucky blue grass. Preference tests showed highly preferred plants to be buffalo grass, sorghum, foxtail millet and sudan grass while non-preferred species were Japanese millet, western wheat-grass, corn, wheat, barley and oats (Capinera 1978). While there is a significant rank correlation between preference and performance, these data highlight many of the problems of experimental diet studies. The most preferred species, buffalo grass, did not prove especially suitable (in the short-term of this experiment). One of the least-preferred species, corn, when fed to larvae produced significant weight increases. It is quite possible, of course, that preferences evolved over long periods in the field are really much better indicators than any short-term measures of performance. Buffalo-grass may provide dietary requirements that are only evident in the long-term; corn may give rise to long-term dietary disorders despite the short-term weight gain observed in these experiments.

3.5.5   Within-plant microhabitat selection

Herbivores are just as selective of the different parts of a plant as they are of one species over another. Microhabitat selection in oviposition site is demonstrated by the five members of the leaf-hopper guild on the grasses *Holcus mollis* and *H. lanatus*; the insects prefer one of three vertical zones and, where two species share a zone, one prefers to lay its eggs on leaf and the other on stem (P. Thompson 1978; see also section 5.7.2). Animals also show seasonal patterns of exploitation of different parts of the same plant species; for example, the beach vole *Microtus breweri* feeds mainly on dune grass *Ammophila breviligulata*, taking the leaves in spring and early summer (the breeding season) when they are richest in protein, taking roots in late summer and autumn, and stems through the winter and early spring. Changes in preferences were demonstrated in feeding trials and were shown to be correlated with changes in the fibre, water and phosphorus contents of the tissues (Goldberg *et al.* 1980).

The gall-forming aphid *Pemphigus betae* produces a single gall on the leaf of *Populus angustifolia* in which all the progeny of one stem mother are contained. It is thus straightforward to estimate the fecundity of each immigrant female. The winged females show microhabitat selection both between and within leaves, preferring large leaves to small, and positions close to the petiole to those near the leaf tip. The number of progeny per gall is higher on large leaves than on small, and higher in galls near the petiole than in those near the tip. For leaves of a given size, average fecundity declines with increases in the number of competing galls (Fig. 3.25).

Under the ideal free distribution (see section 3.5.1) the settling densities should be adjusted such that the average fitness in areas of differing competitor density will be identical. The dispersing aphids do appear to have distributed themselves over leaves of different qualities (sizes) and varying competitor densities in such a way that average fitness is equal at each density of competitors (the heavy line in Fig. 3.25 shows that the average fecundity is more or less independent of the number of galls per leaf).

The agreement should not be accepted uncritically. First, the females are territorial during the establishment phase and aggressively exclude other individuals from favoured parts of the leaf. Second, the environment within

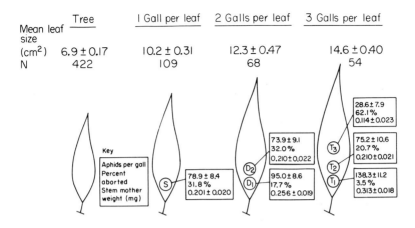

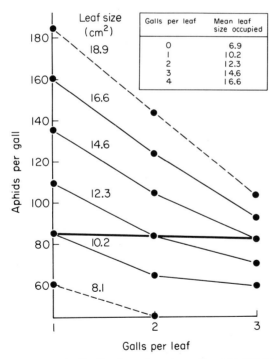

Fig. 3.25 Attack distribution of poplar gall aphid (Whitham 1980). S = single gall per leaf; D = double galled; T = triple galled; N = sample size. Means ± s.e. The heavy line joins the average fitness per aphid for one, two and three stem mothers per leaf. See text for details.

the leaf is spatially heterogeneous. Both these factors contravene assumptions of the ideal free distribution (Whitham 1980).

3.5.6   Switching

Switching occurs when a previously avoided food becomes a preferred food as a result of a change in its abundance or relative abundance (Fig. 3.26). Instead of being under-represented in the diet, it becomes over-represented in relation to its availability.

Switching due to change in the proportional abundance of a food is detected by graphs like those used to detect preference; Fig. 3.26(a) illustrates the principle. When the plant makes up less than 30% of the available food, it is avoided and when it is more common than this, it is preferred. Switching due to absolute increase in abundance is detected from functional response curves; the rate at which a food is taken rises more steeply than linearly, producing an S-shaped functional response curve (Fig. 3.26(b) and see section 3.1.2).

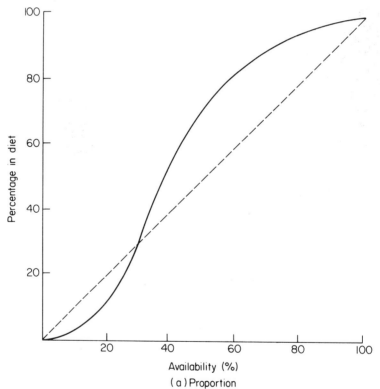

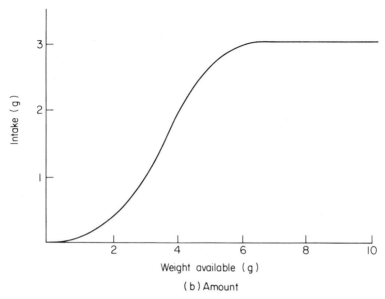

(b) Amount

Fig. 3.26 Switching. (a) Switching on altered proportion. When the food comprises more than 30% of the total available it is preferred and when less, it is
avoided. (b) Switching on altered abundance. When less than $3\,\mathrm{g\,m^{-2}}$ is available, the percentage taken increases with availability. Above $5\,\mathrm{g\,m^{-2}}$ satiation
effects make the functional response negatively density dependent. In many cases
it may be impossible to predict the amount eaten from a given species without a
knowledge of the abundance of *more preferred* foods (see text).

Herbivore switching is most often caused by habitat selection in a patchy
environment; it might also be due to the formation of specific search images
for relatively abundant plants, to improvements in handling efficiency of
commonly handled foods or to more subtle processes such as increasing the
efficiency of exploitation with increased frequency of ingestion. Miller
(1975), for example, found increases in the efficiency of utilization of
abundant food eaten by mallard ducks due to alterations of their digestive
physiology.

Murton (1971) was able to detect switching in wood-pigeons feeding on
maple peas and tick beans laid on an 84 m grid at different densities. The
beans were drugged, so that he could catch the birds after feeding and
examine the contents of their crops. When the two kinds of seeds were
equally abundant, the birds showed a slight preference for maple peas. With
an extreme ratio of 82% tick beans, however, the birds switched to the more
abundant food and showed an average of 90.7% tick beans in their diets.

Interestingly, two pigeons out of the 81 investigated, showed an extreme preference for the rare maple peas and their crops contained 0% and 4.9% tick beans (Murdoch & Oaten 1975).

The food preferences of grey and fox squirrels (*Sciurus carolinensis* and *S. niger*) are determined by both the speed with which the animals can ingest the energy and by the digestibility of the food consumed. However, their preferences are not fixed and they will switch to different food depending on their availability. Bakken (in Smith & Follmer 1972) found that both species concentrated on the acorns of black oak, *Quercus velutina*, during a winter when they were more abundant than acorns of white oak *Q. alba*. The following winter, when white oak acorns were more abundant than black, they formed the squirrels' staple food.

Switching is likely to be most prevalent in spatially complex habitats. If preferred and less preferred species occur in some parts of the habitat in dense patches and in other parts in low-density mixtures, differential exploitation can occur. The high-density patches of the most preferred plants are very attractive and are likely to be depleted rapidly. Once this has happened, the remaining individuals of the most preferred species occur in low-density mixtures. It is quite possible that dense patches of a less-preferred plant species are now sufficiently attractive that the spaced-out individuals of the most preferred species are ignored; switching will have occurred. In this case, the first species is protected not by the density, but by the spatial pattern of the less-preferred plant; for example, grey squirrels, *Sciurus carolinensis*, selected patches yielding high rates of energy intake even though these patches were not sites of concentration of *Quercus rubra* acorns, the preferred food, but contained high densities of less-preferred fruits of species like *Q. prinus, Q. alba* and *Carya glabra* (Lewis 1980).

The importance of switching in population dynamics lies in the fact that it creates a refuge for the plant from over-exploitation. It therefore tends to stabilize plant–herbivore interactions and can lead to increases in species richness in model plant communities (Murdoch 1969, May 1977, Hassell & Comins 1978).

3.6   FOOD QUALITY

Food quality can only be defined strictly in terms of herbivore fitness. The fitness of a given herbivore genotype is measured by its rate of increase when fed a certain type of food. A plant tissue is of high quality when r is large and of low quality when *r* is small. A non-nutritious tissue gives a small, negative value of *r* and a toxic tissue, a large negative value.

Food quality is an extremely complex and elusive thing to measure, compounded of how much nutrient the herbivore can obtain per unit weight of food ingested (e.g. its nitrogen content), the accessibility of the nutrients (digestibility, fibre content, the concentration of complexing agents, etc.) and the concentration of chemical attractants, phagostimulants, repellants and toxins.

It is by no means obvious whether the best measure of food quality should stress the presence of positive factors in the food (desirable amino acids, feeding stimulants, digestible energy) or the absence of negative factors (toxins, resins or deterrents). Studies with domestic livestock have tended to concentrate on the former, since few toxins are abundant in forage crops, and most agriculturalists use organic matter digestibility as an index of food quality (Blaxter *et al.* 1961, Raymond 1969). Some entomologists and wildlife biologists, on the other hand, consider that the absence of toxins or the low concentration of digestibility-reducing substances like tannins and resins are better indicators of high food quality (Feeny 1976, Rhoades & Cates 1976, Bryant & Kuropat 1980).

It is extremely difficult, therefore, (and probably not sensible or necessary) to define a general index of food quality; one man's meat is another man's poison. The nutritional properties of plant chemicals are by no means rigidly determined. If we define a nutrient as a compound required for normal growth, maintenance and reproduction, then clearly that which is a nutrient for one herbivore may not be a nutrient for another. A compound that is a nutrient at one concentration may be a toxin at higher concentrations. A compound may be a nutrient to an animal that has lost its gut flora but not to one whose gut flora can synthesize it (see section 3.6.1). A compound that is repellant to one herbivore may be a feeding stimulant to another. A chemical may be required by the juvenile stages of an animal but not by the adult, and so on (Chapman 1974, Maxwell & Jennings 1980).

Any detailed description of plant food quality is bound, therefore, to be complex, variable and expensive to obtain. The simplest measures are likely to have the double advantages of being cheap to determine and relatively unambiguous to interpret. It must be stressed, however, that chemical measurements are only of value in describing the quality of plant foods that a herbivore habitually eats. Novel food plants may contain extremely 'desirable' levels of soluble nitrogen, but be quite inedible because of physically repellant surfaces, lack of feeding stimulants, presence of digestion inhibiting substances or simply because they taste bad.

3.6.1   The gut microflora

Cellulose is a major component of food intake for almost all the non-sucking, terrestrial herbivores. The bulk of this potential energy is wasted because very few herbivores possess the cellulase enzymes necessary for its digestion, so most herbivores have to defecate most of the cellulose they ingest. Species that rely on high cellulose diets foster a gut flora of microorganisms which is capable of producing cellulases (McBee 1971).

Large vertebrate herbivores are all incapable of cellulase production and rely heavily upon microfloral decomposition in their complex digestive tracts (Hungate 1975). The oesophagous of animals like cattle leads to a large sack (the rumen) where bacteria and protozoa ferment the plant food to produce simpler carbohydrates (like acetic, butyric and proprionic acids) and gases (methane and CO_2). The carbohydrates are absorbed and the gases are belched out. The tough plant tissues are regurgitated to be masticated again; this process of chewing the cud is known as rumination and the animals that do it are known as ruminants. Up to 70% of the energy requirement of a cow is obtained from the acids absorbed from the rumen and about 10% of the digestible energy in the food is lost as belched-out methane (Leng 1970).

Non-ruminant animals also make use of microbial fermentation. In species like rabbits and horses, the fermentation bag is known as the caecum and lies off the lower intestine. This arrangement is apparently less efficient because there can be no rumination (compare the large fragments of plant material in horse dung with the amorphous faeces of cattle), and no recycling of urea nitrogen (McDonald *et al.* 1981). Rabbits, hares and many rodents make up for being unable to ruminate by recycling the contents of their caeca. The caecum produces light-coloured faecal pellets that are eaten direct from the anus and pass to a part of the stomach known as the fundus where fermentation occurs. Other species such as the quokka (a rabbit-like marsupial), langur monkey, sloth and many non-ruminant ungulates have multi-compartmented stomachs in which a large, anterior sack functions as a rumen. A large fraction of total energy requirement is obtained by the absorption of the microbial fermentation products that are formed there.

Herbivorous birds with high-cellulose diets like the willow ptarmigan *Lagopus lagopus* which survives the winter on buds and twigs, have gizzards to grind the food and two large caeca in which fermentation can occur. The grouse is thought to obtain 30% of its energy from fermentation products.

Insects also foster complex gut floras to aid in cellulose digestion (Waldbauer 1968).

In addition to allowing herbivores to extract energy from otherwise indigestible substrates, the gut flora serves two important functions. First, it breaks down and renders harmless, many of the secondary plant substances that would otherwise poison or debilitate the herbivore. In Australia, for example, sheep are occasionally forced to eat *Heliotropium europaeum* when other foods become scarce, and thus to ingest hepatotoxic pyrrolizidine alkaloids like heliotrine and lasiocarpine. Fortunately for the sheep, their rumen microflora is able to detoxify these alkaloids to harmless 1-methyl derivates (Lanigan & Smith 1970).

Second, by synthesis of amino acids or vitamins, the gut flora may rectify dietary imbalances that would otherwise impair growth, survival or reproduction. Sucking insects like aphids, scales and whitefly which live on nitrogen-poor diets all possess complex gut floras of yeasts and bacteria-like organisms which are capable of synthesizing vitamins, amino acids and proteins using recycled or even atmospheric nitrogen (Buchner 1965). The gut flora, therefore, protects the herbivore from the vagaries of food quality.

The gut flora is also remarkably adaptable. If sheep are fed large doses of the non-protein amino acid mimosine, they suffer decreased weight gain, loss of hair and infertility, but if they are given mimosine in small doses at first, then in progressively higher concentrations,their gut flora is able to develop the ability to detoxify and degrade it (Fowden *et al.* 1967). Thus the experience of the gut flora is important in predicting the response of an animal to the ingestion of toxic plants.

Many herbivores have very low requirements for specific amino acids. Ruminants, for instance, are almost independent of the specific amino acid make up of their diets because their prolific gut microflora can synthesize most of their needs; they are affected by the amount of protein in their food but not by its quality (with the possible exception of certain toxic amino acids). Aphids can also synthesize many amino acids via their gut flora, but feeding in *Myzus persicae* is completely halted if methionine is omitted from artificial diets (Dadd & Krieger 1968), presumably because their gut flora is unable to synthesize it (Strong & Sakamoto 1963).

Mechanisms exist for the conservation of nitrogen. In habitats where plant food is typically low in proteins, the herbivores excrete far less nitrogen in their faeces and urine (Schmidt-Nielsen 1979). In ruminants, the main mechanism is for urea to be secreted into the rumen where it can be used by the gut microflora as a substrate for amino acid synthesis; the ruminant then absorbs the amino acids from which new proteins can be built (Houpt 1959). Dairy farmers have long benefited from this process by feeding urea to their cattle as a low-cost alternative to protein supplementation.

3.6.2 Nitrogen

The rate at which a herbivore is able to extract nitrogen from its food is a major factor determining its fitness; this rate is affected by attributes of the animal and of the plant food (McNeill & Southwood 1978). Almost every facet of the herbivore's numerical response is affected by the concentration of available nitrogen in its diet (see section 3.2); for example, Klein (1970) found that in deer, food quality affects growth rate, dispersal, fawn survival, productivity of young and the sex ratio at maturity. Declining food quality (amongst other factors) is responsible for morphological switches from wingless to winged, and from asexual to sexual forms in aphids (Mittler & Sutherland 1969).

The nitrogen content of plant tissues varies greatly (Fig. 3.27) even within a species; in *Quercus coccinea*, for example, roots contain 2%, wood 1%, bark and branch wood 3%, current twigs 4.6%, leaves 8%, flowers 14.5% and fruits 5% nitrogen in dry weight (Woodwell *et al.* 1975).

The minimum nitrogen content of food that will keep body nitrogen levels stable is known for a few species. Ungulates such as deer and cows require at least 1%, while butterfly larvae and grasshopper nymphs need at least 3% (Mattson 1980). The specialist *Eucalyptus* leaf beetle *Paropsis* can survive with only 1% nitrogen (Fox & Macauley 1977).

Herbivores and carnivores face fundamentally different nutritional problems. Whereas the calorific value and nitrogen content of food and body are roughly equal for carnivorous animals, the herbivore must increase calorific value 1.5-fold (from 4 to 6 kcal g^{-1}) and nitrogen content 2.5-fold (from 4 to 10% dry weight) in converting food to body tissue (Southwood 1973). On this simple criterion alone, it is obvious that nitrogen availability is going to limit herbivore growth more often than is a shortage of carbohydrate, and that carbon to nitrogen ratios are likely to form an integral part of most measures of food quality for herbivores.

It is to be expected that sucking and chewing insects will show different responses to total plant nitrogen. For phloem-sucking species, almost all nitrogen compounds will be nutritious, and total nitrogen is likely to be a useful measure of food quality. For the chewing insects, however, the

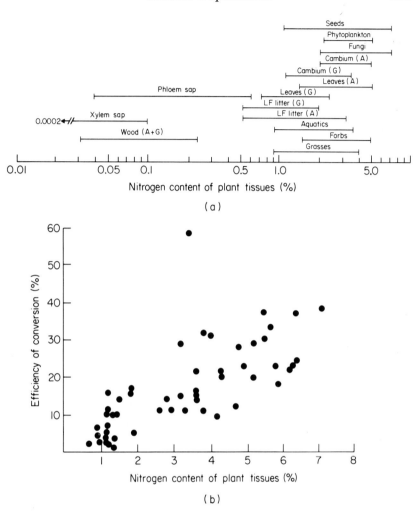

Fig. 3.27 (a) Nitrogen content of different plant tissues and (b) the efficiency of conversion of nitrogen in food into nitrogen in body tissues (Mattson 1980). Low-nitrogen diets are used inefficiently because the rate of feeding and the rate of throughput in the gut are increased to compensate for low food quality; digestion is, therefore, less efficient. The data relate to invertebrate herbivores. G = gymnosperms; A = angiosperms.

presence of protein-complexing agents, non-protein and toxic amino acids, and other non-nutritious forms of nitrogen will mean that total nitrogen is a poor index of food quality (Bernays 1981, Scriber & Slansky 1981). The proportions of different amino acids in the diet can also be important. The amino acids can be divided into categories by their effects on insect development; some are almost always beneficial or necessary (amide (asparagine and glutanine), proline, aspartic and glutamic acids) and others are usually harmful or even toxic (alanine, glycine, L dopa and y-aminobutyric acid; Rosenthal & Bell 1979). The ratio between these 'good' and 'bad' amino acids has been used as an index of food quality in studies of grassland leaf-hoppers; the higher the proportion of 'good' amino acids, the higher the rate of increase of the hoppers (Prestidge & McNeill 1983).

3.6.3 Leaf age and food quality

Leaf age has a profound effect on its suitability as food for both chewing and sucking herbivores. Often young leaves are of the highest quality (e.g. Fenny 1970), but this is by no means always so (Rhoades & Cates 1976).

Trees of *Crescentia alata* artificially defoliated by Rockwood (1974) produced a flush of regrowth leaves in mid season. These young leaves were heavily attacked and eaten by adult beetles of the genus *Oedionychus*. Three such trees lost 100% of their regrowth leaves, while control trees with mature leaves were not attacked. Following beetle defoliation, the trees produced a third flush of leaves. Beetles damaged 82% of these compared to only 27% of the old leaves on control trees.

Dry matter digestibility of pasture grasses tends to decline dramatically with the onset of flowering, as leaf protein levels decline, and fibre and tannin contents rise (Raymond 1969). The digestibility of the diet obtained by sheep can be maintained beyond this time, however, both by selective grazing of the most digestible components of the pasture, and by the effect of grazing in delaying flowering and encouraging the regrowth of young, nutritious leaves (see section 2.7.5). Similar declines in quality with leaf age are known from tree-feeding Lepidoptera (Fenny 1970), leaf-cutting ants (Rockwood 1974) and certain aphids (van Emden *et al.* 1969).

The decline is by no means general. Certain insects will only attack mature leaves (e.g. the cynipid gall wasps *Neuroterus* (Darlington 1974) and some of the miners (Opler (1974) of oak leaves). Also, leaves of the same age produced in different places on the plant, or from shoots with different histories of defoliation, will differ in their quality as food (Bryant & Kuropat 1980; Haukioja 1980). The concentration of secondary plant chemicals often varies with the age of the leaf (see section 3.7.2), with important effects on food quality.

3.6.4 Plant stress and food quality

T.C.R. White (1969, 1974) has gathered a good deal of evidence to suggest
that drought-induced water stress in trees is a potentially important factor in
causing outbreaks of defoliating insects. He proposes that the increase in
amino acids caused by drought-induced proteolysis (Barnett & Naylor 1966,
Gates 1968) represents a dramatic increase in food quality for the insects.
Their enhanced rate of increase then allows them to increase to high
densities. This may be due to release from the control of their natural
enemies or simply a response to an increase in the carrying capacity of
herbivores that their host plant will support (see section 4.2.3).

The effects of water stress on plant food quality are not straightforward,
however. Wearing (1972) found that intermittent water stress was beneficial
to reproduction and survival in the aphids *Brevicoryne brassicae* and *Myzus
persicae* feeding on brussels sprouts, but that continuous water stress was
detrimental. The dominant beneficial effect of intermittent water stress on
diet quality is due to reduced protein synthesis and increased hydrolysis
increasing the availability of amino acids. The main detrimental effect of
continued water stress is through lower plant turgor pressure and increased
sap viscosity which combine to reduce the rate of aphid feeding. Wearing
(1972) suggests that irrigation during dry weather, by keeping food quality
low, may be effective in delaying or preventing aphid outbreaks on these
plants.

Other stresses such as physical damage, disease, or high levels of air
pollution may weaken the plant to the extent that its ability to ward off
herbivore attack is impaired (see section 3.7.2). The grasshopper *Melanoplus
differentialis*, for example, prefers wilted or diseased foliage of sunflowers
over normal, healthy leaves (Lewis 1979), and porcupines increase in
abundance most rapidly when the root fungus *Leptographium* causes an
increase in the sugar content of the wood of their food plants (Spencer 1964).

3.6.5 Food quality and fertilizer application

The principal effect of habitat on average food quality is through soil
nutrient levels (Rodriguez 1960, van Emden 1966, Singh 1970). Thus in
communities where the soil is nitrogen-poor one would expect to find plants
with low foliar nitrogen concentrations and herbivores adapted to low-
nitrogen diets. The herbivores might show very slow development (like the
17-year periodic cicadas which feed on nitrogen-poor xylem fluids) or very

high feeding rates to compensate for the low nutritional quality of their food (Fox & Macauley 1977).

Some plants in nitrogen-poor habitats maintain nitrogen-rich foliage by hosting symbiotic nitrogen-fixing bacteria; others adopt a carnivorous habit and extract nitrogen from insects they catch in pitchers or in sticky traps (e.g. *Drosera, Sarracenia*). Typically, however, plants of these poor soils have low tissue miosture contents, are highly lignified and contain high concentrations of (mostly non-nitrogen based) secondary chemicals (Mattson 1980).

Given the importance of nitrogen availability to the population dynamics of herbivores (see section 3.6.2) it would be expected that the application of nitrogen fertilizer would lead to enhanced rates of increase amongst the animals as well as to improved plant growth.

Leaf-feeding insects removed four times as much foliage from mangrove *Rhizophora mangle* on a high-fertility site enriched by the droppings from a large colony of egrets and pelicans, than from a nearby low-fertility site. Leaf nitrogen levels were 33% higher and the plants on the high-fertility site produced 15% more leaf biomass than on the nutrient-poor site. It is estimated that the excess would have been at least 37% in the absence of the insects (Onuf *et al.* 1977). Other enrichment studies have produced rather different effects (see Table 3.1) and there is clearly no universal response to the addition of nitrogen; for example, in grasses, nitrogen fertilization increases the nitrogen concentration in the phloem sap so that sucking species like aphids and hoppers experience an increase in food quality and exhibit higher rates of fecundity. In such cases, fertilization increases herbivore density (Prestidge & McNeill 1983). On woody shrubs like *Ulex*

Table 3.1 A comparison of the responses to nutrient enrichment in different systems. Entries in the table correspond to whether the difference between treatments (high-nutrient and low-nutrient) are significant and positive (+) non-significant (0), significant and negative (neg), or not reported (?) (Onuf *et al.* 1977).

Localities	Primary production	%N	Losses to herbivores or herbivore production	Herbivore production ÷ primary production
Red mangroves, Florida	+	+	+	+
Experimental ponds, New York	+	?	+	neg
Old field, New York, 6yr	+	0	0	neg
17 yr	0	0	+	+
Salt marsh, Massachusetts	+	+	+	+

and *Erica*, in contrast, the plant responds to fertilization by a rapid burst of growth, but there is no increase in the nitrogen content of the plants' tissues. The herbivores on these shrubs show no significant increase in abundance (Hill & McNeill, pers. comm.).

Application of fertilizer to the host plants did not increase either the growth rates or the nitrogen accumulation rates of two species of hispine beetles feeding on *Heliconia* spp. (Auerbach & Strong 1981).

Fertilizing grasslands is a widely practiced method of improving livestock production. The total nitrogen content of British pasture grasses declines from about 5% dry weight in young leaves to about 1.5% in the mature flowering plant. Defoliation increases the proportion of young leaves and thus maintains a high average nitrogen content. Fertilization can raise nitrogen content to 6% two weeks after application and this declines to 2.5% over the next four weeks (Holmes 1980). This can increase grassland productivity dramatically; with no fertilizer, grass swards yield about 2.5 t dry matter ha^{-1} yr^{-1} and grass/legume swards about $6\,t\,ha^{-1}$ yr^{-1}. Adding $400\,kg\,ha^{-1}$ nitrogen increases the yield of both types of sward to about $10\,t\,ha^{-1}\,yr^{-1}$, although clover survival is poor above application rates of $200\,kg\,ha^{-1}$ nitrogen. Peak grass yields of $12\,t\,ha^{-1}$ yr^{-1} require $625\,kg\,ha^{-1}$ nitrogen but peak profitability occurs at an application rate of $400\,kg\,ha^{-1}$ (Jackson & Williams 1979).

Fertilizer experiments with wild vertebrate herbivores have rarely been performed. Hoffmann (1958) added ammonium nitrate and phosphorus to grassland inhabited by voles *Microtus montanus*. Control populations declined through the summer, while populations on the fertilized plot remained constant then declined in the winter. Fertilization may have delayed the population decline, but it did not prevent it. The difficulty of doing really informative fertilizer trials with mobile vertebrate herbivores is exemplified by Schultz' (1969) work on lemmings. He raised the protein and nutrient levels of about 2.5 ha of tundra vegetation by applying fertilizer but noticed few significant effects on lemming numbers. This study was too short term and on too small a scale to contribute to our understanding of food quality in lemming population dynamics. In work on dune grasslands in Wales, however, it was found that fertilized plots suffered greater rabbit grazing than control plots (Watkinson *et al.* 1979).

The main difficulty in interpreting the results of these fertilizer experiments is that the application causes changes in both the nutrient content of the plant tissues and in the size, shape and phenology of the individual plants. Thus food quality and vegetation structure ('plant architecture') effects are inextricably confounded and observed increases in total number or species richness of insects cannot be unequivocally attributed to improved food quality.

Fertilizing is also used as a means of improving the species composition of pastures (see section 5.1.2). Applications of fertilizer to upland British pastures can lead to a decrease in the abundance of undesirable grasses like *Nardus stricta*. It is unlikely that increased nutrient levels themselves account for the reduction of *Nardus*, but rather that they make it more palatable to sheep, and increase the competitive ability of other, more desirable grasses (Chadwick 1960).

3.6.6   Diet supplementation

Herbivore diets are often deficient in a few crucial minerals, vitamins or trace elements; for example, the caribou *Rangifer tarandus* inhabits arctic tundra and the northern parts of the boreal forest, performing annual migrations between widely separated summer and winter rangelands. In summer they graze on green vascular plants and build up energy and nutrient reserves for the winter. Caribou survive the winter on a diet of lichen which is rich in carbohydrates but poor in nitrogen and minerals. Mineral deficiency develops through the winter and leads to 'salt hunger' in spring which causes the animals to drink sea-water, visit mineral licks, eat urine-contaminated snow and gnaw shed antlers (Staaland *et al.* 1980). Moose *Alces alces* supplement their dietary sodium by eating aquatic plants during the summer (Belovsky 1981).

Grass-feeding ungulates often supplement their diets with woody browse during the dry season in African savannas, when nitrogen levels in the grass fall below the threshold necessary for maintenance (Field 1976). Voles eat flowerheads much richer in nitrogen and other nutrients than their normal diet of leaves, and their reproduction may be stimulated by plant oestrogens (Tast & Kalela 1971).

High-quality foods are available for only two and a half months for arctic populations of beaver *Castor canadensis*. The animals spend most of the year in their lodges feeding on the bark of saplings of willow and alder which they have stored below the ice. This food provides only 8 mg protein cal^{-1} of food ingested. To make up their protein requirements the beavers feed almost exclusively on willow leaves (40 mg protein cal^{-1}), for the short season from June to August that they are available (Aleksiuk 1970).

Diet supplementation in herbivorous birds is frequent during the period when the females are producing eggs and when young are being fed at the nest. Many species are insectivorous at these stages, including foliage feeders (Butterfield & Coulson 1975), seed eaters (Morse 1975, Ward 1965) and fruit feeders (Snow 1971, Morton 1973).

Bizarre forms of nitrogen supplementation are found amongst the Lepidoptera. Several kinds of butterflies drink from urine puddles and others from the rancid flesh of mammal carcasses. Many adults take nectar from flowers and some, like many *Heliconius* spp., feed on pollen. There are sub-tropical moths like *Lobocraspis griseifusca* which take matter and lacrymal secretions from the eyes of large ungulates (Waage 1980). Even the protein from male spermatophores may be an important form of nitrogen supplementation for some female butterflies which do not feed on pollen as adults (Boggs & Gilbert 1979).

Some caterpillars are predatory, too. The dun bar moth, *Cosmia trapezina*, has larvae which feed on the foliage of oak, elm and other trees but are also voracious predators of other caterpillars.

3.6.7   The effect of herbivore feeding on food quality

Just as plant food affects the performance of herbivores, so herbivore feeding affects the food quality of plants. Food quality may improve or deteriorate under herbivore exploitation. In the case of livestock on moderately grazed grassland, feeding tends to improve food quality, because leaf senescence is prevented and the age structure of the leaves shifted in favour of younger, more nutritious regrowth foliage. More commonly, feeding reduces sub-sequent food quality as when regrowth leaves are higher in resins (Bryant & Kuropat 1980), smaller and more fibrous (Benz 1974) or generally less nutritious (Haukioja 1980), or more toxic (Thiegles 1968).

Aphids influence the chemistry of their host plants by injecting saliva, forming stylet tubes, hastening senescence, modifying the balance of carbohydrate sink strengths and by transmitting viruses (Kennedy 1951). Herbivore species that prevent flowering or fruit maturation may cause a build-up of photosynthate in the leaves and stem bases if the plant is sink-limited; this may affect the feeding success of leaf-feeding species.

Actively dividing meristematic tissues tend to be higher in nitrogen content; for example, the beech scale insect *Cryptococcus fagisuga* tends to be more abundant on the callus tissue encircling wounds, than on the normal bark of *Fagus sylvatica* (Wainhouse & Yates, pers. comm.). Cell division may continue longer in tissues attacked by herbivores; galls may, therefore, offer higher quality food to their occupants as well as protection from parasites and predators (McNeill & Southwood 1978).

Changes in food quality determined by the feeding activities of the herbivores are implicated in the population cycles of larch budmoth *Zeiraphera diniana* (Benz 1974; see also section 4.2.4). Following severe defoliation the needles of the larch tree are shorter and their 'raw fibre' content is higher. This increase is correlated with increased mortality (especially amongst the larger larvae, whose survival rate on normal needles is very high) and reduced fecundity (Fig. 3.28). The fact that the tree recovers its leaf quality only after a substantial time-lag may be an important factor contributing to the generation and maintenance of the 9 year cycles in moth abundance (Fischlin & Baltensweiler 1979).

The regrowth shoots of birch and willow are higher in resins than are young leaves on plants unattacked by snowshoe hares. Here too, Bryant (1980) implicates the induced changes in food quality as a cause of the cycles in hare numbers (Vaughan & Keith 1981). Haukioja and Niemela (1977) suggest that cycles in the numbers of *Oporinia autumnata* moths on birch in Scandinavia may also be promoted by chemical changes in the birch foliage following defoliation (see section 3.7.2).

The reverse effect, however, occurs in *Eucalyptus blakelyi* defoliated by

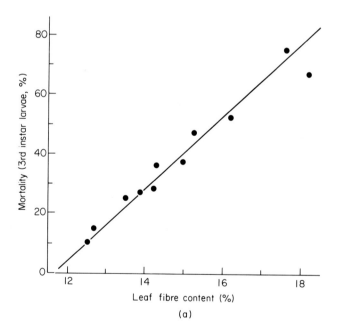

(a)

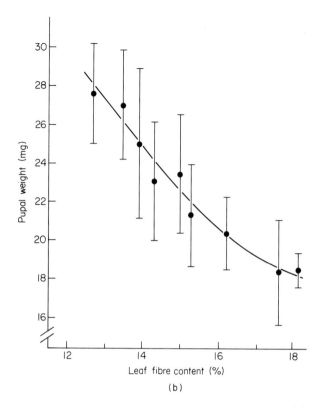

Fig. 3.28 Food quality and demography. (a) Mortality rate of 3rd instar larch budmoth as a function of leaf fibre content; (b) pupal weight as a function of leaf fibre content (Benz 1974). These data are of particular interest because larval feeding in one generation causes increases in fibre content of the foliage in the next (see text).

the sawfly *Perga affinis affinis*; the regrowth foliage is *more* attractive to sawflies in the following generation. Thus, once severely defoliated, the trees tend to suffer chronic insect attack (Carne 1965). Similarly, the psyllid *Acizzia russellae* is 10 times more abundant on young, regrowth foliage on *Acacia karroo* in summertime where the trees had been pruned back in spring, than on normal, mature foliage (Webb & Moran 1978).

Two of the most fascinating examples of reciprocal effects are 'green islands' and 'distant tissue death'. Green islands are areas of green leaf surrounding the mines of certain diptera and micro-lepidoptera in the otherwise brown autumnal leaves of deciduous trees. The tissues around the mine stay rich in proteins long after the rest of leaf has been drained of its nitrogen-rich compounds. It is not clear whether the miner has stopped the plant remobilizing the food it requires for full development by severing phloem vessels or by injecting certain chemicals, or whether the plant, in attempting to isolate the damage done by the miner, has sacrificed the ability to retrieve all its proteins (Spencer 1973).

The phenomenon of 'distant tissue death' is a puzzling effect noted in the attack of cochineal insects *Dactylopius* on prickly pear *Opuntia* in South Africa. The cochineal insects are quite immobile as adults and their feeding inevitably kills the cactus pad they are living on. The pad dies, however, in places where there are no cochineals and stays green until the last, in those places where the insects are anchored (Moran, pers. comm.). It is not known how this is brought about, but it is almost certainly influenced by chemicals injected into the pad by the feeding cochineals.

3.7   PLANT DEFENCES AGAINST HERBIVORES

Most parts of most plants are inedible to most herbivores. If this were not so, it is most unlikely that the perennial plants could ever have come to dominate the land as they do. The plants are defended against herbivory in three main ways; by covering themselves with tough, spiny or inedible surfaces; by suffusing their tissues with chemical deterrents, toxins or digestibility-reducing compounds; or by employing the services of animals like ants to ward off the herbivores.

3.7.1   Surface defences

The commonest surface defences are those in which the epidermis is suffused with lignin, silica, cork or wax so that the tissue cannot be bitten or chewed. The tissues of horsetails *Equisetum* spp. are so rich in silica, for example, that they were used as pan scourers, and there was an international trade in Dutch rush *E. hyemale* for this purpose.

Other plants defend their epidermis by clothing it in various kinds of hairs ('trichomes'); these may take the form of short, sharp hooks or long, matted threads (Levin 1973, Rathcke & Poole 1975, Pillemer & Tingey 1976). The complexity of surface defences can be gauged by the number of names that taxonomic botanists have had to devise to describe the different kinds of hairiness, reminiscent of the 100 or so words that the eskimos have for snow (Mowat 1975). Leaves and stems are described as being hispid, glandular, lanate, villous, arachnoid, ciliate, floccose, hirsute, pilose, pubescent, strigose, tomentose, etc. The effectiveness of surface hairs as a defence was illustrated by Singh *et al.* (1971) working with soybean; densely hairy genotypes had only 2.7 leaf hoppers per plant on average while glabrous plants supported 30.

A different kind of defence against specialist herbivores that search for their food by sight is to have look-alike leaves that mimic the shape of the abundant, non-food, plant species. Gilbert (1975) suggests that visual selection of oviposition sites by adult *Heliconius* butterflies has influenced the evolution of leaf shape in their passion-flower hosts. The hypothesis is that plants bearing leaves that look like the background of abundant, inedible foliage will flourish. The list of specific names of *Passiflora* is striking evidence of the convergence of leaf shape to the abundant forest tree genera that are inedible to *Heliconius*; *discoreaefolia, morifolia, tiliaefolia, laurifolia* and several others.

A particularly subtle form of surface defence is shown by those plants that produce 'egg mimics'. These deter ovipositing insects by fooling them into the belief that an egg has already been laid on that part of the plant. Insects that search visually for oviposition sites, and which suffer cannibalism when more than one egg is laid on each feeding site, are the targets against which this defence might work (Shapiro 1981). *Heliconius* butterflies avoid plants on which an egg has already been laid (in order to minimise intraspecific competition). The plants have modified stipules which bear outgrowths resembling butterfly eggs in shape, size, colour and characteristic position on the tip of the stipule (Gilbert 1975).

3.7.2 Chemical defences

Defensive chemicals have long been known through the widespread occurrence of natural insecticides in plants such as nicotine, derris, pyrethrum and quassia (Jacobson & Crosby 1971). These toxins (and other so-called 'secondary plant chemicals'; Stumpf & Conn 1982) are widely employed in attempts to breed plants which are resistant to insect pests (Beck 1965, Maxwell & Jennings 1980; see also Fig. 3.29).

The vast literature on the relationships between herbivores and plant toxins is accessible through several review volumes (Harborne 1972, Wallace & Mansell 1976, Rosenthal & Janzen 1979) and through reviews of specific chemical groups like phenolics (Levin 1971), cyanogenic glycosides (Jones 1972) and tannins (Bernays 1978, 1981).

The fundamental problem facing a plant which attempts to defend itself

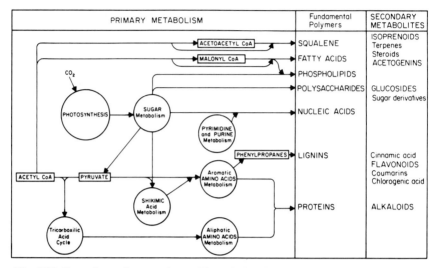

Fig. 3.29 Secondary plant products. The relationship between primary plant metabolism, fundamental polymers and the main groups of so-called secondary metabolites (Kogan 1977). These secondary chemicals were initially thought to be plant waste products; after Fraenkel (1959) they came to be thought of as defences against herbivores. They probably have a variety of functions in defence, metabolism and waste disposal (see text).

from herbivores by producing toxins, is that it is trying to poison physiological systems essentially similar to its own. Also, if the plant can devise ways of making the toxins harmless to itself, then natural selection can probably shape similar systems of tolerance in the herbivores. To protect vulnerable sites of high physiological activity, plant toxins are generally localized in glands, latex or resin systems, or in vacuoles. Other compounds may be stored as inactive precursors and only become toxic when the tissue is damaged; for example, cyanogenic glycosides occur simultaneously with glycosidases which act to release free hydrogen cyanide when the leaf is broken (Jones 1973). Other plant products only become toxic once they are metabolized by the herbivore; fluoracetate in certain Dichapetalaceae, for example, is metabolized by their herbivores to fluorocitrate, a potent inhibitor of Krebs cycle reactions (McKey 1979).

Chemical defence can take many forms; it can act to stop the animal from feeding, or it can act after ingestion. Plants that lack a certain chemical may escape by failing to attract their specialist herbivores. Other plants produce chemicals which deter animals from feeding. Some chemicals interfere with digestion (e.g. substances like tannins which act as protein-complexing agents) and others act as toxins which poison specific metabolic processes (e.g. alkaloids and cyanogenic glycosides).

We would expect pre-ingestion defences to be the rule against mobile generalist herbivores; unless the animal can cue the discomfort it feels to the plant species and tissue from which it fed, the defence is unlikely to be effective. The plant only benefits if the chemicals reduce the rate at which the herbivore returns to feed from the plant.

For specialist herbivores, and for sedentary generalists which spend one or many generations on the same individual plant, feeding deterrence is all but irrelevant (indeed the specialists may be using these same chemicals as attractants or feeding stimulants). In this case, a mixture of post-ingestion toxins and digestion inhibitors is likely to be the plant's most effective means of defence. Since many plants are attacked both by mobile generalists and by sedentary specialists, they defy simple classification by possessing both pre- and post-ingestion defensive chemicals.

The defensive role of secondary chemicals is neatly demonstrated in an experiment by Rehr *et al.* 1973. They fed the polyphagous moth, *Prodensia eridania*, on diets including freeze-dried leaves from three species of acacia trees. One species is naturally protected from herbivores by ants, another is sometimes defended by ants, and the third in unprotected. They found that the foliage of the non-ant acacias was toxic to the larvae, while diets enriched with the foliage of ant-acacias supported normal growth.

Similar work provides strong evidence for the defensive function of such chemicals as sinigrin and other mustard oils in cruciferous plants. Larvae of the black swallow-tail, *Papilio polyxenes*, reared on celery leaves grew normally and showed high survival rates, whereas larvae fed on leaves from celery plants that had been cultured in solution of sinigrin showed much slower growth and greatly increased mortality (Erickson & Feeny 1974).

Some chemicals which are toxic to generalist insects actually increase the growth rate of adapted specialist species when added to artificial diets (e.g. furanocoumarins for the umbellifer-feeding *Papilio machaon*; Berenbaum 1981). Van Emden (1978) measured the feeding rate of the aphid *Brevicoryne brassicae*, a specialist on crucifers, and *Myzus persicae* a generalist, on brussels sprouts leaves differing in their concentration of the mustard oil glycoside allylisothiocyanate. Increasing the concentration of this compound decreased the feeding rate of the generalist aphid as expected, but stimulated the feeding rate of the specialist.

The defensive chemistry of almost all plant tissues changes as they age, presumably reflecting changes in the value of the tissue to the plant (see section 2.3.1). Associated with this is a tendency for tissues of different ages to support different kinds of herbivore species.

Performance of winter moth larvae *Operophtera brumata* declined on oak leaves as the tannin content increased with age (Feeny 1970). Lawton (1978)

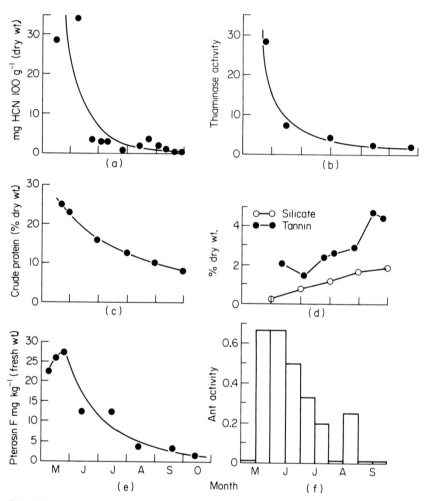

Fig. 3.30 Leaf age and chemical defence. Several defences are most prevalent in young bracken fronds including (a) cyanide, (b) thiaminase, (e) an antifeedant, pterosin F, and (f) the activity of ants attracted by nectar secretions form glands on the unfurling frond. Other defences like (d) silicate and tannin increase in concentration as the frond ages (Lawton 1978). The qualitative defences peak when protein levels (c) are at their highest while the quantitative defences (d) are at low levels in young leaves.

showed that while tannin and silicate increased with the age of the bracken frond, a number of other chemicals, including cyanide, thiaminase and the antifeedant sesquiterpene, pterosin, decline in concentration (Fig. 3.30). Entomologists have long known that cultures of many insects can only be reared on mature grass leaves and perform badly or even starve if fed very young foliage. Alkaloids are commonly found in young grass leaves and the antifeedant halostachine occurs in high concentrations in juvenile *Lolium perenne* foliage. Locusts will only take *Lolium* leaves after they have begun to mature and their alkaloid concentrations have declined. In substrate choice experiments, a variety of herbivorous insects including aphids (*Rhopalosiphum maidis*), bugs (*Leptopterna dolobrata*) and butterflies (*Pararge aegeria*) accumulated on mature grass foliage and ignored young and presumably unpalatable leaves (Bernays & Chapman 1976). Gramine and hordenine are alkaloids found in the seedling leaves of *Hordeum* species which act as antifeedants on the acrid *Melanoplus bivittatus* (Harley & Thorsteinson 1967).

Wheat plants are most vulnerable to being killed by clipping during the first 3 to 4 weeks after sowing, and it is during this period that alkaloids are present in their highest concentrations. It is not surprising that grass plants should protect themselves for the brief period when they are vulnerable to defoliation because they are dependent on the seed and lack the carbohydrate necessary for regrowth. The generalist invertebrate herbivores of creosote bush, *Larrea tridentata* (including three species of grasshopper, a katydid and a geometrid moth larva) all preferred the older leaves which were less resinous than the new shoots (Rhoades & Cates 1976). The tannin and cyanogenic glycoside concentrations of the leaves of the chaparral shrub *Heteromeles arbutifolia* are greatest at leaf initiation, and the concentration of nitrogen-containing glycosides varied seasonally with nitrogen availability.

Fruits containing unripe seeds are often toxic whereas the ripe fruit is attractive to the herbivores that disperse the seeds; *H. arbutifolia* protects its immature fruits with high concentrations of tannins and cyanogenic glycosides in the pulp. On maturation, the concentration of these compounds in the pulp declines rapidly and the glycosides are concentrated in the seeds. Once ripe, the fruits are avidly removed by birds (Dement & Mooney 1974).

The young leaves and twigs of the shrub *Amelanchier alnifolia* are of low quality to browsing vertebrates because of high levels of prunasin, a cyanogenic glycoside. There is a dramatic decline in the concentration of prunasin from 3% of dry wt. after growth initiation in May to 0.5% by June, whereafter levels remain uniformly low through the summer (Majak *et al.* 1980).

One- or two-year-old heather plants are less preferred by grouse than would be predicted on the basis of their nitrogen and phosphorus concentrations; this may provide an explanation of the delay in the increase of grouse numbers following heather burning (Moss *et al.* 1972).

Thus while food quality often declines with leaf age for domestic livestock and many leaf-chewing insects, the effect is by no means universal. In general it is the *kind* of chemical defence that changes with tissue age, rather than the degree of defence; young tissues tend to be defended by repellants and poisons, and older tissues with substances like tannins and lignin.

It is possible that plant tissues might defend themselves by being so low in nutrient content that herbivores could not survive by feeding upon them; for instance, low nitrogen content may protect *Heliconia imbricata* from some herbivory (Auerbach & Strong 1981).

Caswell *et al.* (1973) suggested that plants utilizing the C_4 pathways of photosynthesis suffered less from herbivores than did C_3 plants in the same community. However, Boutton *et al.* (1978) found no significant differences in insect feeding on C_4 and C_3 plants of coastal prairie in southern Texas despite the fact that the C_3 grasses had significantly higher level of nitrogen (1.62% dry weight) than the C_4 grasses (1.34%) throughout the growing season.

The major drawback to reduced nutrient content as a defence is that the herbivores may compensate for low food quality by feeding at a greater rate (see section 3.1.3) so, in the long run, the plant may actually suffer more damage rather than less (Moran & Hamilton 1980).

3.7.3   Protection by animals

Some plants pay protection money to ants in the form of nectar secreted from extra-floral glands. The ants feed on the sugary liquid and, in return, either kill or ward off the plants' herbivorous enemies.

African *Barteria* and neotropical *Cercropia* trees have hollow stems where ants maintain populations of scale insects and other Homoptera. These animals feed on the plant and provide a major food source for the ants with their bodies and their honeydew. The ants are obligate occupiers of the trees and, in return for food and shelter, protect the trees from herbivores and from encroaching vines (Janzen 1979).

The impact of ant defenders on seed production can be dramatic. In a study of *Costus woodsonii* in central Panama, Schemske (1980) found that seed production was reduced by 66% on plants from which the ants had been excluded. The damage was caused by the larvae of an otidid fly, *Euxesta* sp. The adult lays its eggs below the bracts of the inflorescence and the larva kills the floral parts. Ants, especially *Wasmannia auropunctata*, are attracted to the strip of extra-floral nectary on the back of the bract and ward of the ovipositing adult flies. Ant attendance in *Costus* also improves seed dispersal, because fly feeding strips the oily aril which attracts birds to eat the fruit and disperse the seeds. Nectar is only produced, and hence the bract only protected, for the short time that the flower is vulnerable to fly attack. A similar timing effect is shown by the black cherry, *Prunus serotina*. The tree is defoliated by tent caterpillars, *Malacosoma americanum*, and has extra-floral nectaries at which the ant *Formica obscuripes* feeds. The nectar is produced over a three-week period after bud burst during which the tent caterpillars are vulnerable to predation by the ants. Once the caterpillars are too large to be eaten, the nectar is no longer produced (Tilman 1978).

Oak trees over which the wood ant *Formica rufa* foraged suffered a loss of 1% of their leaf area to lepidopterous caterpillars, while trees from which the ants were excluded lost almost 8% (Skinner & Whittaker 1981).

Not all ant activity is beneficial to the plant. Numerous ants are themselves damaging herbivores (leaf cutters or seed feeders) and others encourage sap-sucking herbivores by tending them and protecting them from their natural enemies. The black bean aphid, *Aphis fabae*, is attended by *Lasius niger*; the ant lines up the aphids, spacing them out, and strokes the aphids to encourage them to produce more honeydew. Aphid feeding rate is substantially increased by ant attendance, and while aphid-free bean plants produced 56 seeds and ant-free, aphid-infested plants produced 17, plants with ant-attended aphids set only 8 seeds (Banks & Macaulay 1967).

3.7.4   Facultative defences

Instead of having the tissues permanently suffused with defensive chemicals, an alternative, and possibly less costly option, is to produce the chemicals only as they are needed. Many plants respond to fungal attack by producing antifungal compounds known collectively as 'phytoalexins'; fungi are killed or fail to grow in plant tissues where previous attack has induced the production of phytoalexins. Post-infection increases in secondary compounds have been noted in attacks of fungi and bacteria (Hare 1966, Kuc 1972), nematodes (Krusberg 1963) and parasitic vascular plants (Khanna *et al.* 1968).

Insect feeding also stimulates the production of defensive chemicals by plants. *Pinus radiata* attacked by the wood wasp, *Sirex noctilio*, and *P. sylvestris* attacked by pine sawfly, *Neodiprion sertifer*, both show altered phenol metabolism and the appearance of novel secondary compounds (Thiegles 1968).

Increases in the concentration of secondary compounds following defoliation by herbivores or by artificial means have been noted frequently. Rhoades (1979) removed 50% of the leaves of ragwort, *Senecio jacobaea*, and detected a 45% increase in leaf alkaloids and N-oxidases in the undamaged leaves. Mechanical defoliation brought about a 35% increase in phenolics and a 45% increase in proanthocyanidins in the undamaged foliage of *Carex aquatilis* (Rhoades 1979). Artificially wounded tomato and potato plants produced increased levels of protease inhibitors (Green & Ryan 1972).

Caterpillars of the moth *Oporinia autumnata* grew less rapidly on damaged birch leaves attached to the tree than on isolated damaged leaves, showing that the defensive chemical is probably produced by the tree and transported to the site of damage. The depressant effects of leaf damage on subsequent larval growth are extremely long-lasting and localized. Three years after damage to one birch tree branch, the leaves from that branch were still of poor quality for *Oporinia*, while leaves from an undamaged branch on the same tree remained nutritious for the whole period (Haukioja 1980).

Gypsy moth, *Lymantria dispar*, produced smaller pupae, less fecund adults and suffered increased larval mortality on *Betula populifolia* trees which had suffered two previous defoliations (Wallner & Walton 1979; see also Werner 1979). In each of these cases, defoliation has caused a reduction in the subsequent rate of increase of the herbivore.

In a similar way, beet fly mortality increased by 29% 24 days after beet plants were attacked by *Pegomyia betae* (Rottger & Klingauf 1976). Sheep fertility was reduced 6 weeks after aphid attack on alfalfa had increased the production of its oestrogen mimic, coumestrol (Shutt 1976). Decreased larval growth and survival, and avoidance of trees by ovipositing adults of the sitka spruce weevil, were observed two years after *Picea sitchensis* had been attacked by weevils (Overhulser *et al.* 1972). The effects of larch budmoth, *Zeiraphera diniana*, defoliation in lowering the survival and adult fecundity of moths were still noticeable 4 to 5 years after the attack, due to a combination of delayed leaf production, tougher leaves, higher fibre and resin concentration and lower nitrogen levels (Baltensweiler *et al.* 1977; and see section 3.6.7).

Vertebrates also induce facultative responses in plant chemistry. The snowshoe hare *Lepus americanus* shows all the signs of high-density stress (Christian *et al.* 1965) when fed the leaves of birch which has regenerated from damaged stools (Bryant & Kuropat 1980). Moose feed preferentially on the crown twigs of felled mature trees and break the stems of large saplings to feed on the leading shoot, even though younger plants of the same species are more available. They do this because regrowth tissue is higher in resins than similar-aged twigs from mature plant. Plants are well defended against browsers when they are vulnerable (not yet grown into the canopy) but defend their tissues less when they are out of reach (Bryant & Kuropat 1980).

Facultative defences are not confined to chemicals; for instance, the prickles of *Rubus* on cattle-grazed plants were longer and sharper than those on ungrazed individuals nearby (Abrahamson 1975). In Sudan, goat-browsed individuals of *Acacia raddiana* produced more stunted branches with stiffer thorns than unbrowsed plants. *A. senegal* has no such defences, and is replaced by *A. raddiana* where goat grazing is intense, especially around villages (Seif el Din & Obeid 1971). A similar effect is visible on holly trees *Ilex aquifolium*, where the lower leaves are heavily armed by spines, while leaves on branches in the upper crown have almost smooth edges.

The reciprocal relationship between feeder and fed-upon is illustrated by the woolly aphid, *Adelges piceae*, feeding on the bark of *Abies alba* and *A. balsamea*. The aphids suck the contents of parenchyma cells and appear to stimulate the metabolism of these normally dormant cells, and to increase their soluble-nitrogen content, probably via the action of their saliva. Thus enriched, the diet of the aphids allows rapid population growth. *Abies alba* responds to high densities of the insects by producing a new cambial layer below the feeding sites, laying down a layer of corky tissue and eventually sloughing off the aphid feeding sites as bark flakes. In Europe, therefore, where the tree and the insect are indigenous, fluctuating endemic popu-lations of aphids are maintained without evident damage to the trees. In North America, however, *Adelges piceae* was accidentally introduced and has become a severe pest of *A. balsamea*. The tree seems to be unable to defend itself by sloughing off infested bark, so the aphid population increases until the tree is killed (Kloft 1957, Balch *et al.* 1958).

3.7.5   Polymorphism in chemical defences

Defensive chemistry would be of minor interest in the study of plant–herbivore dynamics if the chemistry of the plants was constant; the level of secondary compounds would merely determine the average rates of fecundity and mortality of the herbivore. However, each time the chemistry of individual plants has been compared, significant differences have been uncovered; some individuals are more prone to herbivore damage than others (Maxwell *et al.* 1972).

Polymorphism in toxicity is widespread. The legumes *Trifolium repens* and *Lotus corniculatus* possess cyanogenic and acyanogenic morphs which coexist in the same grasslands (Jones 1966, Scriber 1978). There is a 20-fold variation in cyanogenic glycoside concentration from foliage of different *Acacia farnesiana* trees within a few hundred metres of one another (Seigler & Conn, in Janzen 1979) and wild ginger, *Asarum caudatum*, varies widely in its palatability to slugs (Cates 1975).

In most stands of bracken *Pteridium aquilinum* in Richmond Park, London, 96% of the individuals contained both the cyanogenic glycoside prunasin and the hydrolase enzyme necessary to release HCN and so make the plants toxic. In a few populations, however, 98% of the individuals were acyanogenic (85% of these lacked both the enzyme and the substrate, while the other 15% contained prunasin without the enzyme). Bracken fronds from these acyanogenic populations were heavily grazed by both deer and sheep. Cooper-Driver and Swain (1976) conclude that cyanogenic polymorphism in bracken has a positive role in determining the degree of herbivore feeding and that there is selection in favour of acyanogenic forms under natural conditions which counters the cost of increased herbivore damage.

The black pineleaf scale insect, *Nuculaspis californica*, feeds on the foliage of ponderosa pine, *Pinus ponderosa*. Edmunds and Alstad (1978) noted vast differences in scale populations on trees of different ages and in different places. By transplanting insects they discovered that survival is much poorer after transfer to another tree than after transfer to another part of the same tree (the trees differ). Second, they showed that what appeared to constitute a 'good tree' for scales (one that supported high natural populations), was not necessarily good for all scales, and insects transferred from another tree often performed badly on it (the insects differ).

These data are strongly suggestive that the trees differ in the kind of chemicals they contain (probably the kind of monoterpenes) and that relatively discrete races of scale insects evolved on individual trees. Since the tree lives for about 200 insect generations and the scales are relatively immobile, there does seem to be scope for selection to operate in this way. As the trees breed by out-crossing, a variety of chemicals would tend to be preserved, while the sedentary and probably inbreeding insects adapt to the specific chemistry of individual plants (Edmunds & Alstad 1978).

Similar polymorphism is evident from transplanting experiments with beech scale, *Cryptococcus fagisuga*, on the bark of different individuals of *Fagus sylvatica* (Wainhouse & Yates, pers. comm.).

Lupins (*Lupinus bakeri, L. caudatus* and *L. floribundus*) are attacked by the larvae of the butterfly *Glaucopsyche lygdamus* which eat the inflorescences, reducing seed production by up to 10% in two species and by up to 100% in *L. floribundus*. Heavily attacked plants are found to contain alkaloids in identical proportions. Lupins in the same population that escape attack or are only lightly infested, have subtly different alkaloid chemistry with altered proportions of four isomers of lupanine and related compounds. The polymorphism in defensive chemistry is maintained, presumably, by greater fitness of the standard chemistry when the butterflies are scarce (Dolinger *et al.* 1973). It must be borne in mind that field observations like these are not proof of the existence of chemical polymorphism; the differences in chemistry could be caused by the herbivores, and the observed pattern of attack could reflect nothing more than the non-random colonization of plants.

Vertebrates are also sensitive to genetic differences in food quality. The relative palatability of Douglas fir, *Pseudotsuga menziesii*, to snowshoe hares, *Lepus americanus*, is strongly inherited, additive and predictable (Dimock *et al.* 1976; Silen & Dimock 1978). The low palatability strains contain increased concentrations of total resins, oxygenated monoterpenes and phenolics.

English oaks (*Quercus robur* and *Q. petraea*) also appear to be polymorphic in the palatability of their acorns. Jones (1959) points out that the differences between the two species are far less than the differences within species. It is possible to find a tree whose acorns have been ignored by squirrels growing alongside another whose acorns have been greedily devoured.

Plants can be polymorphic in physical defences; for example, the tussock-forming grass *Deschampsia caespitosa* is unpalatable to cattle in low pastures but is taken readily in upland swards. The increased palatability is thought to be due to reduced levels of silica in the leaves of upland strains (Davy 1980). It is difficult to interpret this kind of result unequivocally, because alternative grass species are likely to be relatively less attractive in an upland than in a lowland sward; thus what appears to be a reduction in defence might be nothing more than an increase in relative palatability.

3.7.6   Life history and chemical defences

There have been several recent attempts to classify plants in terms of the chemical defences they possess. Feeny (1976) and Rhoades and Cates (1976) suggest that *r*- and *K*-strategic plants ought to have broadly different defences. *K*-strategic plants, being abundant and long-lived are prone to attack by both specialist and generalist herbivores and should possess complex and expensive chemical defences. Since *r*-strategic plants are uncommon in time or in space, they have few monophagous enemies, but must protect themselves from generalist herbivores. The idea is that *K*-strategic plants are mainly defended by dosage-dependent, 'quantitative' compounds while *r* strategists rely on cheaper 'qualitative' toxins. Thus trees nearly all contain digestibility-reducing and protein-complexing compounds like tannins, whereas weedy plants contain toxins like alkaloids or glucosinolates.

These authors have been responsible for promoting a rather unfortunate jargon, however. They call *r*-strategic plants 'unapparent' and *K*-strategic plants 'apparent'. The idea is that plants which are 'easy to find' in evolutionary time will spend more on chemical defence, and will tend to employ chemical defences against which it is difficult for their herbivores to evolve immunity. Hence the preponderance of digestibility-reducing rather than poisonous chemicals in the foliage of trees. The terms are unhelpful, however, because apparency is immeasurable. Whether or not a plant is hard to find depends as much on the sensory acuity and powers of dispersal of the herbivore as on the abundance of spatial distribution of the plant. A specialist herbivore of a rare plant species is *bound* to be good at finding hosts: to say that the plants are 'unapparent' adds nothing to our understanding. Again, we do not need new jargon to appreciate that when plants are abundant both in time (because they are long-lived, or replace one another predictably) and in space, they should support a greater variety of herbivores.

A good deal of attention has been given to the possibility that early and late successional plants might be differentially palatable to generalist herbivores. Cates and Orians (1975) set out to test this hypothesis by feeding leaves from plants of different successional status to slugs. Not surprisingly, they found that early successional plants were more attractive to these early successional animals! Quite the reverse is observed when plants are offered to late successional generalist herbivores; for instance, gypsy moth, *Lymantria dispar*, caterpillars prefer the leaves of trees and other late successional plants to those of weeds and herbaceous perennials (Gornitz, in Bernays 1981).

The fact that there is no such thing as a typical generalist herbivore means that correlations between feeding preferences and plant successional status are likely to be tenuous, at best. Apparent exceptions to the hypothesis are as common as examples supporting it, for instance, three species of *Schistocerca* grasshoppers accepted a greater proportion of later than earlier successional plants (Otte 1975) and tree-feeding Lepidoptera are more polyphagous than the species which feed on herbaceous plants (Futuyma 1976). Plants toxic to generalist vertebrate herbivores like the cow come from all successional stages (e.g. ragwort, *Senecio jacobaea*, from early, wood nightshade, *Solanum dulcamara*, from mid, and yew, *Taxus baccata*, from late successional communities; Forsyth 1968).

It is because generalist vertebrate herbivores like rabbits, sheep and cattle eat late successional plants that grasslands stay as grasslands, rather than undergoing succession to woodland.

3.7.7 The costs of chemical defence

The costs of chemical defence are threefold; first, there is the cost of manufacturing, translocating and storing the chemicals themselves; second, there is the cost of maintaining and running the machinery of facultative defence and repair; third, there is the opportunity cost, which is the loss in productivity and competitive ability compared to genotypes that do not possess the defensive chemicals.

The costs have not been determined directly, but Cates (1975) measured seed production in two morphs of wild ginger, *Asarum caudatum*, which differed in their palatability to the slug *Agriolimax columbianus*. The plant which was unpalatable to slugs and was presumed to contain a defensive chemical produced fewer seeds in the absence of the slug grazing. When the two morphs were grown together in the absence of grazing, the palatable form produced 1.36 times as many seeds, whereas under slug grazing it produced only half as many seeds as the unpalatable morph.

In the absence of herbivores, acyanogenic morphs of *Trifolium repens* have higher rates of sexual and vegetative reproduction than cyanogenic morphs (Foulds & Grime 1972). The growth rate of herbivore-free *Pinus monticola* is negatively correlated with the total monoterpene content of the resin (Hanover 1966), and numerous pest-resistant crop plant varieties yield less than their susceptible counterparts in the absence of pests (Pimentel 1976).

3.7.8   Plant chemicals and herbivore defence

Many specialist herbivores turn the plants' chemicals to their own defence, accumulating secondary compounds from their host plants which render them distasteful to their predators (Levin 1971). The best-known example is the monarch butterfly, *Danaus plexippus*, which feeds on milkweeds, *Asclepias* spp. The larvae concentrate large amounts of cardiac glycosides from the plants which make them and the adult butterflies repellant to birds. Jays learn to avoid the butterflies after one or two trials have made them sick (Brower 1969). This trait may interfere with biological control when beneficial predators and parasites are deterred (Campbell & Duffy 1979).

In addition to using alkaloids for defence against predators, some herbivores use alkaloids in the synthesis of their pheromones. The males of danaid butterflies which feed on plants containing pyrrolizidine alkaloids produce pheromones containing dihydropyrrolizidine derivatives (Edgar *et al.* 1974).

3.7.9   Defensive chemicals and plant fitness

We must be wary in generalizing about the defensive role of secondary compounds. It is just as misguided to assume that all secondary compounds are defensive in function as it was to assume they were all metabolic waste products (see Fraenkel 1959). Many of these products exist in dynamic equilibrium with rapid turnover rates, involving cycles that include primary products such as sugars and amino acids. These secondary chemicals may well have their role in primary metabolic processes as well as in defence (Seigler & Price 1976). They are certainly not mere static doses of poison.

We should also remember that no defence is absolute. All plants, no matter what their secondary chemicals, are attacked by some herbivores in their native habitats; for example, the toxicity of cyanide to insects is evident from its long use by entomologists in killing jars; yet cyanide resistance has been found in the larvae of the common blue butterfly, *Polyommatus icarus*, and the weevil, *Hypera plantaginis*, both of which feed on *Lotus corniculatus*, a plant with cyanogenic morphs (Parsons & Rothschild 1964). The speed at which insects evolve resistance to man-made insecticides demonstrates the difficulty facing the plant that attempts to defend itself with toxic chemicals.

As an alien in Europe *Eucalyptus* trees appear to be very well defended against insect herbivores and suffer almost no attack (in common with many other garden and ornamental aliens). They have very low levels of foliar nitrogen (about 1%) and very high levels of phenolics (up to 25% dry weight). Yet in Australia, native herbivores like sawflies and psyllids regularly defoliate the trees. The average annual rate of defoliation of natural *Eucalyptus* forests is substantially greater than that of temperate European woodlands; the adapted herbivores are abundant and feed at a great rate to compensate for the low quality of their food (Fox & Macauley 1977).

It seems likely that chemicals only provide long-term protection from herbivores that are limited by factors other than food availability. When herbivores are food limited, reductions in food quality will, in the long run, lead to increased rather than decreased leaf consumption, as the herbivores compensate by feeding longer or more rapidly (Moran & Hamilton 1980).

On the other hand, chemical defences, once gained, are difficult to lose. A mutant plant which did not produce secondary substances would be at a strong selective disadvantage in times of high herbivore abundance. The plant attacked by food-limited herbivores is on the horns of a dilemma. If it becomes more toxic, it will gain in the short-term over its less toxic fellows, yet in the long-term it will suffer doubly; it will have to put a large fraction of its net production into defence and the herbivores will eat more of its foliage to compensate for its low quality.

3.7.10   Detoxification of chemical defences by herbivores

Detoxification occurs by one (or several) of four main chemical pathways; oxidation, reduction, hydrolysis and conjugation (Smith 1962). Conjugation is often the crucial step, in which two harmful elements are united into one inactive and readily excreted product. Oxidation in mammals occurs in the liver and in invertebrates in the wall of the midgut. It is brought about by a group of enzymes known collectively as 'mixed function oxidases' (MFO's; Brattsten *et al.* 1977).

The ability to detoxify secondary compounds differs from one individual to another as well as from species to species. Thus some rabbits can detoxify atropine but others cannot (Levin 1976). Polyphagous herbivores would be expected to be able to detoxify a wider range of chemicals than specialists.

Krieger *et al.* (1971) tested this idea with the larvae of 35 species of Lepidoptera. The activity of their oxidases was measured by the rate at which aldrin was converted to dieldrin by homogenized preparations of the caterpillars' guts. Average activity of the polyphagous larvae was 15 times greater than the average for the monophagous species, and oligophagous larvae showed 4 times the rate of oxidation of the monophages. On the other hand, specialists feeding on heavily defended plants would be expected to show highly adapted detoxification physiology (e.g. *Eucalyptus* feeders like koala; Degabriele 1980).

Many species of *Hypericum* possess glands on their leaves and flowers that produce a phenolic quinone known as hypericin. This compound makes *Hypericum* unattractive to insect and vertebrate herbivores alike, and is a factor enhancing the spread of *H. perforatum* as a weed in heavily grazed grasslands. The two *Chrysolina* beetles employed in the biological control of this plant can detoxify hypericin and *C. brunsvicensis* uses hypericin as a feeding stimulant, possibly even storing it as a defence against predators (Levin 1971).

The specialist bruchid *Caryedes brasiliensis* feeds on the seeds of the tropical tree *Dioclea megacarpa*. Its food contains levels of the toxic amino acid *L*-canavanine that are lethal to almost all other herbivores (more than 13% by dry weight). *Caryedes* survives because it possesses enzymes that can degrade canavanine to use it as a source of nitrogen. Also, because it possesses an enzyme that can discriminate between canavanine and arginine, the insect does not incorporate the toxic amino acid into its proteins (Rosenthal *et al.* 1976). The cost of this enzyme machinery must be more than matched by the benefits of feeding on a reliable food source which is not exploited by any other herbivore.

Obviously, a post-ingestion toxin can only work if it can be absorbed through the gut wall. It can harm the animal by interfering with its gut flora, but it can only work directly if its activity is not impaired in the process of digestion. Many of the plant substances that mimic insect hormones (like ecdysone or juvenile hormone) cannot affect animal development when ingested as they cannot pass unaltered through the gut wall. Where such compounds do have a defensive role, it is deterrence or antifeeding activity from their volatile or taste properties that is important. A fine example is provided by Australian pines; when attacked by the termite *Nasutitermes exitiosus* they give off a chemical which mimics the insect's alarm pheromone (Moore 1965).

The detoxification of secondary chemicals by the gut microflora of ruminants was discussed in section 3.6.1.

CHAPTER 4
PLANT – HERBIVORE DYNAMICS

There is no such thing as a typical pattern of plant–herbivore dynamics. The idea that plant and herbivore numbers should follow 'classical predator–prey cycles' (e.g. Lack 1954) stems from a misunderstanding of the artificial and completely neutral cyclic behaviour of Lotka and Volterra's celebrated model (May 1973).

First of all, fluctuations in plant numbers may have nothing to do with herbivore feeding; they may be caused by factors like weather, fire, landslide, competition with other plants and so on. Similarly, herbivore numbers may be determined by factors like severe weather, natural enemies or shortage of shelter and breeding sites. Thus there is no *necessary* link between fluctuations in numbers in the two trophic levels.

When there is a mutual dependence of plant and herbivore numbers, there are several theoretically possible outcomes. The interaction may be quite unstable, leading to the extinction of the plant, of the herbivore or of both populations. The interaction may be stable, returning exponentially to equilibrium following perturbation, or returning via a series of damped oscillations. The system may show stable limit-cycle behaviour where the populations return to their cyclic trajectories following perturbation (unlike the Lotka–Volterra cycles that once perturbed, stay perturbed). The interaction may lead to unstable equilibrium with oscillations of increasing amplitude following perturbation. Finally, there may be apparently chaotic, yet completely deterministic, fluctuations of irregular amplitude and frequency (May 1976).

The outcome depends on the demographic parameters of the plant and herbivore populations and on the timing, kind and degree of density dependence that they exhibit. Animals from permanent habitats, for example, tend to show exactly- or under-compensating mortalities, whereas animals of temporary habitats tend to have very small, under-compensating mortalities at low population densities, rising sharply to over-compensating density-dependent losses as numbers increase (Stubbs 1977).

Long-term field studies provide examples of plant and herbivore populations that follow several of these patterns. Ito (1980) gives a comprehensive review of fluctuation in populations of lower invertebrates, insects, birds, fish and mammals (see also Williamson 1972, Whittaker 1975). As might be expected, there is a continuum of patterns from remarkable stability, through irregular, moderate variability (probably the commonest pattern), to very wide amplitude, more or less regular cyclic fluctuation. Miyashita (1963) reviewed the dynamics of seven herbivorous pest insects in Japan; all showed irregular fluctuations in abundance, and weather was the principal cause of the outbreaks in each case.

It is important that we distinguish those factors that cause the largest variation in population density from year to year (the 'key factors') from those factors that determine the existence and stability of population equilibria (the 'regulating factors'; see Varley *et al.* 1973). In seasonal environments, most deaths and depressions of the birth rate will be due directly or indirectly to abiotic factors like temperature and rainfall. In constant environments, most losses will be due to biotic interactions like parasitism, disease, predation and competition.

Key factors are usually not density dependent and the key factor may vary from year to year; one season drought may cause the largest change in numbers, the next year it may be frosts, the next floods, and so on. While all regulating factors must be density dependent by definition, not all density dependence is regulating, and inverse density dependence or delayed density dependence can be strongly destabilizing (Hassell 1978).

The stability of many plant–herbivore interactions is due to the existence of a refuge in which the plant can escape over-exploitation by the herbivore. The refuge may be literal or figurative. The plant may grow in habitats that are avoided by the herbivores; for example, *Hypericum perforatum* in woodland is not attacked by shade-avoiding *Chrysolina quadrigemina* (Clark 1953), and forbs growing within spiny shrubs are not attacked by rabbits (Jaksic & Fuentes 1980). A certain proportion of the plant population may escape attack because of the distributional behaviour of the herbivore. When animals aggregate in regions of high plant density, isolated plants and low-density patches form a refuge. Equally, when large, isolated plants are preferred, the smaller plants in high-density patches ensure the future of the plant populations. When herbivore numbers are predator regulated or limited by other habitat factors, plant production will satiate the herbivores and the excess plant production constitutes a refuge. Similarly, when part of the plant is ungrazable (e.g. woody rootstocks, larger twigs and branches), a refuge of carbohydrate and non-vulnerable meristems ensures survival of the plant. Perhaps the most important stabilizing factor in the dynamics of short-lived plants is the existence of long-lived seed banks in the soil or the immigration of seed from other habitats (see section 2.5). Seed dynamics are likely to be of great importance in those systems where the storage organs themselves form the food supply, as in the *Scirpus robustus* salt marshes grazed by geese (Smith & Odum 1981).

High equilibrium herbivore density will tend to be correlated with reduced stability for two main reasons. First, when the herbivores treat the plant population in a fine-grained way (see section 3.3), resource exploitation is high and there is no refuge in which plants can escape attack. Second, the lower the equilibrium plant population relative to its carrying capacity, the lower the stability, because the plant population is subject to less intense density-dependent constraints on its growth (see section 4.3.1).

Thus factors tending to make the herbivore treat the plant population in a coarse-grained way (herbivore aggregation, ungrazable plant reserves, habitat selection, predator avoidance, etc.), and factors tending to increase equilibrium plant abundance (reduced herbivore longevity, reduced herbivore reproductive efficiency) will tend to increase plant–herbivore stability.

To search for single-factor explanations of herbivore dynamics will often be futile. Even when one factor is of overriding importance, it may be extremely difficult to establish this unambiguously (Watson & Moss 1970); for example, consider the case of a strongly territorial species of bird inhabiting open scrub. Since each territory requires a look-out point and adequate winter food, the number of territories is fixed, and sets an upper limit to population density. Every year, all surplus birds are driven out by territory holders to the marginal habitat that surrounds the area. The excluded birds are eaten up by generalist predators that prowl the periphery.

That the population is strongly regulated is undeniable; the maximum number of breeding pairs is fixed by the number of territories. In poor years the population may fall below this number but it can never rise above it. So what regulates the population?

The immediate cause of bird losses is predation and the predation rate is strongly density dependent; the more birds there are, the greater the percentage of them eaten. However, it would be misleading to say that the population was predator regulated, because if predators did not kill the excluded animals then something else might (disease, exposure or starvation, for example). Also, the abundance of the predators is not principally determined by the number of birds. The cause of the birds being eaten is that they were behaviourally excluded by territory-holding individuals. It is territorial behaviour that causes the emigration which leads to density-dependent predation. So what causes territory size to be fixed as it is?

If food quality or quantity determines the size of a territory, clearly the population is food limited. If territory size is determined by the availability of nest sites, cover or look-out positions, the population is habitat limited. If animal behaviour determines territory size, independent of the availability of resources, the population is self regulated (in which case the evolution of territorial behaviour is difficult to explain without resort to group-selectionist arguments).

4.1   PLANT POPULATIONS

Most long-term studies of plant population dynamics have not monitored herbivore numbers or feeding damage (Tamm 1956, 1972, Antonovics 1972, Holt 1972, Went 1973). Others have contrasted the floristics, species richness and relative abundances of plant species after different periods of herbivore exclosure, without following detailed fluxes in plant numbers (Rawes 1981, Watt 1981; see also sections 5.1 and 5.2).

Evidence on the dynamics of long-lived plants is largely anecdotal; the woody species on Isle Royale, for example, are said to have recovered from very low levels after wolves arrived on the island and moose numbers were reduced (Mech 1966, Jordan *et al.* 1971). Countless examples of 'range deterioration' caused by overgrazing by domestic livestock attest to the impact that herbivores can have in shaping the structure as well as the dynamics of plant communities (see section 5.2).

Dempster's (1971, 1975) study of ragwort, *Senecio jacobaea*, and cinnabar moth, *Tyria jacobaeae*, on Weeting Heath is outstanding in that attention was given to the dynamics of both plant and animal populations over an extended period. Larval density fluctuated 324-fold over the course of the study while the number of flowering plants varied 184-fold (Fig. 4.1). Larval numbers were clearly food limited and peak numbers of adult moths lag one year behind peak-flowering plant biomass. Plant numbers, however, were not determined by larval feeding. The density of flowering plants is related to the number of rosette plants the previous year. Defoliation, if anything, tends to increase the number of rosette plants because damaged plants produce new shoots from the crown of the rootstock (see section 2.7.5). The main determinant of rosette density on the light, sandy soils of Weeting is rainfall, and plant recruitment is only high when rainfall is higher than normal. Thus the dynamics of the moth are determined by fluctuations in plant biomass (which are due in considerable measure to feeding by the larvae), but plant numbers are determined largely by weather conditions affecting seedling establishment (the plants are microsite limited). We cannot

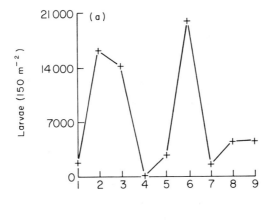

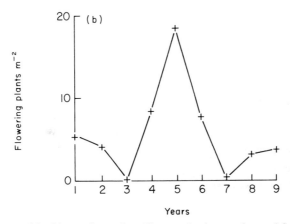

Fig. 4.1 Plant and herbivore dynamics. Changes in the numbers of flowering ragwort plants and cinnabar moth densities at Weeting Heath over a nine year period. The number of larvae emerging in one year (a) is determined largely by the abundance of flowering plants in the previous year (b). Changes in plant abundance, however, are due mainly to changes in germination conditions. The insect is food limited, but the plant is not herbivore limited. (Based on data in Dempster & Lakhani 1979.)

rule out the possibility that plant recruitment may be seed-limited in certain years; if, for example, ideal germination conditions coincided with very heavy larval attack rates, there may be insufficient seed production to exploit all the available microsites (see section 2.6.1).

One of the classic long-term studies of the influence of insect feeding on plant population dynamics was carried out by Waloff (1968) at Silwood Park in southern England. She studied the community of 35 herbivorous insect species, 70 parasitoids and 60 arthropod predators that lives on bushes of broom, *Cytisus scoparius*. As part of this study, one plot of bushes (58×20 m) was sprayed with insecticides three times a year (in May, June and September) over the 11 years from 1966 to 1976 using a mixture of malathion and dimethoate. A similar sized plot was kept as a control.

The insecticide substantially reduced the numbers of the aphid *Acyrthosiphon pisum spartii*, the two psyllids *Arytaina spartii* and *A. genistae*, and the five species of broom mirid bug. The leaf-feeding beetle *Phytodecta olivacea* was reduced by insecticide treatment while another, the weevil *Sitona regensteinensis*, became more abundant in the sprayed area, presumably because its preferred substrate of green shoots was more abundant there. The weevils whose larvae develop inside the seed pod, *Apion fuscirostre* and *Bruchidius ater*, were little affected by insecticide treatment (Waloff & Richards 1977).

Insect feeding on the unsprayed plots reduced the growth of the broom bushes and increased the mortality rate of the plants (Fig. 4.2). The number of pods produced per bush shows a distinct two-year cycle (see section 2.4) and sprayed bushes produced about 2.5 times as many pods as unsprayed bushes over the 11 years of the study. The number of seeds per pod was 1.5 times greater on the sprayed plots (8.88 compared to 5.84 on the unsprayed plot) probably because of feeding by the aphid *Acyrthosiphon pisum spartii* on the green pods. The net effect of increased mortality, reduced growth and depressed fecundity was a 77% reduction in the number of seeds produced over the 11 year period on the unsprayed plot. Unfortunately, the recruitment of seedling broom plants to the two populations was not measured.

In a short-term insecticide experiment, Cantlon (1969) was able to induce a substantial increase in plant numbers following the destruction of the herbivore community. Numbers of the cow wheat, *Melampyrum lineare*, tripled in three years when the katydid *Atlanticus testaceous*, ants and other insects were excluded from woodland plots treated with 2 kg ha^{-1} granular aldrin and occasional foliar sprays of malathion and DDT. *Melampyrum* is hemiparasitic on the roots of jack pine, *Pinus banksiana*, and at the high plant densities brought about by the insecticide, appears to suffer substantially reduced seed production.

Herbivores may exercise their most profound effects on plant numbers by reducing the plants' competitive ability rather than by reducing their fecundity or survival rates directly; for example, writing of the control of

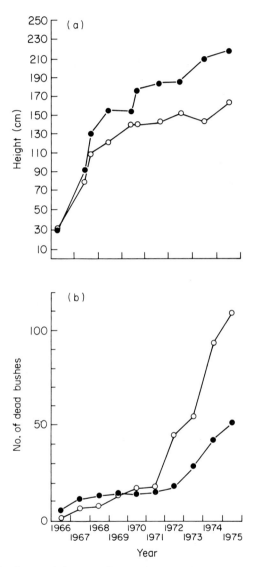

Fig. 4.2 Insect feeding and the growth and survival of perennial plants. The height growth (a) of sprayed (●) and unsprayed (o) bushes of broom and (b) the number of dead bushes over the period 1966–75 (Waloff & Richards 1977).

218   *Chapter 4*

Hypericum perforatum by *Chrysolina* spp. in Victoria, Clark (1953) shows that 'where other wort-controlling agencies are lacking, the insects have caused their host plant to fluctuate violently in density in both space and time without producing a great overall reduction in quantity. By destroying stands of *Hypericum* in areas in which the soils are capable of growing a dense pasture, the insects have paved the way for successful wort control by other factors, e.g. shade-producing herbage and grazing animals. Such areas virtually cease to provide habitats for the insects after the original stand of *Hypericum* is destroyed'.

In the few cases where the numbers of annual plants have been followed over several generations, the populations have shown either exponential damping (*Vulpia fasciculata*; Watkinson & Harper 1978) or damped oscillation (*Agrostemma githago*; Watkinson 1980). The several stabilizing mechanisms of density-dependent fecundity, persistent seed banks or seed immigration that operate in most populations of annuals, suggest that they will show, at worst, damped oscillations, even when their intrinsic rates of increase are very high (Watkinson 1980). The theoretically possible stable cycles and chaos shown by the discrete generation models explored by Hassell *et al.* (1976) are unlikely to be exhibited by these plants.

Some of the best evidence on the impact of herbivore feeding on plant population dynamics comes from successful cases of biological weed control. Where success has been achieved, the weed population has been

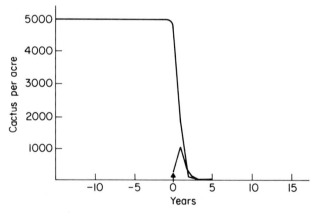

Fig. 4.3 The control of *Opuntia* by *Cactoblastis*. The model of Caughley and Lawton (1981) describes the crash of the cactus population from its pre-release level of 5000 plants per acre to a stable density of only 11 plants per acre within 2 years. Upper curve = cactus abundance; lower curve = moth numbers after introduction at year 0. Parameter values are given in the text.

reduced by the herbivore to a tiny fraction of its former abundance (Wilson 1964, Laing & Hamai 1976), and it is of great interest to know what factors determine the extent to which the stable equilibrium of a plant population can be depressed below its carrying capacity. Beddington *et al.* (1978) compared the average values of depression below carrying capacity in cases of insect pest control with the kinds of depression observed in simple laboratory experiments, and noted a median depression of 99% in the field compared with only 60% in the laboratory. They suggest that field populations can be reduced to much lower stable equilibria because the field is heterogeneous while laboratory populations live in spatially uniform environments; the low density, stable equilibria depend upon the existence of an implicit or explicit refuge for the weeds (see p. 212).

Dodd (1940) describes the greatest success story of weed control in which prickly pear cactuses *Opuntia inermis* and *O. stricta* were virtually eliminated from many parts of Australia. 'Great tracts of country, utterly useless on account of the dense growth of the weed, have been brought into production. The prickly pear territory has been transformed as though by magic from a wilderness to a scene of prosperous endeavour'.

The first insects imported to control the cactus between 1921 and 1925 were not a great success, despite initially promising results in the mid 1920s. There was a moth borer *Olycella*, a plant sucking bug *Chelinidea* and a cochineal *Dactylopius*, among others. Between 1928 and 1930 masses of the moth *Cactoblastis* cactorum collected in South America and reared over the previous 3 years were released. They multiplied rapidly, and by 1932 the original stands of prickly pear had collapsed under the onslaught of *Cactoblastis*; the other insects had been completely eclipsed. In 1932–33 there was a disheartening upsurge of *Opuntia* regrowth as the *Cactoblastis* population crashed following the initial cactus decline. Between 1933 and 1935, however, the *Cactoblastis* recovered and destroyed the regrowth, so that by 1940 there was virtually complete control of the major pest species of *Opuntia*. Forty years later, the moth still maintains the cactus at low, stable equilibrium. The refuge for the weed exists in the highly aggregated egg-laying behaviour of the moth (Monro 1967).

Using data from Dodd (1940) and Monro (1967), Caughley and Lawton (1981) present a model of plant and moth dynamics following the introduction of the insect. The model incorporates logistic plant growth, a Type 2 functional response of the moth to cactus density, exponential herbivore population growth and intense competition between the larvae at high densities (see the models in sections 4.3.3 and 4.3.4 for details).

The parametrized model is:

$$\frac{dV}{dt} = 2.0\,V\left(\frac{5000 - V}{5000}\right) - 6.0N(V/(V + 4))$$

$$\frac{dN}{dt} = 4.0N - 8.92\,N^2/V.$$

Starting with the plant population at its pre-release carrying capacity of 5000 plants per acre, moths are introduced at year 0 and cause a crash to 11 plants per acre within two years (Fig. 4.3). This low equilibrium is stable as a result of the density dependence in larval death rate caused by the egg-clumping behaviour of the adults. The equilibrium herbivore population is 5 egg sticks per acre (Caughley & Lawton 1981). Thus a very simple model provides an extremely precise description of the events.

Can we assume, therefore, that the success of the moth in biocontrol was a predictable consequence of a few simply determined demographic parameters? The answer, of course, is no. If the outcome were predictable, the job of selecting biological control agents would be a great deal more straightforward (Wapshere 1971). The crucial point is that the parameters, especially those of the functional and numerical responses, can only be determined *after* the agent has been released, because their values depend so critically on local conditions. This is why the same pair of weed and herbivore species have produced such different patterns of behaviour in different places (see below).

Some recent successes have been almost as spectacular as the control of *Opuntia*; e.g. the almost total removal of the floating fern *Salvinia molesta* from Lake Moondarra in Australia by the weevil *Cyrtobagous singularis* introduced from Brazil where the fern is endemic (Room *et al.* 1981).

Lantana camara is known to gardeners as a tame and rather tender conservatory plant. In Hawaii, however, it is a pestilential weed that had infested 200 000 ha by 1962. Attempts were made at biological control by searching its native areas of subtropical Central and South America for specialist herbivores. First to be introduced was a Mexican tingid *Teleonemia scrupulosa*, released in 1902 and successful in slowing the rate of spread of *Lantana* (Perkins & Swezey 1924). Later, other herbivores were released which improved control in different habitats. In dry areas, three leaf-feeding moths *Catabena esula*, *Hypena strigata* and *Syngamia haemorrhoidalis* improved control during the 1950s. In wetter parts, two leaf-feeding beetles *Octotoma scabripennis* and *Uroplata giradi* and a stem-boring cerambycid beetle *Plagiohammus spinipennis* have reduced the weed more recently (Andres & Goeden 1971).

Opuntia and *Lantana* provide extremes between which most other successful cases will lie; *Opuntia* was controlled with spectacular success by one herbivore, while a large guild of species is required even to contain the spread of *Lantana*.

Classical success stories of biological control cannot always be repeated; for example, *Cactoblastis* was not very successful in controlling *Opuntia* weeds in South Africa (Pettey 1947). The successful control of St. John's wort (klamath weed), *Hypericum perforatum*, in Californian grasslands by the beetle *Chrysolina quadrigemina* (Huffaker & Kennet 1959, Harris & Peschken 1971), was repeated in Australia (Clark 1953) but not in the drier parts of Canada (Harris 1947b), and the beetle never became abundant after its release in Hawaii (Andres *et al.* 1976).

Dispersal of biological control agents is often extremely slow. The weevil *Rhinocyllus conicus* introduced from France to control nodding thistle, *Carduus nutans*, in Virginia, only spread 1.6 km during the first three years after its release. Clearly, in the early stages of biological control, food is present in excess and any tendency for density dependent emigration would be minimal. Once the thistles began to show signs of damage, the weevils dispersed more rapidly and were found 32 km away from the initial release site after 6 years. At this stage the weevil had reduced thistle density by 95% destroying seeds by eating out the capitulum (Kok & Surles 1975). Other species have even poorer dispersal. The cochineal *Dactylopius* which attacks *Opuntia* in South Africa has sedentary, wingless females. When it kills isolated cactus plants it suffers massive mortality (Zimmermann 1979; see also the model in section 4.3.8).

Once a weed has been reduced in abundance, it is important that cultural practices are revised so that the conditions favouring outbreak of the weed do not recur. In most of the cases, this involves a change in grazing practice to ensure that overgrazing occurs as infrequently and on as limited a scale as possible. Ragwort, St. John's wort and prickly pear all spread rapidly only when competition from grasses and other desirable species is reduced by overgrazing. If overgrazing continues, periodic outbreaks of the weed are likely to recur, and several of the failures of biological control are almost certainly due to failure to alter pasture management to favour the pasture plants and disadvantage the weeds (Cameron 1935, Dodd 1940, Clark 1953).

The successful cases of biological control have several things in common. All the plants were aliens and the vast majority of them were perennials. All the insects were aliens introduced specifically for the purpose of pest control. None of the insects proved damaging to other valuable crop plants thanks to good fortune in the early days, and to careful screening more recently (Wapshere 1971). No native weeds have been controlled by insects. Very few annual weeds have been completely controlled by herbivores, probably because most can produce small amounts of seed even when heavily infested by insects and many support vast banks of seed in the soil. None of the perennial plants successfully controlled by insects was reduced in density by seed-eating herbivores; the successes were due to leaf feeders, stem feeders or flower and seedhead eaters. The weevil *Apion ulicis*, introduced into New Zealand in an attempt to control gorse, *Ulex europaeus*, has become one of the most abundant insects in New Zealand, yet despite eating up to 95% of the gorse seeds every year, has no appreciable impact on the numbers of the plant (Tillyard 1929, Miller 1970).

In summary, while there is no doubt that herbivorous insects can sometimes cause dramatic reductions in plant abundance, it remains to be proved whether the subsequent stable, low-density equilibrium is regulated by herbivore feeding. The changes in land management which typically follow successful weed control may foster the development of a plant community in which the weed is no longer competitive. So while the herbivores caused the crash in weed abundance, it may be interspecific plant competition which regulates the equilibrium, with the weed confined to local and temporary soil disturbances where competition is slight.

Biological control schemes provide priceless opportunities for large scale field experiments in population dynamics. It is unfortunate that funds are so rarely available for detailed pre-release studies of weed dynamics or for long-term follow-up of both the plant and the herbivore populations.

4.2   HERBIVORE POPULATIONS

It has proved possible to predict the population densities of a few herbivore species with some precision. The pest prediction system developed by Way *et al.* (1977) for forecasting outbreaks of the black bean aphid *Aphis fabae* on field beans *Vicia faba* in southern England is notable. By sampling aphid egg density on the twigs of its winter host plant, the spindle bush *Euonymus europaeus*, they can predict the likely summer infestation with remarkable precision (Fig. 4.4(a)). Attempts to predict the winter population on spindle from data on the summer population on beans have not met with such success (Fig. 4.4(b)). Evidently, the factors regulating aphid numbers

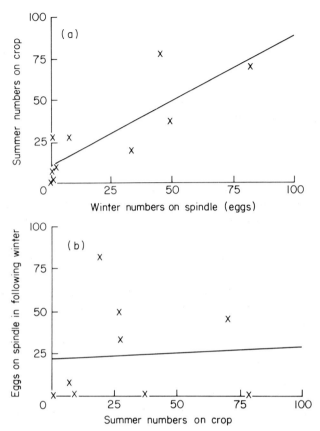

Fig. 4.4 The predictability of population density. The number of *Aphis fabae* on bean crops in summer can be predicted with considerable precision from a count of egg numbers on their winter host *Euonymus europaeus* (a). There is no correlation at all, however, between the number of aphids on the crop in summer and the number on spindle bushes the following winter (b). The key factors affecting population change evidently occur in the period between leaving the crop and returning to the winter host, when the aphids feed from a variety of weed plants and on uncultivated ground (Way & Cammell, unpublished data).

operate between invasion of the bean crop and return to the spindle in autumn; it is equally clear that it may take years of study just to find out which part of the life cycle is important in population regulation!

Continuity of a species in a particular place over a period of many years is not evidence of population regulation. Although few detailed studies have been made, it is clear that many populations persist by extinction followed by re-establishment through immigration. One of three populations of

Euphydryas editha on Jasper Ridge became extinct in 1964 and was re-established by immigration in 1967 (Ehrlich *et al.* 1972). The butterfly inhabits serpentine grasslands on top of a ridge in the Santa Cruz mountains in California, where the larvae feed on the annual plantain *Plantago erecta*. Three sub-populations, more or less isolated by the restricted flight behaviour of the adults, fluctuated quite independently over the study period; in precisely the same weather conditions, one increased more than tenfold, one fluctuated within a narrow range, while the third became extinct (Ehrlich 1965). In the same way, populations of both cinnabar moth and ragwort have been observed to become extinct in Dutch dune communities studied by van der Meijden (1976, 1979).

Our main interest is in the extent to which herbivore dynamics are determined by changes in plant abundance and in the degree to which these changes are brought about by herbivore feeding. A classic example of plant limitation of herbivore numbers occurs in the desert woodrats, *Neotoma lepida*. These animals rely on cholla cactus, *Opuntia bigelovii*, to provide not only food but also water, materials for den construction and shelter from predators. In southern California, 80% of the variation in woodrat density was accounted for by variation in the density of cacti (Brown *et al.* 1972). It is not known whether there is any reciprocal relationship by which the rodents influence the population density of the plants. The woodrats are not absolutely dependent on the cactus, however, for they live in regions where the cactus is absent and in such places rocks or screes provide their cover.

It is almost universally accepted that food availability determines the carrying capacity of a habitat for vertebrates (while recognizing that numbers may be held below the carrying capacity for numerous reasons; see section 3.2). It is by no means clear, however, exactly when or how to measure this food availability, in order to calculate the maximum number of herbivores that may be supported. For vertebrates, it is normally assumed that winter food availability sets the carrying capacity (e.g. Eadie 1970). For roe deer, on the other hand, it seems that summer food is the determinant of the carrying capacity, K (winter K is 1 to 50 times summer K); this is because intense territoriality leads to the emigration of non-territory holders (mainly yearlings) before the onset of winter (Bobek 1977).

4.2.1   Stable herbivore populations

Many herbivore populations show great stability in numbers despite apparently severe fluctuations in environmental conditions (Williamson

1972). Wild white Chillingham cattle in Northumberland have fluctuated in abundance very little over the three centuries for which records exist; there were 28 animals in 1692, the peaks of 80 and 82 were in 1838 and 1913, and the low of 13 occurred after the severe winter of 1947/48. Even though the cattle are provided with a limited amount of winter fodder, variation of less than one order of magnitude is, nonetheless, remarkable (Lack 1954). In Kanha Park in India, annual Forest Department census figures show constancy in population density for four of the wild ungulate species (chital, sambur, blackbuck and gaur) and a steady decline of a fifth, barasingha (Schaller 1967).

Herbivorous bird populations can also show great stability. In a population of Canada geese, *Branta canadensis*, on 21 islands in the Columbia River, Canada, the number of breeding pairs changed only by a factor of 2.1 over a period of 20 years (Hanson & Eberhardt 1971).

Some herbivorous insect populations are also quite constant. Ehrlich and Gilbert (1973) studied the butterfly *Heliconius ethilla* for two years in Trinidad and over this entire period, including two dry seasons, there was no significant difference in adult density from one sampling period to the next (20-day intervals). The adults of *Heliconius* are very long-lived and can lay eggs over a period of up to 6 months. This extended oviposition confers great potential for countering destabilizing fluctuations in larval food availability or predator density (Gilbert 1975). The great longevity of these butterflies is possible because of their ability to nourish themselves on pollen obtained from species of *Anguria* vines. The pollen contains amino acids as well as sugars, and feeding pollen increases egg production fivefold over butterflies fed on sugar alone. Also, the *Anguria* plants flower over a continuous period of up to three years; while individual male flowers only last for one day before dropping off, an individual inflorescence can produce one new flower every two days for between 3 and 12 months. This continuity of pollen in one place, coupled with the spatial constancy and plant constancy of the adult butterflies accounts for their long life and the stability of population density (Gilbert 1975).

Hirose *et al.* (1980) have carried out a detailed survey of the population dynamics of the citrus swallow-tail butterfly, *Papilio xuthus*, living in suburban satsuma groves in Fukuoka, southern Japan. They sampled eggs, larvae, pupae and adults over four generations per year for three years. The density of first instar larvae varied extremely little over 12 generations, while the densities of 5th instars, pupae and adults differed substantially (Fig. 4.4). Two points are clear from this figure. First, an almost constant recruitment

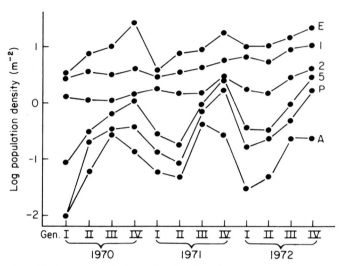

Fig. 4.5 A stable insect population. The density of eggs (E), 1st, 2nd and 5th instar larvae, pupae (P) and adults (A) of citrus swallow-tail, *Papilio xuthus*, over 12 generations and three years (Hirose *et al.* 1980). Despite considerable variation in adult densities, there is almost constant recruitment of first instar larvae.

of first instar larvae occurs despite variability in the number of ovipositing females. Second, variations in density from year to year are caused by factors that act between the 2nd and 5th larval stadia.

The main stabilizing factor in this population is the ovipositional behaviour of the adults coupled with density-dependent dispersal; when many butterflies hatch, a high proportion fly out of the study area (only 10% remained within 400 m of the release point), while a higher proportion of the animals from a low hatch stay to oviposit locally. A further set of stabilizing processes acts on the egg population which is subject to density-dependent parasitism by chalcid wasps like *Trichogramma* spp. The destabilizing factors acting on the larval population were not identified but probably reflected seasonal differences in weather and predation by birds and larger wasps (Hirose *et al.* 1980). This work highlights the important point that the stability one attributes to a population may depend upon which particular phase of the life cycle is studied. On adult counts, the population might be said to be variable, while on first instar larval counts it is highly stable.

Otte and Joern (1975) attribute the stability they observed in the numbers of the grasshopper *Ligurotettix coquilletti* to male territoriality; the insects maintained constant population size over two years despite a high turnover of individual male territory holders.

Dethier and MacArthur (1964) tested the hypothesis that adult butterfly densities were regulated by density-dependent emigration; they released a large number of *Melitaea* and yet there was no increase in the number of eggs over that predicted to be laid by the resident butterfly population. The carrying capacity of these animals is determined by the behaviour of the adults and not, apparently, by the abundance of the *Aster* plants which are their larval food (see section 3.3.1).

Sometimes herbivore stability may be due to habitat limitation. The giant tortoises of Aldabra Island in the Indian Ocean are large, long-lived, sluggish reptiles capable of exerting a great influence on the plant communities in which they live, grazing down the plants to a smooth, lawn-like turf. They are limited, however, by the absolute necessity of shelter to provide shade from the midday sun. Peak tortoise numbers are, therefore, fixed by the number of shading places (which, it must be admitted, tends to decline at high reptile densities due to damage to bushes and tussocks; Coe *et al.* 1979).

4.2.2 Eruptive populations

Population eruption typically follows the introduction of an animal species into an unexploited habitat. An exponential increase in numbers leads to over-exploitation of the plants. A crash in numbers then leads to a lower, more stable plane of herbivore density. This classic pattern was illustrated by the goat-like Himalayan thar, *Hemitragus jemlahicus*, following its introduction into the mountains of New Zealand (Caughley 1970).

The animals were released in 1904 and increased until they occupied over $3600\,km^2$ in the Southern Alps. Increasing populations showed high fecundity, low mortality and high levels of body fat reserves ($r = (a - b) = 0.306 - 0.178 = 0.128$). The population increased until depleted food supplies led to an increase in juvenile mortality. Most mortality occurred in late winter when the short grass was buried under the snow. At this time of year, the snow tussock grasses, *Chionochloa* spp., form the most important food item in maintaining the condition of the animals. It was exploitation of this accessible late-winter feed that caused the decline in the eruption; *Chionochloa* formed 56% cover in areas where thar were increasing but was reduced to about 4% in places where ungulate populations had peaked. The stable equilibrium plant community that ultimately develops is due to modification of the habitat by the grazing animals (Riney 1964, Caughley 1970).

For native species, the traditional explanation of ungulate eruptions is that reduced predation (usually due to the intervention of man) leads to a massive increase in herbivore numbers. Food supplies are then drastically depleted and the ungulates crash to a lower, food-limited equilibrium or become extinct. Caughley (1970) describes how such anecdotes can pass into folklore and then into dogma, in his detailed refutation of the story of the Kaibab deer. In fact, few well-documented cases follow this pattern.

Herds of migratory wildebeest in Serengeti increased in numbers fivefold between 1961 and 1977. There is no evidence that this eruption was due to a reduction in predation. The first phase of the increase is thought to be due to the elimination of rinderpest (a viral disease of ruminants akin to bovine measles) that was an unintentional consequence of the vaccination of domestic livestock in surrounding areas. A second, more recent phase may have been due to increased rainfall in the dry season leading to higher food availability and lower juvenile mortality at this usually critical period (Sinclair 1979).

Klein (1968) describes the population eruption of reindeer, *Rangifer tarandus*, on St. Matthew Island, a small island in the Bering Sea, midway between Russia and Alaska. Twenty-nine animals were introduced to the island in 1944. In the presence of abundant food and in the absence of natural enemies, their numbers increased exponentially to a peak of 6000 in 1963 (an average $r = 0.281$). By this stage the lichens that formed the basis of their winter food supply had been virtually eliminated, and most of the animals were in poor condition because of severe competition for browse in the summer. The winter of 1963/64 was very severe and there was an extremely deep accumulation of snow. The combination of poor condition, scarce and inaccessible food led to mass starvation and the population crashed from 6000 to 50 (more than 99% death rate).

Not all island populations of reindeer crash to extinction, however, as shown by Leader-Williams' (1980) study of the South Georgia populations. These are showing a gradual decline as the density of their preferred winter grass *Poa flabellata* is reduced.

We should be wary, however, of reading too much into the shape of a population trajectory. Fig. 4.6 shows the numbers of coypu, *Myocastor coypus*, in East Anglia. It has all the attributes of a classic population eruption: rapid initial increase; crash; then oscillation to a stable, food-limited equilibrium. In eruption models the herbivores are explicitly food-

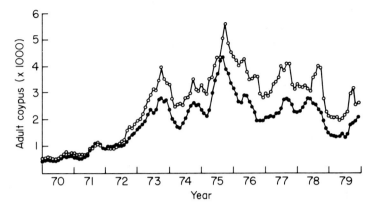

Fig. 4.6 Interpreting population trajectories. Coypu numbers in eastern England appear to follow a classic pattern of herbivore eruption; increase, over-exploitation then crash to a food-limited equilibrium. The actual causes of these changes, however, have little, if anything, to do with food, and are due to severe weather and changes in hunting pressure (see text). Adult female (○) and adult male (●) (Gosling *et al.* 1981).

limited and the decline after the initial peak is due to over-exploitation of food plants. In the case of feral coypu, however, Newson (1966) suggests that its only serious enemies in England are man and bad weather. The rapid increase between 1970 and 1973 occurred when the intensity of pest control was very low and the winters mild. Twelve days of continuous freezing weather and an enlargement of the trapper force were the likely cause of the crash in 1975/76. Subsequent fluctuations are probably due to similar causes; harsh winters in 1977 and 1979 caused steep declines, while low hunting pressure allowed increases in years with mild winters (Gosling *et al.* 1981). Clearly, the pattern of population growth we see in the field cannot be used to infer whether or not the herbivores are food limited.

4.2.3 Outbreak populations of herbivorous insects

Just as with ungulates, eruptions of insect populations can follow introduction into a foreign environment. This can be used to advantage in the biological control of weeds (see section 4.1) but is an embarrassment when the insect is an accidentally introduced pest of crop plants (see section 4.2.5). Outbreaks of native herbivores are also common, especially amongst pest insects (Miyashita 1963); for example, outbreaks of locusts occur when high rainfall allows ideal breeding conditions (Waloff 1976). Many of the best-documented outbreaks, however, occur in forest pests.

Spruce budworm, *Choristoneura fumerana*, has been studied more intensively than any other forest insect since its disastrous outbreak in the 1940's killed most of the mature balsam fir, *Abies balsamea*, and a great deal of the white spruce, *Picea glauca*, over 52 000 km² of Ontario and adjacent Quebec (Morris 1963).

The life cycle of the moth is relatively complex. Eggs are laid in masses of 10 to 50 and the larvae hatch in August and crawl under bark or budscales to hibernate. In spring the larvae emerge from their hibernacula and create mines in old needles. When bud burst occurs, the larvae leave the mines and move into the buds, feeding on the tender, expanding needles. As the shoot elongates, the larva spins the needles together with silk threads to form a shelter in which it feeds until pupation. The adults emerge in July and August, mate then oviposit. There is only one generation per year.

Trees begin to die after five years of defoliation and, after an eight-year outbreak, almost all are dead. Tree mortality continues even after the budworm population has crashed because of fungal infestation and attack by other insects on the weakened trees that survived.

The cause of the outbreak cannot be established in retrospect but it was correlated with a large area of virtual *Abies* monoculture becoming overmature at the same time, and with a run of dry, clear summers. The overmature, low-diversity forests in which the outbreak began grew up because balsam fir regenerates better from seed after fire or logging than do the other conifer species, and the loggers tended to leave the fir trees standing, and only removed the more valuable timber. Spruce budworm tends to do little damage in mixed spruce/fir/hardwood stands where the conifers are overtopped, and under these conditions balsam fir seems to represent rather low-quality food for the insects because growth rates and fecundities are low (Ghent 1958). Alternatively, the microclimate in these mixed woods may hold the intrinsic rate of increase of the budworm at a level low enough for the predators, parasites and diseases to maintain it at a low, stable equilibrium.

The outbreak probably began, therefore, because of a dramatic increase in high-quality food for the budworm (Kimmins 1971). A large number of trees became over-mature and, therefore, more susceptible at the same time, and the run of dry summers further served to increase food quality since the nitrogen availability in drought-stressed plants is higher (White 1974). The budworm does not reach a stable high-density equilibrium because the trees are killed.

Despite the inevitable complexity due to differences in sites, weather and management history, the fundamental dynamics of the system are quite simple. In a uniformly overmature block of forest, an outbreak is almost inevitable, and all the large trees will be killed. In a young forest an outbreak is impossible, and very low, endemic levels of spruce budworm will prevail. In a mixed forest, or a block of medium age, the picture is less clear. Models of spruce budworm dynamics are based on the assumption that natural enemies (mainly birds) are capable of holding the larvae at a low, stable equilibrium (Ludwig *et al.* 1978, Clark & Holling 1979; see also Fig. 4.7). There are three possible equilibria for this system: the low, predator-mediated equilibrium; an unstable, intermediate equilibrium; and a high stable equilibrium where the insects are limited by food availability. Perturbations that moved budworm density from the low equilibrium above the unstable break point would cause the system to flip to the higher,

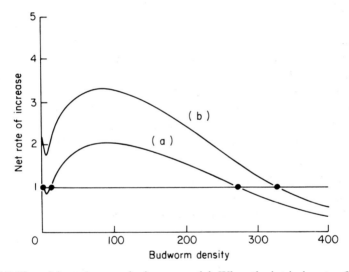

Fig. 4.7 The celebrated spruce budworm model. When the intrinsic rate of increase is low (a) insect numbers have two stable equilibria. The highest equilibrium is food limited and leads to the death of the trees; the equilibrium is locally stable but does not persist because insect numbers crash following the death of the mature trees. The lower equilibrium is predator limited; density-dependent predation by birds and small mammals regulates insect numbers at such a low level that no damage is done to the trees. When the intrinsic rate of increase of the insects is higher (b) because of improved food quality or better weather conditions, the entire graph rises. There is then only one equilibrium, and the system flips to the higher, destructive outbreak equilibrium (Ludwig *et al.* 1978). The evidence for the existence of the lower, predator-mediated equilibrium is extremely tenuous, however (see text).

outbreak state. This state cannot persist for long because successive defoliations kill all the mature trees; the forest then returns to a juvenile state, and budworm numbers crash to low endemic densities.

The 'interesting' dynamics (the existence of multiple stable states, and the catastrophic flips from one state to another) depend upon two assumptions; an S-shaped functional response and constancy in predator numbers. When predator numbers vary as a function of budworm abundance there may be only one equilibrium (see section 4.3.3). More important, the assertion that low-equilibrium budworm populations (endemic levels) are maintained by density-dependent bird predation is plausible but untested. The simpler hypothesis, that the budworm is limited by the availability of susceptible food plants, has never been satisfactorily tested, and there is no critical experimental demonstration of the role of predation in these sparse insect populations. Birds may take 100 to 300 budworm per tree during the course of larval development (Dowden *et al.* 1953), and Mook (1963) showed that certain warblers underwent a numerical response when budworm increased in abundance. The existence of a stable equilibrium determined by density-dependent predation on the larvae however, is still more an article of faith than a proven fact (Morris *et al.* 1958, Buckner 1966).

The other classical eruptive forest insects are the bark beetles. One species *Dendroctonus rufipennis* laid waste almost 20% of the Englemann spruce in Colorado in the mid and late 1940's, killing all the mature trees in the outbreak areas. The trees are killed by a combination of girdling from the egg-galleries and larval mines of the beetle, and fungal attack that blocks the phloem. They usually die in the year they are attacked.

Bark beetles do not kill healthy mature trees under normal conditions because defences such as resinosis prevent successful beetle breeding. The equilibrium density of bark beetles is set not by predators and disease (although these may account for a substantial number of insect deaths; Shook & Baldwin 1970) but by food availability. In a natural forest with a relatively stable age structure, weak and dying trees are scarce. These trees are attacked by the beetles but adult densities never become sufficiently high that mass attack can render healthy trees vulnerable (see section 3.1.5). Bark beetle numbers are restricted by a shortage of susceptible hosts (Coulson 1979).

When perturbation (like windthrow, fire damage or the over-maturity of a single-aged stand) increases the number of susceptibles, large and rapid increases in beetle numbers allow healthy trees to be overcome. An outbreak then occurs which is only terminated when food availability becomes limiting once again; this is usually not until all the full-grown trees have been killed. The Colorado eruption is thought to have been started by a violent windstorm in June 1939 that laid flat large areas of spruce. These dead and dying trees were ideal breeding grounds for the beetles which increased to very high densities over the following three years and were then able to attack healthy trees.

These bark beetle outbreaks are unusual in the sense that, once the outbreak is initiated, the rate of beetle population growth is positively density dependent (see section 3.1.5); the more beetles there are, the faster the population expands. There is no other outcome for these dynamics than 100% tree mortality followed by a crash in beetle numbers. To prevent outbreaks like this, foresters practice 'sanitation felling' to remove senescent trees, extract all wind-blown and fire-damaged timber as quickly as possible, and attempt to maintain forests of a stable age structure. Some dead timber may be left as a source of breeding material for bark beetle predators and parasites.

Outbreaks of spruce beetles, *Dendroctonus rufipennis*, on *Picea engelmanni* probably die out because of a scarcity of sufficiently large trees. The picture is not entirely straightforward, however, since even populations of beetles in test logs (dead wood that could not react to beetle attack) showed reduced fecundity as the outbreak declined. The causes of this are unknown (McCambridge & Knight 1972).

Finally, it must be pointed out that some outbreaks have no local demographic cause at all, and are simply driven by immigration from distant parts. The British Isles are home for several species of Lepidoptera whose populations are occasionally boosted by large immigrations from continental Europe. The large white, *Pieris brassicae*, is a good example, and vast flights of this butterfly have been encountered by ships in mid channel making towards the English coast. Other species of butterfly maintain viable populations entirely by immigration into Britain. Species like the red admiral, *Vanessa atalanta*, and the painted lady, *V. cardui*, are the commonest of these. The point here is that unless the details of immigration are known, it is all too easy to assume that spectacular population changes are due to internally generated demographic processes.

4.2.4   Population cycles

The existence of regular, high-amplitude cycles in animal abundance was first pointed out by Elton (1942) in his analysis of the trading records of the

Hudsons Bay Company. Cycles have since been recorded in most of the vole populations that have been studied in detail (Krebs & Myers 1974, but see Krebs 1979) and in certain forest Lepidoptera (Baltensweiler *et al.* 1977). Cole (1951) sounded a cautionary note by pointing out that the appropriate null hypothesis for detecting regular cycles was not numerical constancy, but a random series of turning points (peaks and troughs). However, the amplitude of 4-year cycles in arctic lemming abundance (200-fold) and 10-year cycles in boreal forest populations of snowshoe hare (20- to 40-fold) leave us in no doubt about the existence of regular cycles (Fig. 4.8).

Cycles may be driven by periodic fluctuations in environmental conditions. Alternatively, they may be caused, even in a constant environment, by internally generated demographic processes due to time-lags or over-compensating density dependence (e.g. Nisbet & Gurney 1982). In the field, any tendency towards demographically generated cycles will sometimes be

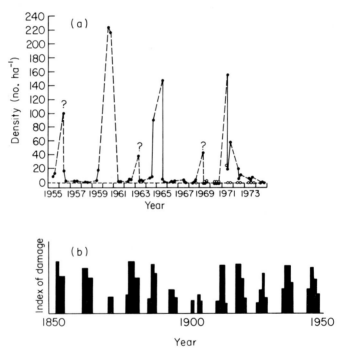

Fig. 4.8 Population cycles in herbivores. (a) Brown lemmings at Point Barrow, Alaska (Batzli 1981); (b) larch budmoth in the Engadin, Switzerland (Baltensweiler 1964). These long-term, periodic fluctuations in abundance (3–4 and 9–10 years respectively) are driven by internal demographic processes rather than by external periodicity in weather.

accentuated and sometimes opposed by fluctuating conditions of weather and food. We shall only expect to observe long-term cycles in natural populations, therefore, when the demographic processes tending to produce them are especially pronounced.

The first requirement for cycles is a high intrinsic rate of increase. The amplitude must be large enough for the cycles to be recognized as such against the background of random, aperiodic fluctuations in weather. Thus, in a cycle consisting of three years increase and one year of decline, r must be large enough to give changes in density somewhere between 100- and 1000-fold; this means that r must be 1.54 to 2.30 at least.

Even with an r-value large enough to produce 4-year cycles, cyclic behaviour is by no means inevitable. If density-dependent reductions in the rate of increase begin at moderate densities and act gradually, the herbivore population will increase smoothly to a stable equilibrium. If, however, density-dependent factors only become important at high densities, and act suddenly and intensely, cycles will be almost bound to occur. Many populations with r-values and density-dependent processes such that relatively low amplitude cycles would occur in constant environments will never have their cyclic behaviour detected in the field against the background 'noise'.

The best-documented cycles occur in three high-latitude vegetation zones of the northern hemisphere. On open tundra heaths, both lemmings (*Lemmus* and *Dicrostonyx* spp.), voles (*Microtus*) and willow ptarmigan (*Lagopus lagopus*) undergo a 4-year cycle. In the transitional birch forest between the tundra and the boreal forest snowshoe hare, *Lepus americanus*, also follows a 4-year cycle. In the boreal forest itself, herbivore species like snowshoe hare, ruffed grouse, *Bonasa umbellus*, and spruce grouse, *Canachites canadensis*, follow a 10-year cycle (Keith 1963).

The cause of cycles has been ascribed to a variety of more or less plausible factors including sunspot cycles, cold winters, wet summers, disease, ectoparasites, predation, food depletion, cover depletion, human hunting and alternating selection. It was originally thought that it was only the periodic declines in density that required explanation, since the potential for exponential growth is a fundamental attribute of all populations (e.g. Dymond 1947). Recent work on the theory of cycles has shown that while over-exploitation due to time-lags is perhaps the most frequent cause of stable limit cycles, the magnitude of the intrinsic rate of increase affects both the existence and the frequency of the cyclic behaviour exhibited (May 1973).

Lemmings and voles

Krebs and Myers (1974) have reviewed the population dynamics of microtine rodents and conclude that their numbers are generally cyclic within a period of 3 or 4 years, and that population cycles should be considered the norm for these animals. They find no evidence to confirm Lack's (1954) suggestion that vole and lemming cycles are more pronounced in the arctic than further south.

The cycles are typically asymmetrical, numbers declining at a greater rate than they increase. Also, during an oscillating decline, summer breeding success may equal or exceed that observed during the increase phase of the cycle. High population density is not sufficient to produce a decline, nor is low population density sufficient to stop one (Tapper 1976).

After discounting changes in litter size and pregnancy rate as causes of cycles, Krebs and Myers point out that the duration of the breeding season of most voles is very flexible, and increases in the length of the breeding season are probably the driving force bringing about the increase phase. Breeding below the snow in winter occurs during the increase phase but not in the winter following a peak. Also, summer breeding starts earlier and continues longer during the increase phase than in the years of population decline. Another important demographic parameter associated with cycles is the age of sexual maturity. Changes in maturation rate are very important in microtines and young Norwegian lemmings of just 20 days of age were found to be pregnant during an increase phase, whereas the maturation rate of summer-born young is strongly density dependent, and at high densities most do not mature until the following year (Fig. 3.9).

Survival rates also change over the course of a cycle. Adult mortality rates are low in the increase and peak phases but adult survival is poor during the decline and when density is low. Juvenile losses are high in the peak phase and during the decline, but there is no cyclic trend in prenatal mortality. Trends in body weight (see section 3.2.1) and emigration rate (see section 3.3.1) during cycles have already been described.

These attributes of cycling populations are widely accepted. There is still no general agreement on how the cycles are caused. The difficulty, as always, lies in separating symptoms and causes. Given that a peak population is bound, by definition, to be followed by a fall in numbers, then all symptoms exhibited by animals at high densities are certain to be correlated with subsequent population decline.

One of the difficulties of fathoming the cause of arctic lemming cycles is the difficulty of doing experiments on the populations beneath the snow (Fuller 1967). It is puzzling, for example, that the high arctic species *Dicrostonyx groenlandicus* appears to decline in abundance every summer, even during the increase phase. If this is general, then winter breeding below the snow must be the norm for this species (Fuller *et al.* 1975). This behaviour is slightly less odd when one remembers that 85 to 98% of the live biomass of high arctic graminoid plants consists of underground parts (Shaver & Billings 1975).

The importance of vegetation cover to small mammals as protection from predators and in ameliorating the environment underneath the winter blanket of snow is stressed by Birney *et al.* (1976), who consider that vole populations are only likely to show cycles where plant cover is high.

Another arctic vole *Microtus oeconomus* feeds mainly on cotton grass, *Eriophorum angustifolium*, in the fell districts of Finnish Lapland. Cotton grass, in common with many arctic plants, takes several years to build up sufficient carbohydrate and nutrient reserves for flowering (the growing season is only 60 days long). Both rodent numbers and cotton grass flowering are cyclic. After a summer of prolific flowering, the plants produce new rhizomes which give rise to a mass of young green shoots the following spring. These new tillers replace the shoots that died after flowering.

The voles feed on the leaves of cotton grass throughout the summer and feed voraciously on the flower buds, flowers and ripe fruits when they are available. In the winter, the voles feed mainly on the rhizomes, preferring firm young shoot bases to the older tissues. When cotton grass flowers, vole reproduction is rapid because of the improved nitrogen quality of their diet. Since rhizome production follows flowering, new shoots are abundant the next year, so rodent survival is high and numbers increase. Numbers rise at a slower rate as food quality declines and emigration eventually causes a population crash. Numbers stay low until the *Eriophorum* flowers again. The hypothesis here is that cycles in plant reproduction drive the lemming cycles (Tast & Kalela 1971). On the other hand, Tikhomirov (1959) found that the voles inhibited cotton grass flowering by consuming most of the buds, and thus may have themselves been a cause of the fluctuations in plant flowering.

Another explanation based on food quality was put forward by Freeland (1974). He argued that the proportion of 'toxic' plants eaten will increase as preferred foods are exploited and that this could drive cycles in fecundity and mortality. Batzli and Pitelka (1975) found no evidence for such a shift in diet composition, but this does not rule out the possibility that more subtle changes in food quality are important; for instance, regrowth twigs of arctic shrubs grazed by snowshoe hare are higher in resins and, therefore, less palatable. Thus high exploitation means that a high proportion of the food plant population will be of low quality at some time in the future (Bryant & Kuropat 1980). This lag could well generate cycles (see also the model in section 4.3.6).

Predation almost certainly does not cause the decline in microtine densities. It is true that large numbers of predators immigrate when vole numbers are high (snowy owl, pomerine jaeger and least weasel in the tundra; racoons, feral cats, short-eared owls and kites in California) but they are eating animals that would have died anyway. Predators may have some role in accentuating some cycles, however, especially if there is a delay in their leaving after rodent numbers have been reduced, and they maintain low animal numbers during a period of plant regrowth (Pearson 1966). This is unlikely to be generally important, however, because the predators would normally emigrate when prey became scarce.

The main argument against cycles being due to herbivore–plant interaction rests on the apparent lack of impact of rodent feeding on vegetation. In data synthesized by Krebs and Myers (1974) from 13 estimates in the literature, only 2 show voles consuming more than 15% of the 'available net primary production'. This would be striking evidence indeed if we could be confident that they had really measured food availability (see 3.1.4). Where food and habitat species are one and the same, as in the arctic lemming systems studied by Thompson (1955) and Pitelka (1957) the animals certainly do appear to over-exploit their food before the population decline. It may well be that in the structurally more complex grasslands at lower latitudes, changes in food quantity and quality are sufficient to cause population cycles, but the detection of these changes requires that very detailed attention be given to the plant populations. Where critical experiments have been done, significant food depletion has been observed. Batzli and Pitelka (1970) in vole exclusion experiments noted an 85% reduction in the volume of the major food species of *Microtus californicus* by grazing of a peak population ($400 \, \text{ha}^{-1}$); this level of grazing intensity was sufficient to reduce seed fall of the preferred grasses by 70%.

A non-cyclic population of *Microtus townsendii* was studied by Krebs (1979) in coastal British Columbia. It has the same reproductive rate as shown by species that cycle, and changes in reproduction, mortality and growth are also similar. Krebs (1979) considers that there is no straight-forward demographic explanation as to why *M. townsendii* does not show cycles. Instead, he favours the hypothesis that cycles are due to changes in spacing behaviour by the adult females caused by high heritability of spacing behaviour. He suggests a genetic continuum of heritability from high values where the animals show strong cycles, to low values in the non-cyclic cases.

Krebs' (1979) suggestion that intraspecific aggression might be the cause of the cycles is interesting in that all other studies of interference (e.g. Rogers & Hassell 1974) have shown it to be strongly stabilizing, damping out, rather than creating cycles (see section 4.3.4). Unfortunately, Krebs presents no model to show how inherited spacing behaviour might produce the cycles.

Other authors argue that cycles have many causes (food, weather, predators, nutrients, etc.) and that different peaks may be due to different processes (Batzli 1981).

Snowshoe hares

Snowshoe hares, *Lepus americanus*, show 10-year cycles of abundance during which their population density varies 20- to 40-fold (Keith & Windberg 1978).

During the population peak, adult survival rate was only slightly reduced by winter food shortage, but juvenile survival was markedly lower. Onset of breeding was delayed and pregnancy rates, ovulation and implantation rates were all reduced following winters when food was scarce (the number of young per female that survive to the end of the breeding season declined from 18.4 to 8.5). Vaughan and Keith (1981) conclude that the cyclic changes in demography of the hares are triggered by a 'delayed density dependent nutritional problem'.

Changes in food quality are implicated in generating the cycles according to Bryant (1980); the regrowth tissues of food plants are richer in resins than normal young shoots (see sections 3.6.7 and 3.7.4).

Absolute food shortage, rather than changes in food quality is the mechanism favoured by Pease *et al.* (1979). Based on an estimated food requirement per hare of 300 g d^{-1} of woody browse, then food was insufficient during a cyclic peak and for one to two winters afterwards. There was a 50% reduction in available food, in hare-browsed as compared to unbrowsed exclosures, and possibly a much greater reduction in the smaller (< 4mm) twigs richer in protein that are preferred by the hares. Preferred browse species like scrub birch, *Betula glandulosa*, rose, *Rosa acicularis*, and saskatoon, *Amelanchier alnifolia*, were much more heavily depleted than were plants like black spruce, *Picea mariana*, and bush cranberry, *Viburnum edule*. Although Pease *et al.* (1979) report changes in the nutrient content of the browse, they contend that these changes are not correlated with demographic changes in the hare population. During winters of peak abundance, the hares browsed some habitat patches so heavily that 100% of the twigs were eaten back to a diameter of 10 cm or more (well above their preferred diameters of 4 mm or less; Woloff 1980).

Larch budmoth

A long-term study of the outbreaks of larch budmoth, *Zeiraphera diniana*, on *Larix decidua* in the Upper Engadin in Switzerland by Baltensweiler and his colleagues, has shown that the population cycles with a mean period of 9.2 years and an amplitude of 0.049 to 237.11 larvae kg^{-1} of branches (representing a 4800-fold fluctuation in density). The cycles are not correlated with weather patterns (Auer 1961, Baltensweiler 1964) and their cause has been the subject of numerous hypotheses. Baltensweiler investigated the role of parasitoids (1964), polymorphism in the insect population (1977), and migration (Baltensweiler & Fischlin 1979). Others have investigated how virus diseases (Auer 1961, Anderson & May 1980) and plant–herbivore interactions (Benz 1974, Fischlin & Baltensweiler 1979) might cause persistent, regular cycles in insect population density.

I shall concentrate on plant–herbivore interactions, not because they are the only, or even the most important cause, but simply because they are the most relevant in our present context. It is unfortunate that so little critical information is available on the changes in food quality (needle length and raw fibre content) that are correlated with increased mortality and reduced fecundity in the moths (Benz 1974; see also Fig. 3.28).

After severe defoliation the tree is 'physiologically weakened' due to lost primary production and subsequent new sprouting in August, and this leads to an increase in raw fibre content of the needles the following year. If there is no significant defoliation, a weakened tree will begin to 'recover' and the raw fibre content falls; if defoliation persists, the fibre content remains high and survival and fecundity amongst the moths remains low. This hypothesis is readily checked by experiment; the fibre content of needles on insecticide treated trees should be significantly lower in post outbreak years than on the infested trees.

It is the putative induced changes in food quality following defoliation that cause the cycles shown by the model produced by Fischlin and Baltensweiler (1979). Larval feeding is assumed to have brought about an increase in raw fibre content from 12% in 1962 to 18% in 1964 (see Fig. 3.28).

4.2.5   The dynamics of exploited herbivore populations

Much of our knowledge of herbivore dynamics comes from the study of exploited populations, in which man culls animals for gain (for food, for skins or for sport), or to rid himself of pests. The herbivores harvested for gain tend to be ungulates, while most of the effort in pest control is directed against insects.

(a) Harvesting ungulates

The first, and perhaps the most important, decision made in cropping an ungulate population is the equilibrium density at which the population is to be maintained, as this will determine the productivity of the herd. If the object is to maintain a high density (as a tourist attraction, for example), net productivity will be low and few animals can be harvested. If meat production is the object, the manager will aim to maintain population density at a level that may be well below maximum. Wildlife managers must choose between large numbers of animals and high productivity; the two are quite incompatible as aims of management (Gross 1969).

It is extremely difficult in practice to know what population density will give maximum productivity. Ecological theory (such as it is on this point) tells us that the density should be below the carrying capacity; perhaps somewhere close to half the carrying capacity when the per capita rate of increase falls linearly with population density (e.g. in the logistic model), or much closer to the carrying capacity if, as in at least some vertebrates, density dependence is markedly non-linear (Fowler 1981; see also Figs 3.11 and 3.14). With strongly non-linear density dependence the population giving maximum productivity will be very close to K and almost any level of harvesting will lead to reduced population equilibrium and lower subsequent yields.

It would take an extremely long time to determine the optimal stocking rate by trial and error, because changes in stocking affect the vegetation, and a judgement of the value of a particular stocking rate should not be made until the age and size structures of both the plant and animal populations are stable. The manager skirts these issues by making a sensible guess and then sticking to it; he defines a target population density which he hopes is close to the optimal, then kills animals each year in such numbers as to reduce stock to this level.

If it can be safely assumed that the population dynamics of the herbivore population are described by a logistic equation (and this may be a big if for

Table 4.1. Choice of model and the behaviour of exploited populations. Three models are fitted to the same data (Fig. 4.9) assuming that the population showed reproduction before harvest, after harvest, or that harvesting relaxed the intensity of density dependence. Each model requires different transformations for the x and y axes and leads to different estimates for r and K. Also, each model demands a different formula for estimating the harvesting rate given the values of r and K. For the relaxed density dependence model, the harvesting formula must be solved numerically for C (see equation under Model A). No matter what model is fit to the data in Fig. 4.9 regression leads to an overestimate of r and an underestimate of K. Thus the harvesting rates are too high and the populations are driven to extinction whether there is relaxed density dependence (Model A) or post-reproduction culling (Model B). For the 'real population' where $r = 0.3$ and $K = 1000$, an annual harvest of 80 animals will drive the population to extinction if there is relaxed density dependence (Model A) but will lead to a gradual increase in numbers if the population shows no relaxed density dependence and harvesting follows reproduction (Model B). The knife-edge of fixed-harvest culling is demonstrated by Model A; taking 80 animals per year leads to extinction, while taking 79 allows the population to increase.

	Reproduction before harvest	Reproduction after harvest	Density dependence after harvest	Real population Model A	Real population Model B
y axis	$\ln((N_{t+1}+C)/N_t)$	$\ln(N_{t+1}/(N_t-C))$	$\ln(N_{t+1}/(N_t-C))$	—	—
x axis	N_t	N_t	$N_t - C$	—	—
estimated r	0.3762	0.4567	0.4076	0.3	0.3
estimated K	946.65	931.29	831.30	1000	1000
estimated equilibrium	473	466	416	500	500
cull formula	$\frac{K}{2}(e^{r/2}-1)$	$\frac{K}{2}(1-e^{-r/2})$	transcendental: solve for C in Model A.	$\frac{K}{2}=\left(\frac{K}{2}-C\right)\exp\left(\frac{r}{K}\left(\frac{K}{2}+C\right)\right)$	$\frac{K}{2}(e^{r/2}-1)$
annual cull	98	95	91	79 80	81
fate of the exploited population A	extinct in 11 years	extinct in 12 years	extinct in 11 years	population drifts upward to 582 after 50 years / population extinct in 85 years	—
fate of exploited population B	extinct in 16 years	extinct in 17 years	extinct in 16 years	—	numbers all but constant; 495 after 100 years

many vertebrates; Fowler 1981), then various simple methods exist for determining the harvesting rate which will allow the maximum sustained yield (see Caughley 1977). The logistic requires a knowledge of the rate of increase, r, and the carrying capacity, K. In the unlikely event that both of these are known, the population density is simply reduced to $K/2$ and harvested at an instantaneous rate of $r/2$ which will yield $rK/4$ animals per year (see Fig. 4.26). If the animals are to be killed in a single brief hunting season, then the percentage kill is computed from the appropriate formula in Table 4.1. If, for example, we want to harvest a population growing at $r = 0.3$ with reproduction occurring after harvest, we would cull $100(1 - \exp(-0.15))\%$ or 13.9% of the animals present at the beginning of the hunting season.

As a rough rule of thumb, therefore, managers aim to harvest at about $r_{max}./2$; thus, for North American ungulates, managers tend to harvest between 10 and 20% of the animals per year. The peccary, *Peccari tajuca*, population of Arizona, New Mexico, for example, is harvested at an annual rate of 15%, white-tailed deer *Odocoileus virginianus* at about 13%, mule deer, *O. hemionus*, at 17% and pronghorn, *Antilocarpa americana*, at 17%. The desert bighorn sheep, however, is harvested at an annual rate of less than 2% (van Dyne *et al.* 1980).

In the more usual case when neither r nor K is known, then data from the first few years of harvesting can be used to obtain estimates. A fixed number of animals (C) is culled each year and the rate of population decline is obtained from the annual head count. Then r is calculated by regression from one of the discrete generation forms of the logistic.

In discrete form, the logistic can be structured in several different ways, each leading to different estimates of r and K from the same data; for example reproduction may occur before or after harvesting:

$$N_{t+1} = N_t \exp(r(K - N_t)/K) - C \qquad \text{(before)}$$

or

$$N_{t+1} = (N_t - C)\exp(r(K - N_t)/K). \qquad \text{(after)}$$

Similarly, density dependence may occur after the harvest, so that culling improves the survival and fecundity of the remaining animals, in which case the appropriate model is

$$N_{t+1} = (N_t - C)\exp(r(K - N_t + C)/K).$$

For each equation, r is estimated by plotting the natural logarithm of

Chapter 4

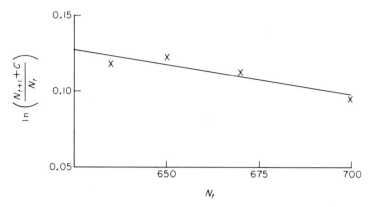

Fig. 4.9 Estimating r and K for harvested populations by regression. If it is assumed that reproduction occurs before harvest, then $\ln((N_{t+1} + C)/N_t)$ should be plotted against N_t (see Table 4.1). The population declined from 700 at the outset to 670, 650, 635 then 615 after the fourth year. The intercept of 0.3762 provides an estimate of r. The slope of the graph is $-r/K$ so the estimate of K is $0.3762/0.000\,397\,4 = 947$. In fact, the population had 'true' values of $r = 0.3$ and $K = 1000$ (see Table 4.1).

some function of the ratio N_{t+1}/N_t against N_t; the detailed transformations are shown in Table 4.1.

An example should make the method clear. Suppose that an initial population of 700 animals was subjected to an annual cull of 100. Numbers fell to 670 in the first year, 650 in the second, 635 in the third and 615 in the fourth. Regression of $\ln((N_{t+1} + C)/N_t)$ against N_t allows an estimate of r (the intercept) and of K from the slope of the line $(-r/K)$ (Fig. 4.9). The estimates of r and K and of the prescribed annual harvest rates and equilibrium population densities for the three models are shown in Table 4.1.

If the estimate of r is lower than the population's actual rate of increase, then numbers will gradually build up to a higher stable equilibrium. If, as here, the estimate of r is too high, then harvesting at a fixed annual kill of C will lead to the rapid extinction of the population. The maximum sustained yield under fixed cull harvesting is extremely unstable; if the real population in our example had density dependence after harvest (Model A), then taking 79 animals per year from a population of 500 would lead to a gentle increase in numbers, while harvesting 80 would drive the population to extinction. Also, real populations are harvested in variable environments, and random year to year changes in r and K exacerbate this instability (see below).

This analysis highlights three major pitfalls. First, it is folly to harvest a constant number of animals from an obviously declining population; either a fixed percentage of the population should be taken each year, or harvesting should be stopped altogether until numbers recover. Second, the regression method for estimating r and K is extremely sensitive to errors in the head count when the number of data points is small. These logistic-based methods are excellent for providing a rough guide to r and K but they should not be taken too literally. The estimates of r and K should be re-estimated each year using all the available information on population densities and harvests (Caughley, 1977). Thirdly it matters greatly what model is used; simply altering the assumption about the timing of density dependence can lead to diametrically opposed predictions (Table 4.1; see also Le Cren & Holgate 1962 pp. 365–8).

In some cases the manager also decides on the age and sex of the animals to be killed at each harvest. The standard approach to problems of this kind is to adopt age-structured population models, such as Leslie matrices (Usher 1972). As before, the size of the harvest is based on the decision as to the equilibrium density at which to maintain the population. The problem consists of deciding what fraction of the cull to take from each age class. As shown by Beddington (1974) the optimal solution is to take animals from only two age classes. All of one age class should be removed (effectively reducing the longevity of the animals), and a proportion of the animals from another age class should be taken. This proportion is determined by the relative reproductive values and relative yields of animals of different ages, and the strategy is robust whether single or two-sex harvesting is practised. A worked example for Scottish red deer is provided by Beddington and Taylor (1973).

The concept of maximum sustained yield has come under severe criticism in recent years (e.g. Larkin 1977). The removal of a fixed number of herbivores (constant yield) is likely to lead to over-exploitation and stock depletion because of random year to year variation in herbivore performance and through errors in data collection leading to the overestimation of population size and intrinsic growth rate, or underestimation of the carrying capacity. As the harvest is increased towards what would be the maximum sustained yield in a constant environment, the return time of the population following disturbance tends to infinity, and the population fluctuations become unboundedly large as time goes on (Beddington & May 1977). Harvesting with constant effort is *less* destabilizing (see section 4.3.9), but the return time still increases with the harvesting rate. The most stable pattern of exploitation is to allow constant escapement; in years when herbivore numbers fall below the defined management threshold ($N_T \approx K/2$), no animals at all are harvested (e.g. Ricker 1958). When numbers are higher, then $N - N_T$ are culled. It is unfortunate that the most stable system is the most difficult to manage; it requires annual pre-harvest census and leads to fluctuating labour requirements and to variable prices.

It should be expected that the herbivore density giving peak productivity will increase gradually over the years as the vegetation recovers after the initial reduction in herbivore numbers from K to $K/2$ (Caughley 1977).

(b) Harvesting herbivorous pests

In harvesting pest herbivores with a view to control, one immediately runs into the problem that as population density is reduced, so food availability and habitat conditions for the survivors improve; for example, after 23 years of hunting large mammals in an attempt to eradicate tsetse fly in Botswana, Child *et al.* (1970) found that for all but three of the species, the numbers shot per year had actually increased! Those that did decline were probably depleted by factors other than hunting. Pest control measures taken at that time of year, or in that part of the life cycle before density dependence operates are likely to be completely ineffective when density dependence is at all strong. The classic study with herbivorous birds was by Murton *et al.* (1974) who showed that no matter how many wood pigeons were shot in autumn, the same number breed the following spring (see section 3.2.4). Similar effects are found when vole populations are experimentally cropped (Krebs 1966, Watts 1970b); over 1700 *Microtus californicus* were removed during one year from an area of less than 1 ha with no depression of population density. Herbivores are replaced by immigration of animals from other habitats, or by improved survival and breeding of the residents due to relaxed density dependence.

Murphy's law operates here; in its ecological guise it states that a population harvested for gain will have demographic properties which make it easy to exterminate, whereas a pest population will yield a high, stable harvest!

The relationship between pest susceptibility and the scale at which a crop is cultivated is a contentious issue in pest control. One school of thought follows Pimentel (1961) in believing that large scale monoculture is dangerous and likely to lead to severe pest infestation (Root 1973, Thresh 1981; see also 'Resource concentration', section 5.4).

The other school of thought follows Darwin (1859) in believing that small patches and isolated plants are most at risk from herbivores. Darwin writes 'anyone who has tried, knows how troublesome it is to get seed from a few wheat or other such plants in a garden; I have in this case lost every single seed. This view of the necessity of a large stock of the same species for its preservation, explains, I believe, some singular facts in nature, such as that of very rare plants being sometimes extremely abundant in the few spots where they do occur'. The idea that pests are proportionately more damaging in small plots is also backed up by the observation that many vegetables suffer greater damage in gardens than in extensive commercial fields (e.g. pea moth, *Cydia nigricana*). Way and Heathcote (1966) found fewer aphids per plant in dense bean crops than in sparse, and in natural communities, isolated plants are attacked by certain Lepidoptera in preference to plants in clumps (Thompson & Price 1977, Wiklund & Ahrberg 1978). Pest damage is often patchy and plants at the field edge are damaged more severely than those in the centre of the crop (see Fig. 3.17). This edge effect is also observed with vertebrate pests like rabbits in cereal crops and brown hares on turnips (Hewson 1977).

When pest numbers are determined by the abundance of the crop, then large monocultures may suffer greater pest problems. However, when pest numbers are determined by factors outside the crop (alternative foods, winter habitat, weather conditions, etc.), small patches may suffer higher damage because of predator-satiation effects in the larger fields.

The real risk of extensive monocultures in temperate latitudes is economic rather than ecological; in years of severe pest outbreak the risks of economic loss are higher, but the rate of damage per plant is likely to be lower in an extensive crop than in a smaller patch. In a tropical intercropping system, however, associational resistance may significantly depress pest numbers in small crop patches (Altieri *et al.* 1977; see also section 5.4).

The vast literature on the control of herbivorous insects by predators and parasites is accessible through comprehensive reviews such as de Bach (1974) and Huffaker (1980), and the dynamics of predatory insects in biological control are discussed by Hassell (1978).

4.3   POPULATION MODELS

Mathematical models are often criticized as being too simple-minded and too ecologically naive to be of any value in understanding how real ecosystems behave. The philosophy followed here is that while it is patently clear that an understanding of these models is not a *sufficient* condition for a full understanding of population behaviour in the field, it is definitely a *necessary* condition for such understanding. Unless we know the consequences of simple assumptions about plant/animal interactions, we are in no position to interpret the complex data we obtain from the field, where interaction effects are compounded by genetic, climatic and spatial variability.

The units in which the plant population is described will depend on the species and on the purposes of the study. For work on the dynamics of annual plants, rosette-forming biennials and most woody plants, it will be appropriate to model individual organisms (what Harper (1977) calls 'genets'). For studies of the dynamics of plants that show extensive lateral growth and where the recognition of genetic individuals is difficult or impossible (as in many grassland species), it will be necessary to deal with a plant unit such as the leaf, the tiller or the rooted node. In agricultural studies of pasture productivity, it may be sufficient to model the dynamics of above-ground, green biomass.

In the models that follow, plant abundance, V, is the number of units per m^2 whether they be genets, tillers or grams of dry weight. It should be borne in mind that since numbers map to biomass in a non-linear way, results from one kind of study will not necessarily apply in another (see Fig. 2.2). Differential equations are employed to described those plant–herbivore interactions that are more less continuous (sheep feeding on pasture, multi-voltine insects feeding on evergreen trees) and difference equations to describe those that are discrete (annual plants). Many real interactions are hybrid with discrete bouts of herbivore attack on more or less continuously growing plants (e.g. univoltine insects on trees).

4.3.1   Lotka – Volterra: the basic model

Models of plant–herbivore dynamics consist of two equations, each of two parts; an equation to describe the rate of change in plant abundance which is given as the gains, A, minus the losses, B, and an equation to describe the rate of change of the herbivores which is their gains, C, minus their losses, D. The two equations are usually linked in that losses to the plant will be gains to the herbivore; C is typically a function of B. We write the models like this:

$$\frac{dV}{dt} = A - B$$

$$\frac{dN}{dt} = C - D.$$

The models are elaborated as the four components, *A, B, C* and *D*, are made progressively more realistic.

The simplest model of this kind was derived independently in the 1920s by Lotka (1925) and Volterra (1926). They assumed that the plants would increase exponentially in the absence of herbivores ($A = aV$); that in the absence of plants, the herbivores would starve and their numbers would decline exponentially ($D = dN$); that herbivores encountered plants at random, and that the depression in plant population growth rate was a linear function of the number of encounters ($B = bNV$); and that herbivore population increase was also a linear function of the number of encounters ($C = cNV$).

The dynamics of this model are of little interest to us, since the 'predator–prey cycles' it produces show neutral stability and are of no value in describing real interactions (May 1973).

4.3.2   Resource-limited plants

The most obvious fault with the Lotka–Volterra model is the assumption that the plants would increase without limit in the absence of herbivores. We can improve this by allowing that when herbivores are absent, the plants would grow to their carrying capacity, K, following an S-shaped curve. The simplest way to model this is to use the logistic curve; component A becomes $aV (K - V)/K$ while the other three elements are unaltered;

$$\frac{dV}{dt} = \frac{aV(K - V)}{K} - bNV$$

$$\frac{dN}{dt} = cNV - dN. \tag{4.1}$$

Such models have a long history (Leslie & Gower 1960, Tanner 1975) and their behaviour can be investigated in two ways. The equations can be solved numerically on a computer, using Runge-Kutta or some related technique, to produce plots of numbers against time as in Fig. 4.10. Alternatively, and more comprehensively, the models can be solved analytically. In equilibrium both dV/dt and dN/dt are zero, so the equilibrium plant abundance $\overset{*}{V}$ and the equilibrium herbivore abundance $\overset{*}{N}$ can be found by setting $A = B$ and $C = D$ and doing a little algebra.

$$\overset{*}{V} = \frac{d}{c} \tag{4.2}$$

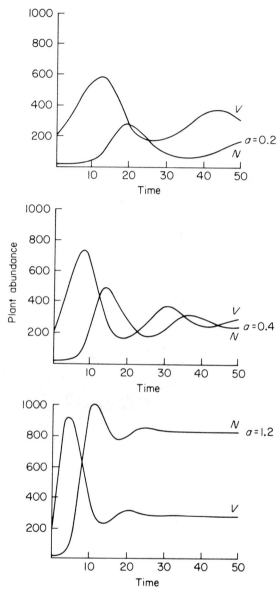

Fig. 4.10 Resource-limited plants. The effect of the plant's intrinsic growth rate *a*; increasing *a* from 0.2 to 1.2 increases stability and increases herbivore equilibrium density, but has no effect on equilibrium plant abundance. Other parameters are $b = 0.001$, $o = 0.001$, $d = 0.3$, $K = 1000$. As in the models that follow, the curve *V* represents plant abundance and the curve *N* represents herbivore numbers (in arbitrary units).

$$\overset{*}{N} = \frac{a(K - d/c)}{bK}.$$ (4.3)

Despite its extreme simplicity, a good deal can be learned about the behaviour of a whole family of models from this example. We shall look at the importance of each of the five parameters in turn.

It is clear from equation 4.2 that the plants' intrinsic rate of increase, a, has no effect on the equilibrium population of plants; the faster the plants grow, the faster the herbivores eat them up. Herbivore population equilibrium does, however, increase as a linear function of a (equation 4.3). The time taken for the plant and herbivore populations to return to equilibrium after perturbation also depends critically upon the plants' rate of increase; Fig. 4.10 shows how increasing a increases herbivore abundance and speeds the return to equilibrium.

The parameter b describes the feeding rate of the herbivores and the response of the plant to feeding. Changes in b do not affect the size of the equilibrium plant population (equation 4.2) but increases in b lead to *lower* herbivore equilibrium. This counter-intuitive result comes about because the more efficient each herbivore is at food gathering, the fewer herbivores can be supported by a given level of plant production. Fig. 4.11 shows the effect of increasing b; herbivore equilibrium is reduced while plant equilibrium and stability are unaffected.

The numerical response of the herbivores is described by c; the higher the value of c the more herbivores are born per unit feeding. It is essentially the efficiency with which herbivores turn food into progeny. The parameters b and c are clearly interrelated since the animal cannot turn food into progeny until that food has been sought-out and consumed. As one would expect, the higher the value of c, the lower the equilibrium population of plants and the higher the equilibrium herbivore density (equation 4.3). Rather less obvious, however, is the effect of c on stability.

In this whole family of models, the overall stability derives from the tight density-dependent control on plant abundance that comes from having a fixed carrying capacity, K. The closer the equilibrium population of plants lies to K, the more highly density dependent are changes in plant (and hence in herbivore) abundance. When plant equilibrium density is low, the plant population is under lax control, and tends to increase exponentially when herbivore numbers are reduced. Thus the lower the equilibrium plant population, the lower the stability of plant and herbivore numbers. This is why stability declines as increasing values of c cause lower plant, and higher herbivore equilibria (Fig. 4.12).

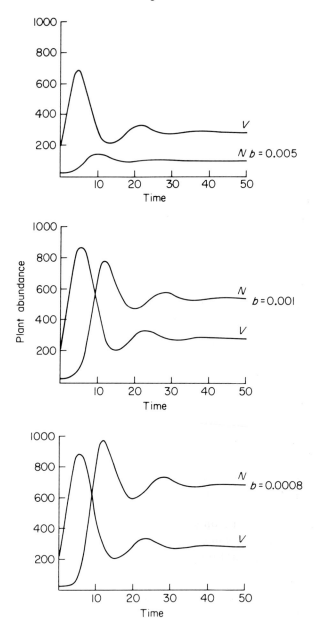

Fig. 4.11 Resource-limited plants. The effect of herbivore searching efficiency *b*. Increasing *b* from 0.0008 to 0.005 *reduces* herbivore equilibrium density but has no effect on stability or on equilibrium plant abundance. Other parameters are $a = 0.5$, $c = 0.001$, $d = 0.3$, $K = 1000$.

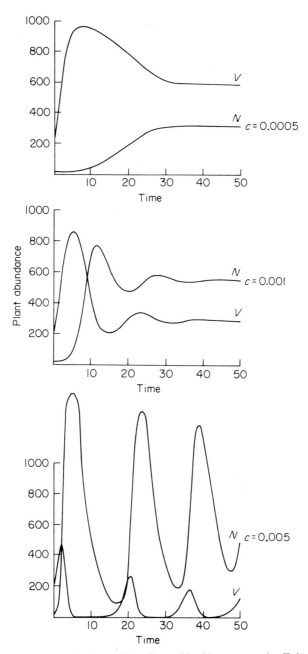

Fig. 4.12 Resource-limited plants. The effect of herbivore growth efficiency c. Increasing c from 0.0005 to 0.005 reduces equilibrium plant abundance and thereby reduces stability but increases equilibrium herbivore numbers. Other parameters are $a = 0.5$, $b = 0.001$, $d = 0.3$, $K = 1000$.

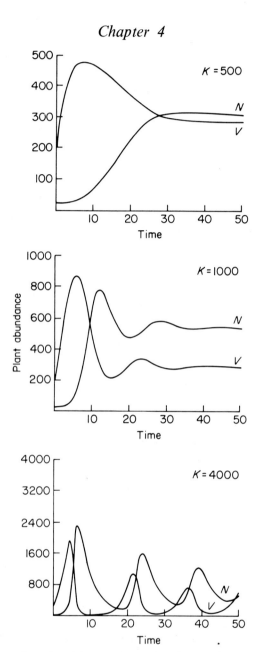

Fig. 4.13 The paradox of enrichment. Increasing the carrying capacity of the environment for plants *K* from 500 to 4000 increases herbivore equilibrium density but has *no* effect on plant equilibrium (note scale changes on y axis). The main effect of increasing *K* is to reduce stability (see text for details). Other parameters are $a = 0.5$, $b = 0.001$, $c = 0.001$, $d = 0.3$.

The effects of herbivore death rate on the behaviour of the model are directly opposite to those of c. Increasing the herbivore death rate increases stability, because it leads to an increase in equilibrium plant numbers. Obviously, the higher the herbivore death rate, the lower the equilibrium abundance of herbivores (see equation 4.3).

Plant carrying capacity, K, is critical to the stability of the model, because the proximity of equilibrium plant abundance to K determines the intensity of density-dependent control. It is clear that an increase in K should lead to an increase in the number of herbivores, since the animals are food limited, but, as we have seen, the equilibrium plant population is determined solely by c and d. Thus increasing K means that the equilibrium plant population is *proportionately* lower and the intensity of density-dependent control on plant numbers is, therefore, reduced. This is why increasing K reduces the stability of the model (Fig. 4.13), and may be the explanation of Rosenzweig's (1971) 'paradox of enrichment'.

4.3.3   Herbivore functional responses

As we have seen (see section 3.1) the feeding rate of each herbivore is limited by its body size and physiological condition, so that there is a maximum rate at which a given number of herbivores can deplete the vegetation. There are two cases of interest; when the feeding rate rises smoothly to an asymptote (a Type 2 functional response) and when feeding rate is density dependent at low levels of plant availability (an S-shaped or Type 3 functional response). For the first, using Ivlev's (1961) formulation, we have

$$\frac{dV}{dt} = \frac{aV(K-V)}{K} - bN\{1 - \exp(-eV)\}$$

$$\frac{dN}{dt} = cN\{1 - \exp(-eV)\} - dN \tag{4.4}$$

which means that when plants are abundant, herbivore depletion is negatively density dependent; increasing the amount of vegetation reduces the percentage that is consumed. We should expect this model to be rather less stable than the last; it has a stable equilibrium that is approached by damped oscillations but, when K is increased, it can demonstrate stable limit cycles (rather than the neutral cycles shown by equation 4.1). (Note that although the parameters b and c have been retained for familiarity, their values will differ from model to model; see figure legends.)

The effects of the parameters, a, b, c, d and K on equilibrium densities and

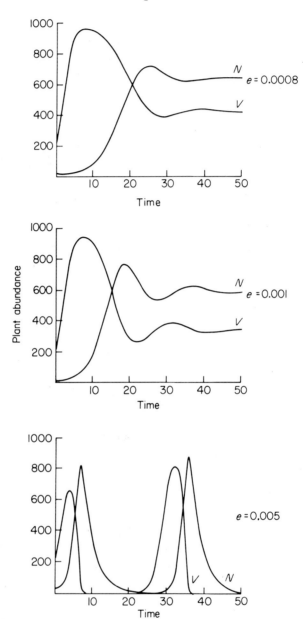

Fig. 4.14 Herbivore functional response. The parameter *e* measure the slope of the functional response. Increasing *e* from 0.0008 to 0.005 reduces equilibrium plant abundance, slightly increases herbivore abundance but substantially reduces stability. Other parameters $a = 0.5$, $b = 0.001$, $c = 0.001$, $d = 0.3$, $K = 1000$.

stability are just the same as in the previous model (equation 4.2). The equilibrium populations of plants and herbivores are

$$\overset{*}{V} = \frac{-\ln(1 - d/c)}{e} \tag{4.5}$$

$$\overset{*}{N} = \frac{aV(K - \overset{*}{V})}{bK(1 - \exp(-e\overset{*}{V}))} \tag{4.6}$$

Clearly d must be less than c, because the herbivore population must be able to increase in size when food is plentiful.

The parameter e of the functional response reflects the searching efficiency of the herbivores. Small values of e mean that food intake per herbivore rises slowly with increasing food availability; food intake rises rapidly to its asymptotic rate when e is large. Broadly, intake is an increasing function of availability when e is small, and is independent of availability when e is large (compare the curves a and c in Fig. 3.1). The effects of increasing e are shown in Fig. 4.14, where the outcome is altered from rapidly damped stable equilibrium to violent cyclic behaviour.

The moral is that a herbivore with high searching efficiency *can* reduce plant abundance to low levels, but only at the cost of reduced stability. Some kind of density-dependent constraint on the herbivores' rate of increase is necessary if low equilibrium plant abundance is to be associated with stable herbivore numbers (see section 4.3.4 on interference; section 4.3.7 on patchiness, etc.).

In the second case we can model sigmoid functional responses by replacing Ivlev's asymptotic function by an S-shaped curve such as

$$\frac{dV}{dt} = \frac{aV(K - V)}{K} - \frac{bNV^2}{e' + V^2}$$

$$\frac{dN}{dt} = \frac{cNV^2}{e' + V^2} - dN. \tag{4.7}$$

Under most conditions an S-shaped functional response model behaves exactly as a Type 2; there is a single stable equilibrium, and increasing b reduces herbivore equilibrium just as in equation 4.2. The fact that herbivore feeding rate increases with plant density when plants are scarce allows at least the possibility of a second, stable plant equilibrium at low biomass levels. Fig. 4.15 shows how this could come about. If herbivore numbers are fixed at, say, 400, then increasing b takes us from a single stable equilibrium at high plant biomass, through a condition where two stable equilibria are

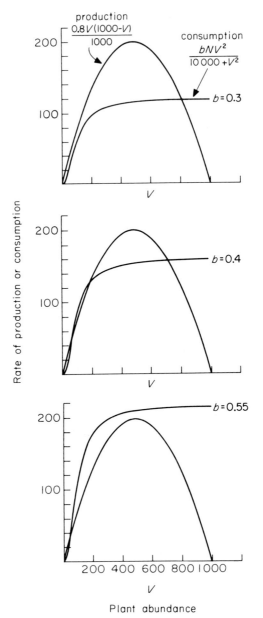

Fig. 4.15 Sigmoid functional responses and multiple equilibria. With constant herbivore numbers the rate of plant production and the rate of plant consumption can show (a) a single, high plant density equilibrium ($b=0.3$); (b) two stable equilibria separated by an unstable equilibrium ($b=0.4$); and (c) a single, stable, low plant density equilibrium ($b=0.55$). A disturbance that increased the value of b from 0.3 to 0.55 would cause a discontinuous crash from the higher to the lower plant equilibrium. Increased searching efficiency could come about through increased temperature, or through the removal of other plant species, rendering the hosts more obvious. Other parameters $a=0.8$, $e'=10\,000$, $N=400$.

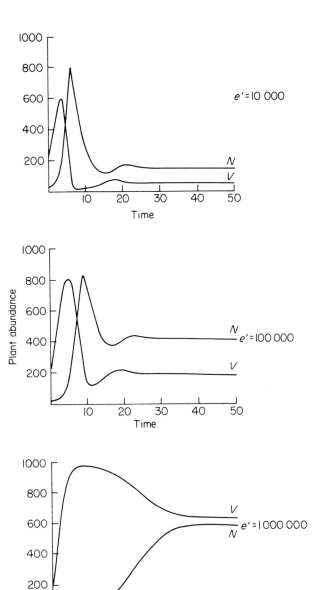

Fig. 4.16 Sigmoid functional responses with variable herbivore numbers. The multiple equilibria of Fig. 4.15 do not exist when herbivore numbers are determined by plant abundance. Increasing e' increases plant equilibrium and increases stability by reducing the slope and removing the destabilizing asymptote of the functional response curve. Other parameters $a = 0.8$, $b = 1.0$, $c = 0.001$, $d = 0.3$, $K = 1000$. Just as with the type II functional response, increasing b from 0.5 to 4.0 reduced herbivore equilibrium without affecting equilibrium plant abundance (as in Fig. 4.11).

separated by a third, unstable equilibrium, to a state where there is only one stable, low-density equilibrium. With the intermediate values of b, the system might be expected to flip from one equilibrium to the other following large perturbations (for examples see Ludwig *et al.* 1978, Clark & Holling 1979, May 1981b). However, in our model herbivore numbers are *not* fixed. When we change b we alter the equilibrium abundance of herbivores, so the whole functional response curve in Fig. 4.15 fluctuates between the three states as numbers vary following perturbation (Fig. 4.16).

With variable herbivore numbers there is only a single equilibrium at

$$\overset{*}{V} = \frac{e'}{c/d - 1} \tag{4.8}$$

$$\overset{*}{N} = \frac{a}{b}\left(\frac{e'}{\overset{*}{V}} - \frac{e'}{K} + \overset{*}{V} - \frac{\overset{*}{V}^{2}}{K} \right) \tag{4.9}$$

and the multiple equilibria only arise when herbivore numbers are fixed (see, for example, Noy Meir 1975 and section 4.2.3).

Since functional responses invariably mean that when plants are abundant the per capita rate of plant depletion declines with increasing plant density, they are always a destabilizing factor in plant–herbivore dynamics. The two models above retain their stability because of the overriding stabilizing effect of density dependence in the plants. Functional responses are complicated, however, when plant abundance and quality are correlated. In grasslands, for example, at high standing crop biomass, a high proportion of the crop consists of older, less nutritious leaves. In contrast, when the forage is grazed down, the regrowth consists entirely of young, high-quality foliage, so that increases in quality compensate for reduced availability at high grazing pressures, and intake actually falls at very high levels of pasture abundance. This effect is strongly destabilizing and tends to produce severe overgrazing unless the plants possess an ungrazable reserve (Noy Meir 1975).

4.3.4   Herbivore numerical responses

It has been assumed in all the models so far that herbivore numbers are a linear function of the amount of food eaten. Clearly this need not be so (see sections 3.2 and 3.3). Maximum herbivore fecundity may be limited by factors other than food availability or, conversely, a threshold

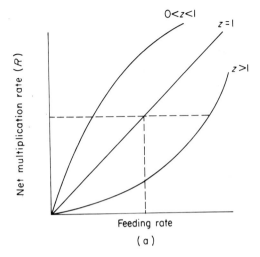

(a)

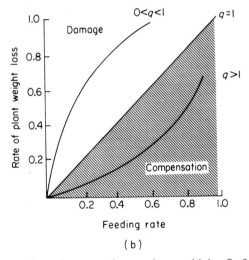

(b)

Fig. 4.17 (a) Starvation tolerance and starvation sensitivity. So far the herbivore's net multiplication rate (R) has been a linear function of the feeding rate $(z = 1)$. Population growth will cease $(R \leq 1)$ at higher feeding rates for starvation-sensitive $(z \geq 1)$ than for starvation-tolerant $(0 \leq z \leq 1)$ species. (b) Plant compensation occurs when the rate of weight loss by the plant is less than the rate of herbivore feeding $(q > 1)$; damage occurs when the rate of weight loss exceeds the rate of feeding $(0 < q < 1)$. See text for equations.

amount of food may be necessary before reproduction is possible (Fig. 4.17). The constant z is used to describe the starvation-sensitivity of the herbivore. Values of z between 0 and 1 describe a starvation-tolerant species (one, for example, that can draw on fat reserves for reproduction), while values of $z > 1$ describe starvation-sensitive species whose net rate of increase falls off steeply with declining food availability.

$$\frac{dV}{dt} = \frac{aV(K - V)}{K} - bNV$$

$$\frac{dN}{dt} = c(NV)^z - dN. \tag{4.10}$$

So far we have assumed a linear relationship between herbivore numerical response and feeding success ($z = 1$) which produces gently damped oscillations with the parameter values employed in Fig. 4.18. A starvation-sensitive herbivore ($z > 1$) suffers a greatly reduced rate of increase at low plant densities and, as would be expected, equilibrium herbivore density is lower. Starvation-tolerant animals, by drawing on their food reserves, maintain their rate of increase at lower plant densities, so their equilibrium abundance is higher.

It might be expected that starvation tolerance would lead to over-exploitation of the plant population and would consequently be destabilizing; this is indeed the case with discrete-time models (Crawley 1975). Here, however, because herbivore numbers are adjusted to plant availability continuously and without a time-lag, over-exploitation is never serious and stability is little affected (Fig. 4.18). In the field, however, over-exploitation by long-lived, starvation-tolerant herbivores can be highly destabilizing because of the time-lag in their numerical responses (e.g. elephants in East African woodlands; Buechner & Dawkins 1961, Glover 1963, Laws *et al.* 1975).

An important numerical response in many herbivores acts through dispersal. We can model this simply by making the loss rate of the herbivores density dependent. If the response acts throught mutual interference (through aggressive behaviour, for example) we would put

$$\frac{dN}{dt} = cNV - dN^m \tag{4.11}$$

where $m > 1$ reflects the intensity of interference (Rogers & Hassell 1974).

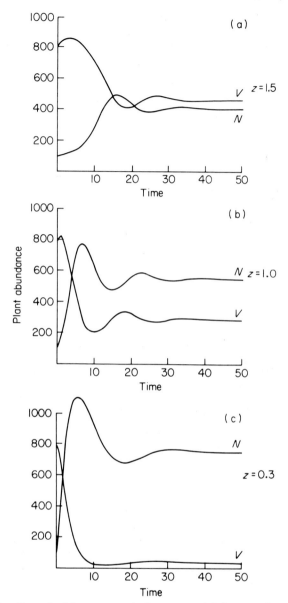

Fig. 4.18 Non-linear herbivore numerical response. (a) A starvation-sensitive herbivore ($z = 1.5$) causes high equilibrium plant abundance but rather low herbivore density. (c) A starvation-tolerant species (with, for example, high body fat reserves) causes a great reduction in equilibrium plant abundance and leads to higher equilibrium herbivore densities. The parameter z has little impact on stability (but see text). Other parameters $a = 0.5$, $b = 0.001$, $c = 500 \times (1/500\,000)^z$, $d = 0.3$, $K = 1000$.

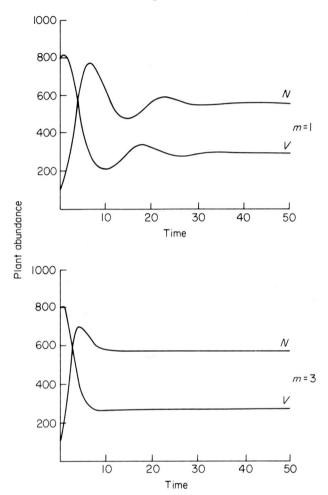

Fig. 4.19 Mutual interference. Interference between herbivores causes a reduction in searching efficiency as herbivore numbers rise. This is strongly stabilizing because it prevents over-exploitation of the plants. Herbivore death rate is independent of herbivore density for $m = 1$ but increases steeply with herbivore density when $m = 3$. Other parameters $a = 0.5$, $b = 0.001$, $c = 0.001$, $d = 180 \times (1/540)^m$, $K = 1000$.

Density dependence in herbivore death rate is strongly stabilizing (Hassell 1978). As the intensity of mutual interference is increased, the model passes from damped oscillations ($m = 1$) to exponential stability ($m \gg 1$, see Fig. 4.19). Since interference slows the rate of increase at high herbivore numbers, plant populations can be reduced to levels that are both low and stable (cf. Fig. 4.14). This model becomes more stable as *m* increases because interference prevents over-exploitation of the plants. If the plants were not subject to their own density-dependent constraints, extreme interference could lead to the plants escaping control.

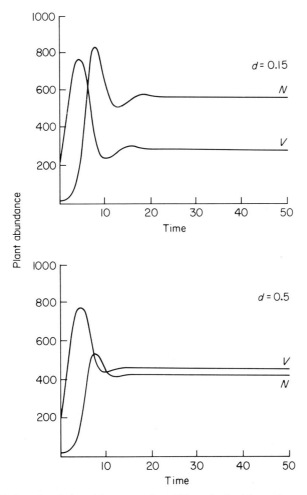

Fig. 4.20 Emigration induced by starvation. When the herbivore loss rate increases linearly with N/V then increasing *d* from 0.15 to 0.5 leads to increased stability, reduced herbivore equilibrium and increased plant abundance. Other parameters $a = 0.6$, $b = 0.001$, $c = 0.001$, $K = 1000$.

Alternatively, the loss rate could be density dependent due to hunger-induced emigration. In this case it is not the absolute number of herbivores that is important, but the amount of food per herbivore, and we would model emigration as

$$\frac{dN}{dt} = cNV - \frac{dN^2}{V} \tag{4.12}$$

This response, too, is strongly stabilizing; as d increases, the return to equilibrium occurs more rapidly (Fig. 4.20; and see Fig. 4.3).

4.3.5   Plant compensation

Plant compensation occurs when the depressive effects of herbivore feeding on plant growth are less than the additive effects of feeding on herbivore increase (see section 2.7.1). We replace the linear damage function $- bNV$ with the non-linear term $- b(NV)^q$. When $q > 1$ the plant compensates (Fig. 4.17) for herbivore feeding and when $0 < q < 1$ herbivore feeding damages plant growth. The model is now

$$\frac{dV}{dt} = \frac{aV(K - V)}{K} - b(NV)^q$$

$$\frac{dN}{dt} = cNV - dN \tag{4.13}$$

Plant compensation affects both stability and population equilibria. When the plant suffers a reduction in growth rate in linear proportion to herbivore feeding we obtain the damped oscillations shown in Fig. 4.21(b). This is the model ($q = 1$) that we have assumed so far. When compensation occurs, the plant suffers proportionately less at low rates of herbivore feeding than at high ($q > 1$). This is strongly stabilizing and leads to an increase in herbivore equilibrium. Plant density is quite unaffected (Fig. 4.21(c)). When herbivore feeding is damaging even at very low levels (either because they feed on particularly vulnerable plant parts like meristems, transmit viruses and so on), the reverse trends occur. For values of q between 0 and 1, mean herbivore numbers are reduced and the model is a good deal less stable (Fig. 4.21 (a)).

This is a paradoxical result because it suggests that compensation is of no benefit to the plant; it simply serves to make the herbivore more abundant. Again, this is an artefact of the continuous time base of the model. If the herbivores only fed on the plant for a limited time each year (like univoltine

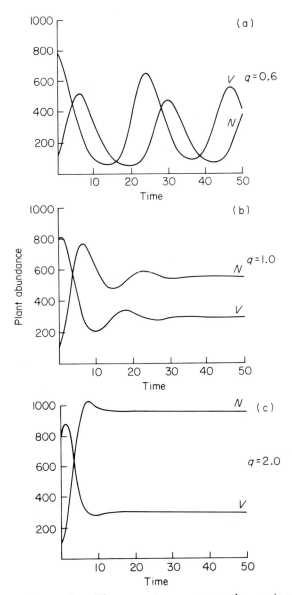

Fig. 4.21 Plant compensation. The parameter q measures the way in which herbivore feeding reduces plant abundance (see Fig. 4.17). We have so far assumed a linear damage function ($q = 1$). When the plant compensates for herbivore feeding, plant weight loss is less than herbivore food removal ($q \geq 1$). When the plant is damaged by herbivore feeding, it loses more weight than the herbivores remove directly by feeding ($0 \leq q \leq 1$). (a) Plant damage ($q = 0.6$) is strongly destabilizing compared with (b) linear damage, while compensation (c) is strongly stabilizing ($q = 2.0$). Compensation of this kind does *not* benefit the plant and serves only to make the herbivore more abundant (see text).

insects on trees), the plant would have scope to compensate for their feeding once the attack had abated (see section 2.7.8). Some plants may even create a herbivore-free period for themselves by shedding their leaves after heavy herbivore attack (like the *Eucalyptus* described in section 3.3.1). The regrowth foliage then experiences a period free of herbivores in the time before re-invasion occurs.

This model tells us that we should not expect to find strong plant compensation in plants that are subject to continuous feeding by food-limited herbivores. Compensation can benefit the plant, of course, when the herbivores are limited by natural enemies or are kept at low densities by bad weather.

Feeding that causes disproportionately high damage $(0 < q < 1)$ is strongly destabilizing because it means that herbivore feeding is negatively density dependent; the rate of damage per herbivore declines with increasing plant density.

4.3.6   Impaired plant regrowth

Regrowth ability may depend upon more than just the residual leaf area, and it is unrealistic to assume that the effects of high herbivore density act through feeding alone. When animal numbers are high relative to food availability, due to herbivore immigration or catastrophic reduction of plant availability through fire or flood, the herbivores damage the plants in numerous ways (trampling, fouling, grubbing-up, etc.). In this way the ability of the plant population to recover from grazing may be seriously impaired, so that following perturbation, the vegetation recovery period is much longer than expected. To model this requires that the rate of regrowth, a, is a decreasing function of grazing pressure, N/V; it is convenient to assume that regrowth declines exponentially with N/V, so that

$$\frac{dV}{dt} = \frac{a\exp(-\varepsilon N/V)(K - V)}{K} - bNV \qquad (4.14)$$

Increasing ε means that the regrowth potential of the plants is strongly depressed by high grazing intensities (N/V). When regrowth depression is slight we obtain the usual output of damped oscillation; as ε is increased we pass from more violent oscillations into cyclic behaviour (Fig. 4.22). As usual, the decrease in stability is due to the depression of plant biomass to very low levels (so that density-dependent regulation on the plants is lax) and the lack of density dependence in the other three compartments of the model.

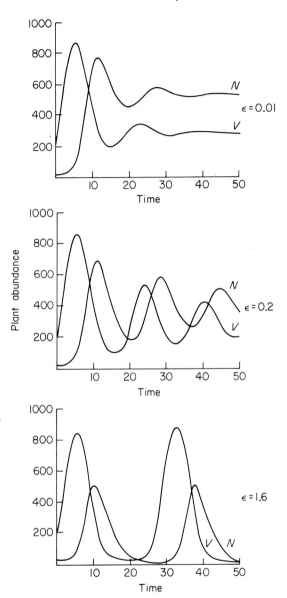

Fig. 4.22 Impaired plant regrowth. When high grazing pressure reduces the intrinsic rate of plant growth then plant and herbivore numbers are likely to fluctuate. Increasing the damage parameter from 0.01 to 1.6 leads from damped oscillations through more persistent damped oscillation to stable limit cycles. This model may help in understanding why some populations of voles undergo cycles of abundance whereas others do not (see text).

This model embodies a good deal of plausible ecological behaviour; for example, it shows that the same model of food limitation can produce cycles or stable equilibrium under different conditions (cf. Krebs 1979). It allows for the existence of cyclic behaviour without reference to unmeasurable time-lags (May 1973), although it must be admitted that the depression of regrowth serves the same purposes as a time-lag by not allowing an increase in plant biomass until grazing pressure has declined. Further, the model does not rely on the existence of saturating functional responses (Batzli *et al.* 1981) or of density dependence in the herbivores' numerical responses (Krebs & Myers 1974) to produce cycles.

This model also describes the kind of cycles demonstrated by larch budmoth (see section 4.2.4). The mechanism by which the cycles are generated in the model mimics the fact that feeding by the moth larvae reduces the subsequent rate of larch growth, because in the years following severe defoliation the needles are shorter (there is less leaf area), they are produced later and elongate more slowly.

However, as with all long-term studies, we are unable to make any concrete statements about the cause of the pattern of dynamic change that is shown. The fact that a model shows the same pattern as the data is no indication that the structure of the model matches the structure of the real plant–herbivore system. These models are descriptive and not explanatory. What we require are detailed, long-term field experiments to fathom the precise mechanisms that affect the rates of increase of the plants and the herbivores (whether caused by induced changes in plant chemistry, long-term damage to regrowth potential, altered nutrient availability, depletion of preferred foods, or whatever).

4.3.7   Pattern of herbivore attack

So far we have assumed that each plant suffers equally under herbivory. Clearly this will not be the case when herbivore attack is aggregated because of limited dispersal, patchy plant distribution, heterogeneity in plant size, differences in food quality and so on. The models that follow apply particularly to sessile herbivorous insects but their spirit applies equally well to mobile, larger herbivores whose attack rate per plant is uneven for one reason or another.

Unfortunately, the assumption that herbivores cause unequal damage to different plants makes the sums a good deal harder. The total number of herbivores, N, is distributed over V plant units; some units are herbivore-free, others have $1, 2, 3, \ldots i$ herbivores per unit, so that

$$N = \sum_{1}^{V} i$$

Herbivores are distributed over the plants according to a probability distribution (like the poisson or the negative binomial) so that the proportion of plants with i herbivores is $p(i)$.

Linear damage function

The simplest assumption is that the plants suffer as a linear function of the number of herbivores on them, i. The number of plants with i herbivores is $p(i)V$ and if we define α as the intrinsic plant loss rate per herbivore (the 'damage rate', analogous to the virulence of a parasite), we predict that the rate of loss of plants with i herbivores will be $\alpha ip(i)V$. Adding up the loss rates for plants with every possible number of herbivores we obtain

$$\alpha V \sum_{i=0}^{\infty} ip(i)$$

which is the rate of herbivore-induced plant mortality (or fecundity depression). No matter what probability distribution is chosen to describe the data, the term $\sum ip(i)$ is simply the mean number of herbivores per plant, N/V; it is the expectation of i, $E(i)$.

The loss rate due to herbivore feeding, therefore, simplifies to αN because the V's cancel. The model for plant population dynamics is just

$$\frac{dV}{dt} = aV - \alpha N \qquad (4.15)$$

To simplify the mathematics, I have ignored the logistic term in plant growth; if the herbivores can regulate plant population without this term, the model would simply be more stable if it were included.

Sessile herbivores suffer losses from four causes:

(1) by natural herbivore mortality (at a rate dN);

(2) by natural mortality of their host plant (at a rate bN);

(3) by their killing the host plant they are on (at a rate αi);

(4) by losses during dispersal in search of another plant.

Only the third category requires further explanation here; we shall deal with dispersal later. Of the $p(i)Vi$ herbivores on plants with i herbivores, a fraction αi is lost because they kill the plant that feeds them; this self-induced

Table 4.2. The expectations of i, i^2 and i^3 for the binomial, poisson and negative binomial distributions (see text).

	Binomial	Poisson	Negative binomial
$E(i)$	$\dfrac{N}{V}$	$\dfrac{N}{V}$	$\dfrac{N}{V}$
$E(i^2)$	$\dfrac{N}{V} + \dfrac{(k'-1)}{k'}\left[\dfrac{N}{V}\right]^2$	$\dfrac{N}{V} + \left[\dfrac{N}{V}\right]^2$	$\dfrac{N}{V} + \dfrac{(k+1)}{k}\left[\dfrac{N}{V}\right]^2$
$E(i^3)$	$\dfrac{N}{V} + 3\dfrac{(k'-1)}{k'}\left[\dfrac{N}{V}\right]^2 + \left[\dfrac{N}{V}\right]^3 \dfrac{(k'-1)(k'-2)}{k'^2}$	$\dfrac{N}{V} + 3\left[\dfrac{N}{V}\right]^2 + \left[\dfrac{N}{V}\right]^3$	$\dfrac{N}{V} + 3\dfrac{(k+1)}{k}\left[\dfrac{N}{V}\right]^2 + \left[\dfrac{N}{V}\right]^3 \dfrac{(k+1)(k+2)}{k^2}$

mortality occurs at a rate $\alpha i \times p(i) V i$ or $\alpha V p(i) i^2$. For the whole plant population we sum over all levels of infestation and obtain

$$\alpha V \sum_{i=0}^{\infty} i^2 p(i).$$

The term $\sum i^2 p(i)$ is the expectation of i^2 and its value depends upon the particular probability distribution used to describe the herbivore attack. Values of $E(i^2)$ are given in Table 4.2 for the binomial, poisson and negative binomial distributions that are commonly used to describe regular, random and aggregated patterns in ecology.

Now losses due to plant death and natural herbivore mortality add to $(d + b)N$ and herbivore-inflicted losses through host plant death are $\alpha V E(i^2)$. Assuming a negative binomial distribution, multiplying out and cancelling the V's, we arrive at the equation for herbivore dynamics:

$$\frac{dN}{dt} = cNV - (d + b)N - \alpha N - \alpha \frac{(k + 1)}{k} \frac{N^2}{V}. \tag{4.16}$$

The two new aspects we investigate with this model are the damage rate of the herbivores, α, and the spatial distribution of herbivore attack as reflected by the binomial (k') or negative binomial (k) parameters that describe the extent of clumping.

Damage rate α affects both the stability and the equilibrium size of both populations. For intermediate values of α plant abundance is both low and stable (Fig. 4.23(a)). Increasing α leads to an increase in plant abundance, a reduction in herbivore numbers, and to a reduction in stability (Fig. 4.23(b)). This counter-intuitive result was discovered by Anderson and May (1978); because the model is only stable when herbivore attack is aggregated (see below), high damage rates lead to a high death rate amongst the herbivores since they perish with their host plants. Plant abundance increases with α because herbivore numbers are reduced; clearly those plants that are attacked by highly damaging herbivores are quickly killed. The greatest depression of equilibrium plant abundance occurs at intermediate levels of herbivore damage rate; this is a result of great interest in biological weed control.

Attack pattern has a profound influence on the behaviour of the model. If a regular pattern of attack is assumed, there is no stable equilibrium. When herbivore attack is random, plant and herbivore numbers cycle at low average densities (Fig. 4.23(d)). If herbivore attack is aggregated, however, so that some plants are heavily attacked and others escape altogether, then a

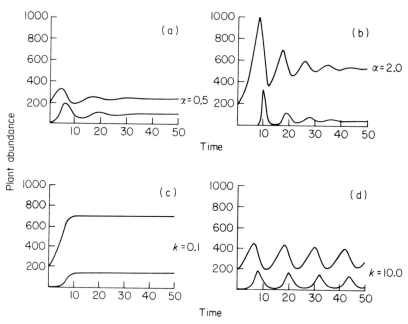

Fig. 4.23 Different plants suffer different rates of herbivore attack. Sedentary invertebrate herbivores are killed when they kill their host plant, so that increasing the virulence of the herbivores *increases* the equilibrium abundance of plants and reduces stability (a) $\alpha = 0.5$, (b)$\alpha = 2.0$. The stability of the model and the equilibrium plant density also depend upon the degree of aggregation of the herbivores. Increasing k of the negative binomial from (c) $k = 0.1$ which represents pronounced aggregation to (d) $k = 10.0$ which is more or less random (poisson) attack brings about a substantial reduction in plant abundance and also a change from exponential stability to persistent cycles. Upper curve = plant abundance; lower curve = herbivore numbers.

stable equilibrium is possible. For highly aggregated attack (k very small) the model demonstrates exponentially stable equilibrium (Fig. 4.23(c)). Since so many plants escape attack when k is small, equilibrium plant density increases as aggregation becomes more intense.

The interesting dynamics of this model centre on the fact that herbivores which kill their hosts also kill themselves. This may or may not be realistic for insects. In some extreme cases, for example, herbivore reproduction will not be successful *unless* the host is killed (e.g. bark beetles which require mass attack to overcome healthy trees, see section 3.1.5). At the other extreme, cochineal insects, which kill an isolated *Opuntia* cactus, are bound to die, so limited is their dispersal ability (Zimmermann 1979). Between these extremes lie a whole range of possibilities where not only plant density but also plant spatial distribution affect the survival rate of the herbivores. If isolated plants are preferentially attacked and the animals are not highly mobile, then killing the plants means almost certain death. We would, therefore, predict that herbivores which preferentially attack isolated plants would possess density-regulating mechanisms which limit herbivore numbers per plant below the level at which the plant dies.

Non-linear damage function

The model is improved by allowing that the loss rate is a non-linear function of the number of herbivores per plant, i. To simplify the mathematics, assume that damage is a function of i^2; this mimics plant compensation. Now the plants with i herbivores are damaged at a rate $\alpha Vp(i)i^2$ and, summing over all levels of infestation we obtain $\alpha V \sum i^2 p(i)$, so

$$\frac{dV}{dt} = aV - \alpha VE(i^2) \tag{4.17}$$

where the expectation of i^2 depends on the distribution of herbivore attack (see Table 4.2).

The herbivores, as before, suffer losses due to natural mortality, natural plant death and herbivore-induced plant death. The first two terms are unaltered but now a fraction αi^2 of the $Vp(i)i$ herbivores succumbs with the plants it kills. So this component of the loss rate is $\alpha Vp(i)i^3$ and, as before, summing over all classes of infestation gives us $\alpha V \sum i^3 p(i)$.

The summation term is the expectation of i^3 (see Table 4.2) so that

$$\frac{dN}{dt} = cNV - (d + b)N - \alpha VE(i^3). \tag{4.18}$$

This model possesses a stable equilibrium for both aggregated and random distributions of herbivores (Fig. 4.24). Thus randomly distributed herbivore attack can regulate plant population density so long as the plants can compensate for low levels of herbivore attack (as in section 4.3.5). If the mean number of herbivores per plant is small, and the degree of under-dispersion not too great, this model can possess a stable equilibrium even with a binomial distribution of animals per plant. The stability conditions and details of the equilibria can be found in Anderson and May (1978).

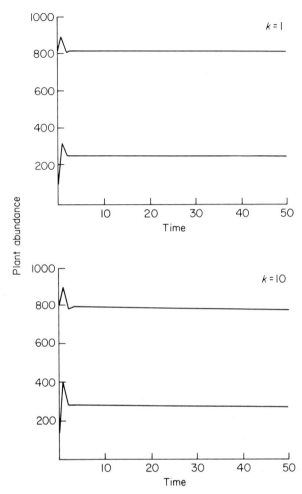

Fig. 4.24 Plant compensation with unequal attack rates. Assuming that plant losses increase with the square of herbivore numbers per plant ($q = 2$) increases stability dramatically. Now random attack distributions ($k = 10$) are just as stable as aggregated attack ($k = 1$) (cf. Fig. 4.23(d)). Upper curve = plant abundance; lower curve = herbivore numbers.

4.3.8   Ineffective herbivore dispersal

As we have seen, many sessile herbivores have very limited powers of dispersal (cochineal insects, scale insects, etc.) and suffer great losses when emigrating from the plant, particularly in low-density plant populations and from isolated plants in aggregated populations (see section 3.3.2). To model

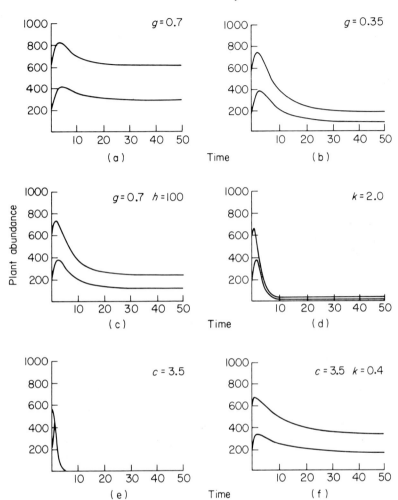

Fig. 4.25 Reproduction within the plant and inefficient plant-to-plant dispersal. Altering the fraction of the animals that disperse in search of new host plants; reducing *g* reduces equilibrium plant and herbivore abundance. (a) *g* = 0.7; (b) *g* = 0.35. Dispersal efficiency; reducing *h* from (a) 250 to (c) 100 leads to a reduction in equilibrium plant abundance but has no effect on stability. Reducing aggregation from (a) *k* = 1.0 to (d) *k* = 2.0 leads to a dramatic reduction in plant abundance, but the equilibrium remains stable. Increasing the rate of reproduction within the plant from (a) *c* = 2.2 to (e) *c* = 3.5 causes the herbivore to eat the plant population to extinction. Increases in herbivore reproductive rate can be accommodated if attack becomes more aggregated; (f) has *c* = 3.5 but *k* reduced to 0.4. A major shortcoming of this model is its assumption that *k* remains constant despite exponential increase in herbivore numbers per plant.

this we can allow that the proportion of emigrants finding a host plant is proportional to $V/(h+V)$; when V is large almost all emigrants survive, but when V is small and h (inefficiency in dispersal) large, only a small proportion finds host plants.

In this model, herbivore numbers increase by reproduction on the host plant and by emigrant animals establishing on new plants. The dynamics are described by

$$\frac{dN}{dt} = (c-g)N + \frac{gNV}{h+V} - dN - \alpha V E(i^2) \qquad (4.19)$$

where g is the emigration rate. As in section 4.3.7, a linear damage function is assumed.

The model is only stable when an aggregated distribution of herbivores is assumed. If herbivores are distributed at random, the model can be stabilized by allowing that reproduction on the plant is density dependent or that the rate of emigration is density dependent. Inefficiency in dispersal (measured by h) has little effect on the dynamics; increasing h merely reduces equilibrium herbivore density without affecting stability (Fig. 4.25). A detailed analysis is given in May and Anderson (1978).

4.3.9 Herbivore numbers not food limited

In many agricultural systems and for the bulk of monophagous herbivores that are predator regulated, the impact of herbivore feeding on the plant is not matched by a reciprocal influence of the plant on animal density. This condition is modelled simply by assuming that there is a fixed amount of herbivore feeding, bN, so that

$$\frac{dV}{dt} = \frac{aV(K-V)}{K} - bN \qquad (4.20)$$

$$\frac{dN}{dt} = 0.$$

When there is no herbivore feeding, the plant increases to its stable equilibrium at K, the carrying capacity. As N is increased, the equilibrium plant population is reduced but remains stable. Once N is increased above the break point of $aK/4$ the plants are driven to extinction. It is important to note that, except at exactly $N = aK/4$, this model has two equilibria, one stable and one unstable (Fig. 4.26(a)). While the system responds to small perturbations in plant abundance by returning to the higher equilibrium, it will crash to extinction if the perturbation takes V below the lower equilibrium (see Harvesting, section 4.2.5).

Non-food-limited herbivores may exert a less profound influence on the plant population if they take a constant fraction of the plant parts available rather than a constant number. For this we put

$$\frac{dV}{dt} = \frac{aV(K-V)}{K} - \beta V \qquad (4.21)$$

where β is the instantaneous rate of harvesting. The dynamic behaviour can be seen from Fig. 4.26(b); the stable equilibrium plant population is $K(1 - \beta/a)$ and maximum sustained yield occurs at $\beta = a/2$ when the plant population is at $K/2$. The yield at this point is $aK/4$. While the plant population is stable no matter what the rate of feeding, it is important to note that the return time to equilibrium following disturbance increases with β; thus in a randomly fluctuating environment, higher exploitation would lead to higher variability in plant abundance (Beddington & May 1977).

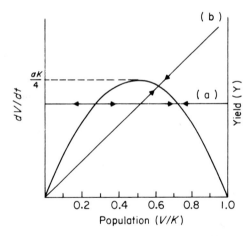

Fig. 4.26 Plant dynamics with fixed herbivore numbers. The curve shows the rate of plant population growth dV/dt as a function of plant density V/K. (a) When each herbivore feeds at a constant rate there is a stable high-density equilibrium and a lower, unstable equilibrium. If plant abundance is reduced below $0.3\,K$, the population crashes to extinction. (b) When each herbivore takes a constant fraction of the available plants there is only one equilibrium and this is always stable. Note, however, that the return time to equilibrium is longer at higher feeding rates (see text).

4.3.10 The role of predators

The earliest attemps at an analytical understanding of the role of predators in regulating herbivore abundance suggested that predators were not capable of maintaining a stable equilibrium population of their prey (e.g. the Lotka–Volterra models of the 1920s; May 1973). The legacy of this work was that predator–prey cycles were for many years considered to be the norm in trophic interactions (Lack 1954). Later, Nicholson and Bailey (1935) developed a model of predator–prey dynamics that incorporated a linear functional response and random search (for details see Hassell 1978). This model, again, produced no stable equilibrium, but led to diverging oscillations and ultimately to the extinction of the prey. Evidently, if predators were to be able to bring about low, stable populations of herbivores they had to exhibit some kind of mechanism that prevented over-exploitation of the prey. The first such mechanism was studied by Hassell and Varley (1969). They showed that if the searching efficiency of the predators declined as predator density rose, then over-exploitation could be prevented and a stable equilibrium could result. They called this process 'mutual interference' and marshalled data from the laboratory and the field to show that it might be of widespread importance amongst insects. The simplest way to ensure that over-exploitation does not occur is to provide a refuge for the prey population. The first analytical investigation of refuges in predator–prey dynamics was by Hassell and May (1974). They suggested that most prey populations are distributed patchily and that predators aggregate in regions of high prey density. Their models produced stable herbivore equilibria because the low-density patches of prey were under-exploited and formed a refuge from predation. The same stabilizing effect can be brought about when a polyphagous predator switches to an alternative food species when the abundance of its preferred food species falls below a threshold level (Murdoch & Oaten 1975). Aggregation in regions of high prey density may lead to depletion of the prey and hence to competition amongst the predators; this 'pseudo-interference' is stabilizing because predator efficiency declines with predator density (Free *et al.* 1977).

It is now well established that predator populations *can* regulate the abundance of their herbivorous prey and that the equilibrium can be stable (May 1981b gives an excellent overview). Clearly, predator-regulated equilibria must be lower than equilibria determined by the availability of plant food. One of the important questions centres on the factors that determine the degree to which a natural enemy can depress the herbivore population below the level it would attain in the absence of predation (Beddington *et al.* 1978); this is discussed more fully in section 4.1.

For simplicity we shall assume that the dynamics of the plant–herbivore interaction can be described by a single, time-delay logistic equation for the growth of the herbivore population. The plants set the carrying capacity for the herbivores, K', and influence stability via the time-lag, T, which reflects the fact that increases in plant abundance only lead to changes in herbivore numbers after a delay roughly equal to the development time of the herbivore. When T is large the herbivore population shows damped or cyclic behaviour; when it is small the population approaches equilibrium smoothly (May 1973; see also Fig. 4.27).

$$\frac{dN(t)}{dt} = rN(t)\frac{K' - N(t-T)}{K'}. \tag{4.22}$$

Let us take the simplest possible model for the dynamics of the predator, P, (the Lotka–Volterra equation)

$$\frac{dP}{dt} = \gamma N(t)P - \delta P \tag{4.23}$$

and for the effect of the predator on herbivore dynamics

$$\frac{dN(t)}{dt} = rN(t)\frac{K' - N(t-T)}{K'} - \varepsilon N(t)P. \tag{4.24}$$

The dynamics of this system have been explored by May (1973; see also Fig. 4.27). Predators can stabilize previously unstable plant–herbivore interactions when the natural time-scales of vegetation recovery (T), herbivore births ($T_1 = 1/r$) and the geometric mean of herbivore birth and predator death rates ($T_2 = 1/(r\delta)^{1/2}$) are related as

$$T_1 < T < T_2.$$

This tells us that with no predators, the time delay in plant regrowth is long compared to herbivore population growth ($T > T_1$), so the feedback is destabilizing. With predators whose natural death rate is such that $T_2 > T$ the time delay in plant regrowth is relatively small so the feedback becomes stabilizing. In other words, long-lived and starvation-tolerant predators are more likely to produce stable plant–herbivore systems.

In a unique natural experiment on Isle Royale in Lake Superior, Michigan, the impact of moose on plant populations with and without vertebrate predators can be seen (Mech 1966). Moose arrived on the island in the early 1900s, probably by swimming from Canada, and wolves arrived by

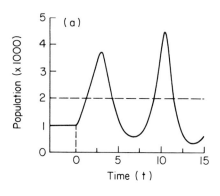

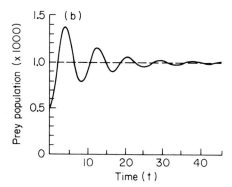

Fig. 4.27 The impact of predators on plant–herbivore dynamics. The upper curve (a) shows herbivore numbers in the absence of natural enemies; the interaction is unstable because the time lag ($T = 3\pi/5$) in the herbivores' response to food availability is long compared with its characteristic time scale ($1/r$). This pattern of herbivore dynamics may describe the moose of Isle Royale prior to the immigration of wolves (see text). The lower curve (b) shows herbivore numbers in the presence of long-lived predators; e.g. moose numbers after the immigration of wolves. The system in (b) is stable because the natural time scale is now longer than the time-lag, thanks to the relatively low death rate of the predators (from May 1973). Under different conditions, addition of a natural enemy could well destabilize a previously stable plant–herbivore interaction (see text).

crossing the frozen lake some time in the late 1940s. We therefore have anecdotal information on about 40 years of plant–herbivore interaction and about 40 years of plant–herbivore–carnivore interaction. The first period was characterized by two large build-ups in moose numbers followed by marked suppression of forest growth and rapid decline in moose density. The second period, during which wolf density stabilized at about 23 animals, has been characterized by greater stability in moose numbers and less dramatic impact on the forest trees. This example does suggest that predators can reduce the likelihood of eruptive outbreaks by vertebrate herbivores (Jordan *et al.* 1971).

A recurrent theme in the literature of herbivory concerns 'predator release'. A period of low, stable numbers is followed by an outbreak of herbivores which severely damages the vegetation. Sometimes numbers crash very soon after the eruption, while in other cases they stay high for a longer period (see section 4.2.3). Natural enemies are assumed to keep the numbers low and stable during the endemic phase. Then, for one reason or another, the intrinsic rate of increase or the carrying capacity of the herbivores is increased. Under these new conditions, natural enemies rise to their limit of abundance and can no longer exert a controlling influence on herbivore numbers. Herbivore density erupts exponentially until limited by food availability or some other density-dependent factor.

The cause of the increase in herbivore r or K is known in a few cases. Increased food quality is often implicated; this may be due to application of fertilizer (Lawton & McNeill 1979), to stress following drought or fire (White 1974) or to senescence of a large proportion of the plants (Morris 1963). Increased food quantity, K, is implicated in the archetypal eruptions of plague locusts, where high rainfall in their breeding areas causes increased grass production (Waloff 1976).

The difficulty is that density-dependent, natural-enemy attack during the endemic phase has been demonstrated unequivocally in very few studies. There are formidable sampling and statistical problems in detecting density dependence in sparse populations. East (1974), however, did detect density-dependent mortality by soil-dwelling carabid beetles on the pupae of several oak-defoliating Lepidoptera of *Erannis* spp. and *Operophtera*. These moths were at approximately one-fifth of outbreak (i.e. defoliating) densities. Predator regulation is more often assumed than critically established (see section 4.2.3).

4.3.11 Annual plants

A different kind of model is needed to describe the impact of grazing on the dynamics of annual plants. Development in annual populations consists of discrete bursts of growth and seed set followed by longer or shorter periods of dormancy. These periods may be regular (as in many dune annuals) or irregular (as in desert plants). It is, therefore, appropriate to represent this system by difference rather than by differential equations. Also, since rather

few species of annual plant have monophagous herbivores, and few
herbivore populations are regulated by the abundance of one species of
annual plant, it is appropriate to assume that herbivore feeding is constant
(i.e. unaffected by changes in plant abundance).

The plant population is made up of two classes; seeds and adults
(although for some species the seeds may be of many ages). Some annuals,
like those inhabiting sand dunes, have short-lived seeds and form no seed
bank; others like many agricultural weeds form vast, long-lived banks with
complex dormancy mechanisms (see section 2.5).

The simplest case is epitomized by the grass *Vulpia fasciculata* studied by
Watkinson and Harper (1978). On sand dunes in north Wales, 90% of the
seed germinated, 69% of the seedlings survived to flower and each individual
produced 1.7 mature seeds on average. Mortality was density independent,
but fecundity was significantly lower at high densities (Fig. 2.10), so that if V_t
plants germinate in the autumn, s survive to reproduce and fecundity is
density dependent we have

$$V_{t+1} = sV_t f\,(V_t).$$

Depending upon the shape of the density-dependent function $f(V_t)$ we obtain
different dynamics (see May 1981a for details).

The plants are eaten by rabbits and are killed when their floral meristems
are cut. Rabbit numbers are not determined by *Vulpia* abundance so we can
represent rabbit numbers as constant at N. The impact of rabbit feeding may
be felt before or after plant density dependence in size has determined seed
set; thus there are two models;

either $$V_{t+1} = sV_t f\,(V_t) - cN$$

or $$V_{t+1} = (sV_t - cN)\,f\,(sV_t - cN).$$

In the first, grazing accentuates the reduction in seed production due to
competition while in the second, grazing relaxes the intensity of density
dependence; the behaviour of the two models is shown in Fig. 4.28 assuming
that seed production per plant declines exponentially with density
$\{f(V_t) = a.\ \exp(-bV_t)\}$.

Three important results emerge from this simple model. First, the timing
of herbivore attack relative to the time in the life cycle when density
dependence operates on plant recruitment can be critical. Grazing that is
destabilizing when it follows density dependence can be readily accom-

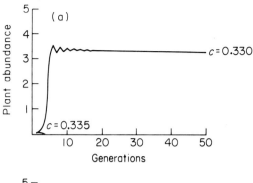

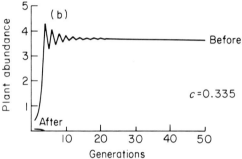

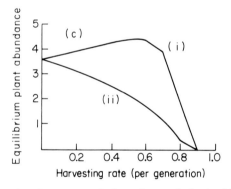

Fig. 4.28 Harvesting from a population of annual plants. (a) Removing a constant number of plants. Tiny changes in the harvest cause a sudden crash from high, stable plant abundance to extinction; increasing *c* from 0.330 to 0.335. (b) Timing of harvest. A harvesting rate that causes extinction when applied after density dependence has determined seed set per plant (*c* = 0.335) can be readily accommodated if harvesting occurs *before* density dependence. (c) Removing a constant proportion of the plants. When plants are harvested before density dependence, increasing the harvesting rate can *increase* equilibrium plant abundance (i); when harvesting occurs after density dependence, there is a monotonic decline in equilibrium plant density with increased harvesting rate (ii). Other parameters *a* = 6.0, *b* = 0.5.

modated if it occurs before seed output per plant is determined. If a constant fraction (dV_t) rather than a constant number (cN) of the plants is removed by herbivores, attack before plant density dependence can actually *increase* equilibrium plant abundance (Fig. 4.28).

Second, harvesting a fixed number of plants can lead to rapid depletion,

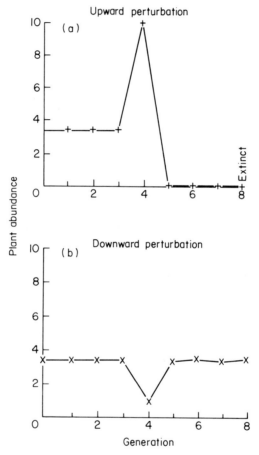

Fig. 4.29 Stability of annual plant populations to large disturbances. When density dependence is intense, upward perturbation (a) may lead to extinction (V increased to $3 \times \dot{V}$ at generation 4). The same system is exponentially stable when perturbed downwards (b) (V decreased to $1/3 \times \dot{V}$). This effect may occur when very high-density populations of seedlings are especially prone to fungal disease, or when high-density plant populations fail to produce seed.

and the crash in vegetation can happen very suddenly with little advance warning through a deterioration in the stock of plants. Maximal harvesting is on a knife-edge of disastrous over-exploitation (see section 4.3.9).

Third, and rather more esoterically, populations of plants can respond in different ways to perturbations of their density upwards or downwards (Fig. 4.29). With a fixed number of plants harvested, perturbing equilibrium plant population upwards can lead to over-exploitation and extinction. Perturbing it downwards relaxes density dependence in plant recruitment and allows a smooth return to equilibrium. Thus global stability may depend not only on the size of the perturbation but also on its direction. Perturbations of plant abundance downwards are likely to be much more common in nature than perturbations upwards (due to immigration or human release of seeds).

4.3.12 Annual plants with a seed bank

Inclusion of a seed bank makes the model a good deal more complex. Assumptions must be made about the fraction germinating per year (a function of the kind and extent of dormancy and of the degree of soil surface disturbance); about the rate of seed loss to above- and below-ground predators (whether the loss rate is a function of seed density, for example); and about the influence of seed density on the germination rate (whether there is density-dependent chemical inhibition, for instance).

The size of the seed bank, B_t, is reduced each season by germination, G, predation and other losses, L, and topped up by current seed production, R.

$$B_{t+1} = B_t - G - L + R.$$

The size of the mature plant population is determined by the germination rate, the size of the seed bank and the impact of foliage-feeding animals. With constant rates of germination and loss we have $G = gB_t$, $L = lB_t$, and, as before, each of the N herbivores feeds at a rate c, then

$$R = sG\, f(sG) - cN$$

or
$$R = (sG - cN)f(sG - cN)$$

depending on the timing of herbivore attack. In the second case the plant may compensate for early herbivore feeding due to relaxed density dependence (depending on the function chosen for fecundity $f(sG)$). Using

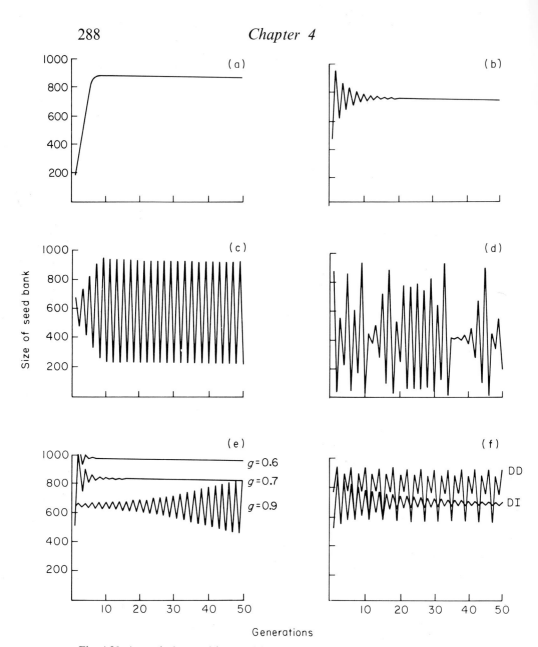

Fig. 4.30 Annual plants with a seed bank. This model can display a wide variety of dynamics — changing survival rate, s: (a) exponential stability $s = 0.1$; (b) damped oscillations $s = 0.3$; (c) stable cycles $s = 0.5$; (d) irregular, aperiodic fluctuations (chaos) $s = 0.85$. The existence of dormant seeds can be highly stabi-

the exponential function, as before, the full model can be written

$$B_{t+1} = B_t(1 - g - l) + asgB_t \exp(-bgB_t) - cN$$

for post density-dependent herbivore attack. Obviously, with so many parameters the behaviour of the model can be immensely complex, and it can display a wide variety of behaviour (Fig. 4.30).

Not until a great deal more is known about the parameter values typical of field populations of annual plants, will we be able to distinguish population fluctuations such as these from changes due to abiotic conditions. What little evidence there is, suggests that annual plant populations in constant environments tend *not* to show complex periodic behaviour, and are characteristically quite stable (see section 4.1).

---

lizing; (e) reducing the germination rate from $g = 0.9$ to $g = 0.6$ changes previously cyclic behaviour ($s = 0.5$) to rapidly damped stable equilibrium. Note that equilibrium plant numbers do not increase as g is reduced since $\overset{*}{V} = \ln$ $(a.s)/b$ when there is no seed predation or defoliation ($l = c = 0$); (f) a counter-intuitive result from this model is that density-dependent seed predators are *destabilizing*. With density-independent seed predation of 50% (DI) plant numbers show damped oscillations, whereas with density-dependent seed losses (DD) the population follows a complex 5-point cycle. This is because the seed bank forms a stabilizing refuge, and density-dependent seed predation destroys the refuge.

CHAPTER 5
COMMUNITY DYNAMICS

Herbivores play a central role in community dynamics. They affect every aspect of the plant community; its species richness, the relative abundance of species within it and its physical, three-dimensional structure. They influence the kind and relative abundance of species of predatory, parasitic and disease organisms. By their activities they create or modify much of the spatial heterogeneity which is so important in the dynamics of all the species.

5.1 HERBIVORES AND PLANT-SPECIES RICHNESS

Herbivores exert direct effects on the number of plant species; adding plants by their mutualistic behaviour and subtracting plants by eating them to extinction. Most widespread, however, are their indirect effects which operate through alterations in the relative competitive abilities of plants under herbivore feeding, and through the creation of competition-free germination microsites.

5.1.1 Direct effects

Plant-species richness is increased directly by the presence of mutualist herbivores like pollinators, dispersers and defenders. Several plant species have obligate pollinators. The agave, *Yucca filamentosa*, relies for pollination on the moth *Tegeticula yuccasella* whose larvae feed on the developing *Yucca* seeds. Wild fig trees, *Ficus carica*, depend entirely for pollination on tiny chalcid wasps *Blastophaga psenes* which develop by feeding on neuter flowers within the flask-shaped receptacle. The male wasps are wingless and never leave the fig in which they develop. On emergence, they bore into an ovary inhabited by a female wasp, mate with her, then die. The winged female then breaks out of the inflorescence and carries pollen to a new generation of young inflorescences. She pollinates the fertile ovaries of these, and another generation of the wasps develops in the neuter flowers (Proctor & Yeo 1973).

Seed germination may require that the seed passes through the digestive tract of a herbivore (see section 2.5.2); germination of numerous species of tree from the Ivory Coast was enhanced when the fruits were eaten by elephants and defecated in their own competition-free and nutrient-rich heap (Alexandre 1978).

Plant-species richness is reduced directly when heavy grazing leads to extinction. This is more likely to be brought about by generalist, or at least oligophagous, herbivores than by specialists, simply because specialists would tend to decrease in abundance as the plant became scarce. The monarch butterfly, *Danaus plexippus*, feeds on two kinds of milkweed, *Asclepias curassavica* and *Calotropis procera*, in Barbados. The butterfly has all but eliminated *A. curassavica* by maintaining its numbers on *C. procera*. This reduction in plant species has also affected the herbivore community. Two milkweed bugs, *Oncopeltus* spp., rely on *Asclepias* and they cannot eat the seeds of *Calotropis* because its pod walls are so thick. Feeding by an oligophagous butterfly has, therefore, caused the demise of two specialist herbivores (Blakley & Dingle 1978). In a similar way, the western pine beetle, *Dendroctonus brevicomis*, tends to exclude *Pinus ponderosa* from the mixed forests on the western slopes of the Californian Sierras (Stark & Dahlsten 1970).

Reductions in plant species due to specialist herbivores are not unknown. The elimination of the Bermuda cedar, *Juniperus bermudiana*, by the introduced scale insect, *Lepidosaphes newsteadi* (Bennett & Hughes 1959), and the several examples from biological weed control (see section 4.1) attest to this.

5.1.2 Indirect effects

Selective feeding by herbivores modifies the competitive relationships between the component plant species. Diversity will increase when selective feeding on a previously dominant species reduces its vigour and allows the spread of less competitive, but more grazing-tolerant (or less heavily grazed) plants. Thus Darwin (1859) writes 'If turf which has long been mown, and the case would be the same with turf closely browsed by quadrupeds, be let to grow, the more vigorous plants gradually kill the less vigorous, though fully grown plants: thus out of twenty species growing on a little plot of turf (three feet by four) nine species perished from the other species being allowed to grow up freely'.

Selective feeding will decrease species richness when a plant species that is normally uncompetitive in the absence of grazing becomes aggressively competitive when its competing plants are eaten but it is avoided. The classic example of this is the spread of the unpalatable grass *Nardus stricta* in British hill pastures (Chadwick 1960, Nicholson *et al*. 1970; see also section 5.2.2).

It is important to stress that non-selective feeding on plants that are equally grazing tolerant cannot affect species richness; for example, while mowing causes spatially uniform defoliation, it is wrong to assume that it is non-selective (e.g. Duffey *et al.* 1974). The mower does not harvest plant species in proportion to their abundance and tall plants suffer disproportionately high rates of defoliation while rosette species escape defoliation altogether. Since mowing is selective it can (and usually does) increase species diversity.

Herbivorous insects may play an important role in determining the species diversity of trees in certain tropical forests, not so much by their effects on the adult trees but by killing their seeds and seedlings. It has been suggested that feeding by herbivores on the foliage of seedlings (Connell 1971) and on the seeds themselves (Janzen 1970) is so intense close to the parent trees (where the specialist herbivores are aggregated), that recruitment close to the parent is unlikely. The hypothesis is that herbivores are important in creating and maintaining the classic rain forest community by increasing spacing between the individuals of each species; diversity is increased because dominance by one species is prevented.

This model has been tested analytically and empirically by Hubbell (1980) and found wanting. First, in many tropical forests the trees do occur in clumps, contrary to the popular view, and recruitment does occur beneath parent trees despite the depredations of herbivorous insects. Second, because of high variability from tree to tree in the size of the seed crop, the predicted spacing of peak recruitment from the parent would vary so much that any regularity in adult spacing would be practically undetectable in real communities, even when underlying heterogeneity was minimal. Finally, if the survival and seed-rain axes of Janzen's (1970) model are scaled realistically, Hubbell suggests that peak recruitment will always occur below the parent rather than at a distance, because the fall in seed density with distance far outweighs any increased likelihood of survival. Only if herbivore pressure was so intense that survival below the parent was zero would Janzen's model be likely to hold. The clumps of juveniles around parent trees are probably the result of predator satiation, particularly when the juveniles are all the same age (Hubbell 1980).

Species richness may vary substantially from place to place within one community under the same grazing regime. Sagar (in Harper 1977) showed how the number of vascular plant species in 30 cm quadrats varied in a ridge and furrow grassland near Oxford. There were 10 species on top of the ridge and 3 in the trough of the furrow. The opposite effect can be seen in British upland pastures where characteristic soil mounds are sometimes found. The mound tops are leached and hence more acid; the upper soil layer is peaty and supports a low-diversity community dominated by *Nardus stricta* which is largely ignored by the sheep. The hollows between the mounds accumulate leachate and most of the sheep droppings tend to roll into them. The soil shows no peat development, is rich in earthworms and is dominated by relatively species-rich, closely grazed *Agrostis/Festuca* grassland (Cotton 1968).

The most common way in which herbivores increase plant-species richness is by providing competition-free germination sites. We have met several examples of this already (e.g. see section 2.6.1). On the disturbed soil around rabbit burrows in chalk grassland grows a characteristic assembly of plants, unpalatable to rabbits but unable to establish in closed grassland, including *Atropa belladonna, Sambucus nigra* and *Solanum dulcamara* (Thomas 1960).

Dominant monocultures occur rather more frequently in nature than we are sometimes led to believe. In Britain, for example, there are vast areas of pure bracken fern, *Pteridium aquilinum,* heather, *Calluna vulgaris,* birch, *Betula pubescens,* reed, *Phragmites communis,* cord grass, *Spartina* spp., marram grass, *Ammophila arenaria,* and numerous others. These examples call into question the generality, if not the spirit, of Ridley's (1930) much quoted remark that, 'Where too many plants of one species are grown together, they are apt to be attacked by some pest, insect or fungus'.

There is no general relationship between grazing intensity and plant species diversity. Heavily grazed pastures are sometimes more diverse than well-managed grasslands because more weed species are added to the community than forage species are eliminated. Lightly grazed pastures are sometimes more diverse when the herbivores feed selectively on grazing-intolerant species; relaxed grazing pressure allows some of these species to survive. More commonly, species richness will be lower in rank pastures because a few species will become dominant and exclude the lower-growing forbs and grasses (Harper 1969).

5.1.3   Alternative host plants and plant diversity

One plant species can limit the distribution of another by increasing the herbivore pressure it suffers; for example, the grasshopper *Hesperotettix viridis* is resident in arid grasslands in New Mexico on the abundant shrub *Gutierrezia sarothrae.* By its feeding it can prevent the establishment of another composite, the forb *Machaeranthera canescens* (Parker & Root 1981).

It must frequently be the case that uncommon species are prevented from increasing when they are fed upon by a generalized herbivore which is maintained at high density by a common species of plant (Futuyma & Wasserman 1980).

It is commonly observed in Chilean matorral that native annual plants are more common beneath bushes than in the open spaces between them. To test whether microclimatic factors (shade, water stress) or grazing by rabbits was the cause of the observed pattern, Jaksic and Fuentes (1980) removed half the canopy from several bushes, and fenced half of these against grazers. Flower production was not depressed by removing the shade of the bushes, but it was significantly depressed by rabbit feeding. Thus the bushes allowed an increase in plant-species richness by protecting vulnerable annuals from defoliation.

5.2   HERBIVORES AND PLANT COMMUNITY STRUCTURE

The most dramatic impression of the effects of grazing on plant communities is obtained where a fence line separates two areas which have a long history of different grazing management. Despite similar soil, aspect, slope and drainage the plant communities are often so distinct that the fence line is evident from many miles away. Where the fence separates grazed and ungrazed areas the contrast is most obvious; the grazed area consists of closely cropped, mainly graminaceous plants while the ungrazed area is dominated by shrubs and tall herbs (left long enough, of course, the ungrazed side would revert to woodland; a process which demonstrates the relative importance of vertebrate as compared to invertebrate herbivores in affecting succession). Where the fence line separates heavily grazed from lightly grazed areas, or cattle-grazed from sheep-grazed land, the effects are more subtle but nonetheless evident. In the Cheviot Hills on the borders of England and Scotland, for example, fence lines often separate heather moor from grassland which, in winter, makes an almost black and white contrast between the *Calluna* and the *Nardus*. The heather flourishes under low grazing with periodic burning and the *Nardus* under rather more intense grazing without burning. Where extensive grazings are separated from more intensively grazed pastures, the fence line is equally obvious, now dividing white *Nardus* ground from bright green *Agrostis/Festuca* grassland.

The influence of domestic livestock on the plant community is so profound that we can sometimes recognize the herbivore from the vegetation it produces. Thus we know horse pastures by the abundance of dock, *Rumex obtusifolius*, they contain, goose grazings by the dominance of silverweed, *Potentilla anserina* (hence also the plant's specific name), and goat-grazed Mediterranean maquis by the abundance of spiny shrubs and aromatic, sticky herbs.

These changes are due mainly to altered plant competitiveness. Interspecific plant competition can be studied either by keeping total plant density constant and varying the proportions of the different plant species, or by keeping the density of one species constant and varying the number of other plants with which it competes (Harper 1977). In weed studies, for instance, the farmer needs to know how yield is affected when different numbers of weeds grow with a fixed number of crop plants. In this kind of experiment, both density and proportion are inextricably compounded. In most experiments, however, total plant density is kept constant and the proportions of seed sown are varied in 'replacement series' or 'de Wit series' (de Wit 1960). Factorial designs where replacement series are observed at a range of total densities can give an indication of the interaction between absolute and relative densities of the competing plants (McGilchrist & Trenbath 1971).

In general, herbivores could affect the outcome of replacement series in four ways. The herbivore could reverse the relative competitive abilities of the plant species. The plant that is most competitive in the absence of grazing may be fed upon preferentially to the extent that it is dominated by the second, less-preferred plant species. Alternatively, the herbivore might prefer the least competitive plant. The plant is now doubly disadvantaged and will suffer greatly reduced abundance or even become extinct. Third, the herbivore may demonstrate switching (see section 3.5.6) and feed preferentially on whichever plant is the most abundant. This will tend to make the outcome of the different relative-density trials more alike. Finally, the herbivore may be entirely neutral in its effect, taking each plant species in proportion to its abundance. The outcome would then depend upon the relative grazing tolerances of the two species, upon their powers of regrowth and their morphological responses to defoliation. Replacement series with and without herbivores are currently being studied for several systems; with slugs (Dirzo & Harper 1980), beetles (Bentley & Whittaker 1979) and grass-feeding aphids (Packham pers. comm.).

The classic field experiments on the impact of vertebrate herbivores on the structure of plant communities were performed by Jones (1933a) at Jealott's Hill in southeast England. Pastures sown with exactly the same seed mix, and treated with the same cultural and manurial regimes, were simply grazed in different ways by sheep.

In one trial where a ryegrass/clover, *Lolium perenne/Trifolium repens*, mixture was sown, changing only the timing of grazing produced an almost pure ryegrass sward in one case (6% clover) and a predominance of clover (62%) in the other. This was brought about by grazing regimes which tipped the competitive balance in favour of the grass in the first case and of the prostrate-growing clover in the second. To produce a ryegrass sward, the animals were kept off the pasture in early spring (March and April) so that the ryegrass tillers produced vigorous root and shoot systems. When the sheep were introduced to the pasture, the grass plants had substantial reserves of root and shoot and were able to recover rapidly from defoliation. Despite the attentions of the sheep, they were able to over-top and out-compete the clover plants.

The clover sward was created by introducing the sheep to the grassland very early in the year (during March and April) so that the first ryegrass leaves were grazed off as they developed. The sheep were removed before they grazed down to the lower clover plants. By the peak growth period in May and June the clover plants were in fine condition, but the ryegrass had poorly developed root and shoot systems, and suffered from the combined effects of defoliation and competition from the clover.

The full impact of grazing preferences is felt when food supply is high; few polyphagous herbivores are as choosy when they are hungry as when they are stated. In another experiment, Jones (1933b) stocked two adjoining grassland plots at about the same average annual grazing intensity. He varied the distribution of defoliation between winter and summer. On the first plot, the number of stock was varied month by month in relation to the amount of food available (there were 20 times as many animals per unit area in May/June as in the depths of winter). The pasture was kept free of sheep during the crucial period of grass shoot establishment in March/April. Since food was never present in excess, the sheep were never able to exercise their preferences to the full, so that after 4 years, the plot still retained its original productive mixture of plant species and was practically free from any weed grasses or thistles.

The second plot was stocked with as many sheep as it could carry through the winter while summer numbers were limited to twice this level. The heavy grazing in winter so weakened the grass plants that weedy species rapidly invaded and spread. In summer there was substantial overproduction of plant biomass so that the sheep grazed selectively on the preferred and productive grass plants, avoiding the weedy, unpalatable species. After 4 years, these valueless plants occupied one-third of the sward. The combination of heavy grazing on the preferred species in the winter, and selective grazing in favour of the weedy species in the summer, meant that the pasture ultimately became worthless for livestock production (see section 5.2.1).

To prevent this downgrading of pasture quality, modern grazing management aims to avoid overgrazing in the critical period of early grass growth, and to prohibit selective feeding when grass production is at its height. This is achieved by keeping sheep off the pasture when it is at its most productive in early summer. Instead, this grass is cut for hay or silage when it is at its peak nutritional level. Overgrazing in winter is avoided by rotational grazing and the animals are seen through the period of March/April when the pasture must be protected, by the stored feedstuffs harvested the previous year.

Long-term exclosure of sheep from upland grasslands in the northern Pennines has been studied by Rawes (1981; see also Hughes & Dale 1970). Species-poor *Festuca* grasslands become dominated by *Deschampsia flexuosa* when sheep are excluded, while the species-rich *Agrostis/Festuca* grasslands develop a tall, patchy community with *D. caespitosa* and *Holcus mollis*, and herbs like *Alchemilla glabra, Rumex acetosa, Galium* spp. and *Urtica dioica*. Where sheep had previously grazed *Juncus squarrosus* grassland, a community of *Eriophorum vaginatum* developed and the *Juncus* became extinct. *Nardus* grasslands showed little change in 24 years of protection from grazing.

Grazing of salt marshes by sheep leads to the replacement of broad-leaved species like *Halimione portulacoides* and *Aster tripolium* by the grass *Puccinellia maritima* (Gray & Scott 1977).

Similar effects occur in semi-arid grasslands. Grazing exclosures set up in the Serengeti by Watson in 1963 showed substantial changes by 1967 (McNaughton 1979b). The grasses *Andropogon greenwayi* and *Sporobolus marginatus* which made up 56 and 20% of the standing crop in the grazed areas were not found in the exclosures at all. The two tall grasses *Pennisetum stramineum* and *P. mezianum* which made up only 5 and 3% outside, accounted for 72 and 26% of standing crop biomass within the fence.

Rabbit grazing on Australian grasslands at densities as low as 40 ha^{-1} can reduce pasture yield by up to 25% and with increasing rabbit numbers there is a decrease in the abundance of useful grasses and a rapid increase in weeds (Myers & Poole 1963). In eastern England, grazing by rabbits can convert *Calluna* heath to grassland (Farrow 1925).

One of the oddest herbivore-created plant communities is found on the island of Aldabra in the Indian Ocean. The island is unique in that its dominant herbivore is a reptile, the giant tortoise, *Geochelone gigantea*. These animals are present at extremely high standing crop biomasses (over $35\,t\,km^{-2}$ compared with less than $20\,t\,km^{-2}$ which is the peak biomass of game in the African savannas; Coe *et al.* 1979) and are responsible for the creation of closely cropped, lawn-like grasslands known as 'tortoise turf'. Other grasslands on Aldabra are rank and tussocky and support very low tortoise numbers because of rough terrain (the sharp, spiky 'champignon limestone') or lack of shade. The animals are so sluggish in their movements that areas more than 300 m from shade cannot be grazed, and a noticeable increase in plant height and changed species composition occurs at this distance from shade-providing shrubs or grass tussocks. Five of the six grasses and two of the three sedges present in tortoise turf are thought to be endemic to the Aldabra group and almost certainly own their origin and their continued survival to tortoise grazing (Merton *et al.* 1976).

Clearance experiments have shown the importance of grazing by limpets, *Patella vulgata*, on the species composition of the plant communities of exposed rocky shores; for example, one winter all the large limpets and most of the small ones were removed from a 10 m wide strip of shore between the high- and low-water marks. 'An initial growth of diatoms, colonial and filamentous green algae, soon gave way to a dense cover of *Porphyra*, *Enteromorpha* and *Ulva* the following spring. Pure stands of fucoids then developed in the second year, and (after three years) the strip was covered by thick growths of *Fucus spiralis*, *F. vesiculosus* and *F. serratus*. With the reappearance of a large population of limpets, the existing cover of fucoids declined ... and no further settlement of algae took place. Five to seven years after clearance, the experimental areas were almost clear of algae' (Southward 1964). Recent studies of the effect of herbivores on the distribution and abundance of other marine algae confirm these trends (Lubchenco 1978, Hay 1981).

Insects also affect the competitive ability of grasses. Italian ryegrass, *Lolium multiflorum*, is one of the most productive agricultural grasses in Britain. It is fed upon by the larvae of the frit fly, *Oscinella frit* (and others), stem borers that burrow into the central shoot and greatly reduce the plant's vigour. Ryegrass losses to *Oscinella* vary between 15 and 30% per year and frit fly attack is a major factor in the competitive replacement of ryegrass by other species in permanent grasslands (Clements 1978). It is noteworthy that close grazing by sheep can bring about a 10-fold increase in frit fly density because it promotes the constant production of new and susceptible tillers (Henderson & Clements 1979). The insect is doubly damaging in that it reduces forage availability directly, and it reduces the longevity of the productive ryegrass leys by causing *Lolium* to be ousted by weeds and less profitable grasses.

Seed predation by insects may alter the relative abundance of seeds and hence affect the proportions that arrive in germination microsites. The legumes *Astragalus cibarius* and *A. utahensis* probably compete for germination microsites where they are sympatric; *A. cibarius* produces more seed but suffers a higher rate of seed predation by insects like the beetle *Acanthoscelides fraterculus* and the chalcid wasp *Bruchophagus mexicanus*, so that the balance is tipped in favour of *A. utahensis* (Green & Palmblad 1975).

Sometimes an increase in the abundance of a wild herbivore will cause an increase in desirable plant species. The moth *Aroga websteri* periodically defoliates big sagebrush, *Artemisia tridentata*, over wide areas of its native range in the Great Basin. In some years it destroys thousands of hectares, allowing the re-establishment of the native grasses exterminated by overgrazing (Ritcher & Dickason in Andres *et al.* 1976).

5.2.1 Overgrazing

Overgrazing is a term used by the managers of wild or domesticated herbivore populations to describe changes in the plant community which they view as deleterious. A grassland is overgrazed when the harvested animal productivity is lower than it would be under less intense grazing or lower than it used to be before irreversible (or only very slowly reversible) changes in plant composition or soil structure occurred.

We do not talk of natural ecosystems as being overgrazed or undergrazed; as we have seen (Table 1.3) the fraction of primary production that is consumed by herbivores varies greatly from one community to the other. The equilibrium vegetation that persists in an ecosystem is both a cause and a consequence of the prevalent rates of feeding; high rates of consumption are not 'unnatural' nor are they associated with 'instability' (see section 4.3).

The short-term consequence of heavy grazing is a reduction in productivity alone; this is readily reversed if grazing pressure is relaxed (see section 2.7.1). If the stocking rate remains high for a long period, the pasture suffers both a reduction in the productivity of forage species and a change in plant-species composition in favour of non-forage woody plants, undesirable grasses or toxic weeds. These longer-term changes may be much more difficult to reverse simply by relieving the grazing pressure. Overgrazing is typically brought about by increased human population (Schaller 1980), overstocking of fenced populations of domestic livestock (Le Baron *et al.* 1979), extermination of natural enemies or reductions in the area of suitable habitat for wild ungulate populations (Laws *et al.* 1975).

The symptoms described by pasture managers as overgrazing are due to selective feeding which alters the competitive relationships and thus the relative abundances of the different plant species. There is an increase in species which the manager regards as undesirable ('invaders' or 'increasers') and a reduction in those which are most palatable and provide the highest yields of animal products ('decreasers'); for example, the palatable grass *Agropyron spicatum* on alluvial fans and lower slopes in Idaho is replaced by the unpalatable *Aristida longiseta*. 'The type of grazing use which has prevailed in the area since settlement includes heavy use in spring of early-growing, palatable grasses like *Agropyron*. Where this use is both heavy and prolonged, regrowth of the grazed plants is inhibited by the dry summer conditions even when the livestock are removed to summer ranges. In the fall heavy use of fall regrowth further depletes plant food reserves and contributes to lower vigour. *Aristida* largely escapes the effects of such use because of (1) lack of growth available to livestock in early spring and sheltered position of the growing point; (2) low palatability, so that even the young leaf growth is grazed less than that of associated species, and later growth stages are virtually ungrazed; and (3) lack of fall regrowth and hence low palatability in the fall' (Evans & Tisdale 1972).

The influence of livestock grazing on succession in true prairie, mixed-grass prairie, desert grasslands, scrub and mountain rangeland is detailed by Ellison (1960). As overgrazing progresses, floristic changes occur which further downgrade the quality of the range. Woody plants like big sagebrush, *Artemisia tridentata*, mesquite, *Prosopis juliflora*, and oak, *Quercus* spp., obstruct the grazing animals; plants like cheatgrass, *Bromus tectorum*, and medusa head, *Elymus caput-medusae*, with low forage value spread and reduce the carrying capacity for livestock; poisonous plants like larkspur, *Delphinium* spp., locoweed, *Oxytropis* spp. and *Astragalus* spp., sneezeweed, *Helenium* spp., milkweed, *Asclepias* spp., and *Halogeton glomeratus*, cause deaths amongst livestock and greatly reduce stocking levels; many of the weedy plants, because they are ignored by the herbivores, leave large quantities of dry, dead organic matter that constitutes a considerable fire hazard (dry cheatgrass, for example, is extremely inflammable).

In a review of various management practices on Western ranges in the United States, Van Poollen and Lacey (1978) found that herbage production averaged 13% higher when livestock were controlled by a specialized grazing system, than when grazing was continuous. However, mean herbage production was increased by 35% when grazing use was reduced from heavy to moderate and by 28% when it was reduced from moderate to light. This suggests strongly that land managers should place more emphasis on proper stocking density and rather less on implementing complex grazing systems (Van Poollen & Lacey 1978).

Where plant growth is water-limited, as in semi-arid savanna, overgrazing first causes a change in the species composition of the perennial grasses, followed by a decline in the perennial grass cover and a rise in the abundance of annual grasses. Eventually, woody species establish and if severe overgrazing continues, a thicket of woody plants develops to the exclusion of almost all the grass (Walker *et al.* 1981). This kind of range deterioration has been described from the United States (Buffington & Herbel 1965, Clark-Martin 1975), Africa (Barnes 1979, Wickens & Whyte 1979) and India (Singh & Joshi 1979). When grazing animals are removed, the range does not return rapidly to its pristine state, and the woody scrub persists for at least 20 years (Walker *et al.* 1981).

5.2.2 Undergrazing

Undergrazing, too, is an economic rather than an ecological process, and is said to occur when greater profit would result from an increase in stocking rate. In the short term, undergrazing can occur in irrigated or fertilized pastures where self-shading of leaves reduces the rate of net productivity (see section 2.7.1). Long-term consequences of undergrazing include a shift in the species composition in favour of plants avoided by livestock (see section 5.3).

The best example of undergrazing comes from upland sheep pastures in Britain which are dominated by the grass *Nardus stricta*. This species is unpalatable to sheep for most of the year, and at most stages of growth. On the other hand, it is intolerant of grazing, so that when it *is* eaten, it becomes uncompetitive with other, more desirable species. The number of sheep on hill farms is limited by the availability of winter feed and shelter, so that unless extra animals are imported from the lowlands, there is a vast excess of primary production in the summertime. Since food is present in excess, the sheep can exercise their preferences to the full; palatable grasses like *Agrostis* and *Festuca* are grazed heavily and *Nardus* is all but ignored. Each year, therefore, *Nardus* tends to increase at the expense of the better grasses. Also, uneaten stems and leaves die in the sward. The accumulated dead material persists through the winter and reduces productivity the following year both by shading young leaves and by reducing the accessibility of the new green shoots. As production and quality decline, the animals spend less and less time on these parts of the hill, and the long-term decline in pasture value continues (Chadwick 1960, Nicholson *et al.* 1970). The changes in species composition brought about by undergrazing are not rapidly reversed. Long-term sheep exclosures in the Pennines have shown very little change in *Nardus* grasslands after 24 years (Rawes 1981); the individual clones of the grass are very long-lived, and their tough, mat-forming shoot bases tend to prevent the establishment of other plants. Only by forcing the sheep to eat the *Nardus* (by maintaining high stocking densities in small, limed and fertilized paddocks) is the grass replaced by more desirable *Agrostis* and *Festuca* species. The expense of the fencing that this demands, coupled with the losses in condition that the animals inevitably suffer in being forced to eat the *Nardus*, are sufficient to deter most farmers from attempting to reverse the downward trend in pasture quality.

5.2.3   Grazing versus mowing

There are a great number of differences between grazing and mowing. The most important distinction is that while mowing is uniform in height and spatially homogeneous, grazing is typically patchy in space, and different plants are grazed to different heights within a patch.

The species composition of grazed and mown grasslands is distinct. Some of the most conspicuous plants of mown verges, like hogweed, *Heracleum sphondylium*, for example, are never found in grazed pastures (Randall 1981). Similarly, the intensity of mowing can have complex effects on plant-species richness. Regular mowing below 0.5 cm will weaken desirable lawn grasses like *Festuca rubra commutata* and allow the ingress of mosses and low-growing weeds. Mowing above 4.0 cm will cause the fescue to be ousted by taller-growing, coarser grasses (Pearson 1981).

Uniform defoliation alters the competitive balance between the two pasture grasses *Lolium perenne* and *Holcus lanatus*. Light or infrequent defoliation allows the taller *Holcus* to dominate, but *Lolium* is the most successful competitor under frequent severe defoliation (Watt & Haggar 1980).

The three major floristic components (grasses, legumes and others) in the long-term Park Grass Experiment at Rothamsted remain in dynamic equilibrium despite the occasional increase of certain species (due to unusually dry weather, for example). These plots also show great consistency in the negative relationship between standing crop and plant-species diversity, suggesting that strict rules govern the number and relative abundance of species under a given regime of cutting and fertilizer application (Silvertown 1980b).

Mowing and grazing can differ markedly in their effects on plant physiology when the grazing animal produces growth-modifying chemicals in its saliva (see section 2.7.8). Mowing machines neither urinate nor defecate.

5.3 HERBIVORES AND PLANT SUCCESSION

It is clear from sections 5.1 and 5.2 that vertebrate herbivores and some invertebrates can have a pronounced impact on plant community structure. Whether these communities are stable or whether they merit the status of 'climaxes' is a moot point. The question to be addressed here is whether herbivore feeding alters the rate at which plant species are added to, or are excluded from, communities during secondary succession. The role of herbivores in the soil-forming processes of primary succession is not considered.

The suppression of trees and the maintenance of grassland by domestic livestock is a clear-cut indication of the potential of vertebrate herbivores in deflecting successions. We should be aware, however, that these effects are often more visible than real. On an acid heath in Surrey where only small, isolated clumps of mature Scots pine trees could be seen, a fence was erected to exclude the cattle. Almost immediately pine saplings sprang up everywhere in abundance. Intrigued by this, Darwin (1859) looked more closely at the 'treeless' heath and 'in one square yard, at a point some hundreds yards distant from one of the old clumps, I counted thirty-two little trees; and one of them, judging from the rings of growth, had during twenty six years tried to raise its head above the stems of the heath, and had failed. No wonder that, as soon as the land was enclosed, it became thickly clothed with vigorously growing young firs'.

There is no clear consensus about how secondary succession works but there are two main groups of hypotheses (Connell & Slatyer 1977, Horn 1981). The first stresses the importance of random processes; this says simply that species arrive at a disturbed site at a rate proportional to their dispersive abilities and to the local abundance and parent plants. The observed pattern is nothing more than the replacement of small, short-lived plants (often with high rates of dispersal or resident in the seed bank) by large, long-lived, usually woody, species (typically with slower rates of dispersal). There is no site preparation by one species for another and no inhibition of establishment, and herbivores are only important in so far as they affect plant species' immigration rates through their seed dispersal activities, or their establishment rates through the creation of suitable microsites. The effect of delayed seed immigration is shown in Holt's (1972) study of carrot, *Daucus carota*, populations in abandoned fields in Michigan. Late seed immigration led to low rates of seedling establishment and reduced survivorship (Fig. 5.1). These effects are due to successional changes in the vegetation leading to germination inhibition and growth suppression, probably coupled with higher rates of seed predation (Holt 1972).

The second group stresses biological interactions. Connell and Slatyer (1977) describe three models of succession; the first, 'facilitation model' appears to operate in primary successions and in heterotrophic succession

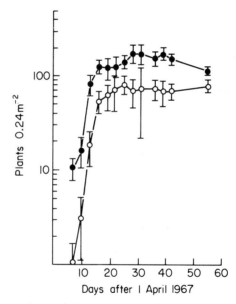

Fig. 5.1 Plant succession and the probability of seedling establishment. Delayed immigration led to reduced numbers of established seedlings, delayed reproduction, and reduced cumulative seed yields of carrot, *Daucus carota*, seed sown in young (●) and old (○) fallow (Holt 1972).

where a plant species is unable to invade a community until a previous species has created the right soil, microclimate and light conditions. Herbivores would generally tend to slow down this kind of succession, because feeding would reduce the rate at which a plant species ameliorated the environment for its successors.

The second, 'tolerance model' suggests that the sequence of species is due to shade-tolerant plants gradually replacing those that can only reproduce in the open. There is little evidence in support of this model. Finally, the 'inhibition model' suggests that established plants exclude invaders until the early species are damaged or die; only then are resources released. When a species is more likely to be replaced by an individual of its own kind the community will come to a stable equilibrium (Connell & Slatyer 1977); when an individual is likely to be replaced by plants of another kind, gap formation in the canopy will lead to the development of a mosaic of patches of different ages and different species compositions (Paine & Levin 1981). In this kind of succession, herbivores would tend to increase the rate of species replacement, because their feeding would reduce the competitive ability of the resident plants and increase the likelihood of invasion and establishment by their successors.

Clearly, these hypotheses are not exclusive and it is quite plausible that the fate of each plant species in succession is governed by a different set of rules. There is unlikely to be 'an effect' of herbivores on succession.

We have seen some examples of how the exclusion of vertebrate herbivores has affected vegetational change. In an unusually long-term study of vegetational change under release from vertebrate grazing, Watt (1981) followed the development of plant communities in rabbit exclosures on the sandy soil of Breckland in east England. Over a period of 38 years after fencing a series of dominants developed (*Festuca ovina, Hieracium pilosella* and *Thymus drucei*) and 21 new plant species were added to the original 11 on the 6 × 6 m plot. Most of these were known to be palatable and sensitive to rabbit grazing. 'Protection from grazing leads to the development of a richer flora, a more continuous cover of vegetation, the provision of greater protection from frost-heaving and erosion … and is likely to lead to the stabilization of the soil surface and the development of a soil profile' (Watt 1981).

Insecticides can be used to exclude herbivorous insects and, while it is admitted that interpretation will be complicated by possible fertilization and phytotoxic effects, the technique holds out great potential for highlighting normal levels of herbivore impact (e.g. Morrow & LaMarche 1978). Some of the problems of the method are highlighted by Shure's (1971) study where the insecticide turned out to be differentially toxic to the seeds and seedlings of different plant species. Increases in plant species diversity following the application of 12 kg ha^{-1} diazion were attributed to the phytotoxic effect of the insecticide on *Convolvulus sepium*, one of the dominant plants. Species normally inhibited by *Convolvulus* flourished on the treated plots.

Defoliation of the dominant trees by Lepidoptera can have an impact on the rate of succession in forests by altering the conditions experienced by the suppressed trees of the understorey. In oak forests in Connecticut defoliated by gypsy moth, *Porthetria dispar*, the understorey is represented by red maple, *Acer rubrum*. Collins (1961) compared the growth rates of maples in open fields and under intact, defoliated and refoliated oaks. Maples in the open field put on 10 times the height growth of those under intact oak canopies, while maples under defoliated oaks grew for two weeks longer than shade trees and put on twice the height. Caterpillars falling from the oak canopy eventually defoliated the maples, too. They refoliated and continued to grow until the refoliated oak closed canopy, one month after larval feeding stopped. Despite being defoliated themselves, the death rate of maples was lower on defoliated plots, presumably due to the increase in their carbohydrate reserves brought about by improved photosynthesis. The frass from the defoliating caterpillars may have a slight fertilizing effect on maple growth in the following year.

The impact of sheep and cattle grazing on the growth and mortality of trees in forest succession is reviewed by Adams (1975). In the New Forest, browsing by deer, ponies and cattle can prevent tree regeneration; bouts of successful regeneration have always coincided with periods of low browsing pressure (Peterken & Tubbs 1965). Long-term exclusion of browsing animals (mainly white-tailed deer, *Odocoileus virginiana borealis*) from 230-year old *Pinus resinosa* forest in Minnesota is described by Ross *et al.* (1970). Regeneration of trees and shrubs outside the exclosure was severely inhibited when deer numbers were high. The relative increase in importance of *Betula papyrifera* over *P. strobus* outside the fence was due both to a preference by the deer for pine over birch, and to the relatively higher susceptibility of the pine to browsing damage. Sugar maple, *Acer saccharum*, rapidly replaces hemlock *Tsuga canadensis* under heavy browsing by white-tail deer, *O. virginianus*, because deer-feeding causes a significant reduction in recruitment of the shade-tolerant hemlock (Anderson & Loucks 1979).

Succession in grassland communities is all too familiar to farmers who manage intensively grazed paddocks sown with highly productive grass strains in temporary leys. After four or five years almost all the species sown in the seed mixture will have disappeared to have been replaced by less-productive, more grazing-tolerant or less-preferred plants. The influence of herbivorous insects on grassland succession has rarely been investigated, but Clements and Henderson (1979) showed that ryegrass swards treated with insecticides over a number of years, retained a much higher proportion of the sown species than did untreated areas, and were far less weedy.

5.4 RESOURCE CONCENTRATION AND ASSOCIATIONAL RESISTANCE

The species richness and physical structure of the plant community can exert a profound effect upon the rate at which a given plant species is attacked by invertebrate herbivores; for example, the numbers of pest insects per unit area and per crop plant have been found to be greater in monocultures than in species mixtures (polycultures) in several studies (Pimentel 1961, Root 1973, Thresh 1981).

Densities of the beetle *Acalymma vittata* were 10 to 30 times greater in monocultures than in polycultures both per unit area and per plant, even when total plant density and host plant density were held constant (Bach 1980b).

Whenever an intercrop contained one non-host plant for a given species of pest chrysomelid beetle, the numbers of that beetle per host plant in the intercrop were reduced compared to a similar monoculture. When both species in the intercrop were suitable as food for the beetles, pest numbers per plant were generally higher than in monoculture for one or both of the crops (Risch 1980).

The interaction between host plant density and relative plant density obviously allows a vast range of different responses. Within the same community, some herbivores may show resource concentration while others do not. We have already seen that certain insects prefer to lay their eggs on isolated plants rather than on plants in high-density patches (see section 3.5.3). Different herbivore species may attack precisely the same distribution of plants in different ways; for example, Cromartie (1975) found that different species peaked in abundance at different densities of *Brassica* plants. Larvae of the butterfly *Pieris rapae* were most abundant on low-density plants, while the beetle *Phyllotreta striolata* was most abundant on intermediate density plots and *P. cruciferae* responded to resource concentration and was most abundant in the densest *Brassica* populations.

Resource concentration suggests that abundant plant species suffer disproportionately high rates of herbivore attack. While this may be true in certain crops, it is not general in natural communities; for example, autumn cankerworm, *Alsophila pometaria*, is a polyphagous moth which defoliates canopy trees in the forests of north-eastern United States. The larvae hatch at budbreak of *Quercus coccinea* and feed on the young, nutritious foliage. *Q. alba* produces leaves 10 days later, and in forests dominated by *Q. coccinea* suffers almost complete defoliation as caterpillars disperse and attack its young, more nutritious foliage. In woods of *Q. alba* the moth is uncommon and concentrated on the scattered, early flushing trees of *Q. coccinea*. While larvae do emigrate to *Q. alba* when they come into leaf, damage is slight because moth numbers are low. In both forests, therefore, the least common species suffers the highest degree of herbivore damage (Futuyma & Wasserman 1980).

The expression of resource concentration by herbivores may be masked by the aggregative behaviour of their enemies, as Shahjahan and Streams (1973) found with the braconid *Leiophron pseudopallipes*. The wasp parasitized a greater proportion of tarnished plant bugs, *Lygus lineolaris*, where their host plants were abundant than where they were scarce.

The interpretation of differences between herbivore population dynamics in monocultures and polycultures is complicated by the fact that the two kinds of plant community differ in so many ways. If the number of host plants is kept constant, adding non-hosts to create the polyculture increases total plant density, the level of competition experienced by the crop plants, the amount of leaf per unit area of ground upon which flying insects can alight, and the number of stems per unit area up which walking insects may climb. If total plant density is kept constant, then there will be more host plants per unit area in the monoculture than in the polyculture, and any differences observed may be pure host-plant density effects rather than responses to the increase in plant-species richness (see section 3.1.2).

Again, where insect numbers per plant are reported we should treat the data with caution. Increases in plant density will usually lead to reductions in mean plant size and hence to a reduction in the surface area per plant upon which insects can live. Similarly, if a fixed number of insects immigrates to a given area of crop, then the average number of insects per plant will decline with increasing plant density. Both these elementary effects mean that reduced insect burdens per plant will be the norm when plant density is increased beyond the point where competition significantly affects plant size.

The reduction in herbivore load in polyculture compared to monoculture has been termed 'associational resistance' (Root 1973, Atsatt & O'Dowd 1976), and several hypotheses have been proposed to explain it. The first is based on the assumption that plants in monoculture are encountered at a higher rate by searching adults because they are, in one sense or another, more conspicuous; this is the 'resource concentration hypothesis' (Root 1973). The second suggests that there is a higher density of predators and parasites in the polyculture because there is a greater variety of nectar sources for the adults and alternative prey for the larvae; this is the 'enemies hypothesis' (Root 1973). The third hypothesis suggests that differences in the structural complexity and microclimate between the two plant communities are sufficient to account for the effect (Tahvanainen 1972, Risch 1980).

5.4.1   The resource concentration hypothesis

The higher numbers of insects in monoculture may simply reflect higher rates of immigration; the rate of arrival of insects per plant is higher in some monocultures (van Emden 1965), but not in all (Bach 1980a). A detailed understanding of the pattern of establishment of immigrant insects rests on a knowledge of flight behaviour and of the relative attractive and repellant properties of different kinds on plants. In the present context, it is important to know whether the probability of an insect alighting per unit area of ground is affected by the total leaf area of plant tissue present, or only by the leaf area of host plant. In the first case, the rate of alighting will be greater in the experimental polycultures (because total plant density is higher) while in the second case the rates of alighting will be the same. Also, we need to know whether the probability of an insect alighting per unit area of host-plant leaf is affected by the presence of other plants. Non-host species may mask the chemical attractants of the host (Tahvanainen & Root 1972), or reduce the contrast between the host and its background (Smith 1976a), or simply hide the plants from view (Rausher 1981b). The probability of alighting on a leaf may also be affected by how many insects are already present; when aggregating hormones are involved the probability may improve with increasing numbers (e.g. the cucumber-feeding beetle, *Diabrotica unde-cimpunctata*; DaCosta & Jones 1971), but if immigrants avoid already heavily infested leaves it may decline (e.g. Whitham 1978).

   Re-take-off may occur at a greater rate from host plants in polyculture than from hosts in monoculture; Bach (1980b), for example, found a higher residence time for cucumber beetles in monocultures than in plant species mixtures, and Lewis and Waloff (1964) describe how individuals of the bug

Orthotylus virescens in the centre of a patch of broom, *Cytisus scoparius*, tend to stay where they are, while individuals on the edge tend to emigrate. Re-take-off rates are usually higher from non-host than from host plants, but the abandonment of apparently suitable hosts often occurs at a high rate (Kennedy *et al.* 1959; see also section 3.5.3). Thus, if differential re-take-off were the only important mechanism, and total plant density in monoculture and polyculture were equal, we would expect higher numbers of insects per unit area in the monoculture, but equal numbers of insects per host plant in both areas.

Similarly, if the number of host plants was constant, and the monoculture and polyculture differed in total plant density, then the number of insects per unit area should be greater in the polyculture but the numbers per host plant should again be the same in both communities.

The residence time of an individual insect on a host plant may be affected by the chemical environment (plant attractants, aggregating pheromones, the masking or repellant chemicals produced by neighbouring plants). Insects will tend to stay longer where attractants are more concentrated (DaCosta & Jones 1971, Tahvanainen & Root 1972). Herbivore population density will also be higher when microclimate favours long residence times. The cucumber beetles *Acalymma thiemei* and *Diabrotica balteata* are more common in monocultures than on cucumber plants shaded by corn, because they prefer sunny leaves and emigrate from shade (Risch 1980). Insects that can obtain all their resources within the monoculture will have higher residence times than those that need to leave to find nectar sources, nitrogen supplements, free water or other requisites outside the crop (Price & Waldbauer 1975). Thus specialist herbivores will form a higher proportion of the insect fauna in the monoculture than on crop plants in the polyculture (Root 1973).

5.4.2   The enemies hypothesis

It has been noted that natural enemy numbers are higher in weedy crops than in weed-free systems (Dempster 1969, Smith 1976b), and the correlation between higher predator densities and lower pest numbers per plant has formed the basis of the 'enemies hypothesis' (Root 1973). Where surveys have attempted to relate pest problems to the availability of non-crop habitats, however, the results have been inconclusive (Lewis 1965, van Emden 1965, Pollard 1971), and most detailed comparisons of monocultures and polycultures have failed to demonstrate that rates of predation and parasitism are significantly higher in the polyculture (Tahvanainen & Root 1972, Root 1973, Bach 1980b). Both Flaherty (1969) and Pollard (1969) did show however that increased natural enemy attack could exert a major influence on pest numbers on crop plants, and the numerous examples of biological pest control on perennial crop plants demonstrate that it would be wrong to discount the impact of natural enemies on herbivore population densities in crops.

The natural enemy response is also varied. In weedy and weed-free Brussels sprouts, for example, Smith (1976b) found that some aphid predators laid more eggs on plants in weedy plots (e.g. *Melanostoma* spp.) while other syrphids (e.g. *Syrphus balteatus*) laid most on weed-free plots where aphid densities per plant were higher. Clearly some species of natural enemy are governed more in their foraging behaviour by prey density while others are more influenced by background plant density; the net effect on prey density will depend upon the abundance and feeding preferences of the different kinds of predator.

5.4.3   The microclimate hypothesis

The microclimate in the monoculture may be conducive to higher fecundity or a lower death rate. The reproduction of herbivorous insects in monoculture and in polyculture has been studied in rather few cases, but there is evidence to suggest that the reproductive rate is significantly lower in the polyculture (Tahvanainen & Root 1972, Altieri *et al.* 1977, Bach 1980b). Whether this is due to microclimatic differences, to lower levels of secondary chemicals (attractants and feeding stimulants), or to differences in food plant quality is not clear. Nor is the effect universal; in her study of fixed-density cucumber plots, Bach (1980a) found no significant difference in the reproductive rate of the beetle *Acalymma vittata* in pure crops or in polycultures of cucumber, corn and broccoli. Increased planting density may contribute to *improved* pest control in certain crop systems where early canopy closure produces microclimatic conditions unfavourable to the pest or more favourable to its enemies (Luginbill & McNeal 1958, A'Brook 1968, Mayse 1978).

The reduction in pest numbers in polycultures is of largely theoretical interest in temperate regions, since the loss in crop yield due to competition with the other plants far outweighs any gains there may be from having a reduced number of pest insects per plant. There is far greater scope for using associational resistance as a practical tool of integrated pest management in tropical intercropping systems (Altieri *et al.* 1977, Thresh 1981).

5.5   HERBIVORE GUILDS

A guild is a set of animal species (usually closely related taxonomically) which feeds on the same plant species or in the same plant community. We refer, for example, to the guild of leaf-mining insects on a particular species of oak (Opler 1974) or to the guild of grass-feeding ungulates in the Serengeti (Sinclair & Norton-Griffiths 1979).

It is of considerable interest to know what factors govern the number and kind of species that make up a guild. What is the role of chance as compared to interactions between species in structuring the group? Do assembly rules exist whereby certain combinations of species occur more often than would be expected by chance alone (Diamond & May 1981), and, if so, how do they operate?

Central to the development of a theory of guild structure is an understanding of the kinds of interactions that exist between the member species. Are they reciprocal, for example, like competition or mutualism, so that the fitness of both species is affected by a change in the abundance of the other? Or are the interactions asymmetrical like amensalism $(0/-)$ and commensalism $(0/+)$, so that one species affects the fitness of the other, but not the reverse (Lawton & Hassell 1981). It is possible, of course, that there are no important interactions of any kind; the guild may simply consist of a set of species thrown together by circumstance and structured by the availability of separate, non-interacting resources, by habitat structure, and by random immigrations and extinctions.

There are no grounds for *assuming* that interspecific competition is a major force in structuring herbivore guilds on host plants (Connor & Simberloff 1979, but see also Grant & Abbott 1980). Structural differences between the plants, in their geographic distributions, histories, chemistries, and neighbouring plant species all affect what herbivores will be found upon them, and in what relative abundances. Chance will play a major role in the structure of the guild on small or temporary patches of plants. Few plant–herbivore systems in temperate or highly seasonal environments may even approach equilibrium configurations of species richness or relative abundance.

A great deal is known about the complex guilds of herbivorous pest insects attacking such crop plants as soybean (Newsom *et al.* 1980) and cotton (Bottrell & Adkisson 1977); rather less is known about the guilds on native plants (but for notable exceptions see Waloff 1968, Lawton 1976, Claridge & Wilson 1981). Few of the 18 species of woody plants studied by Futuyma and Gould (1979) had a highly distinct, specialized fauna. Despite the limited number of detailed case studies, certain generalizations seem sound. Plants with wide geographic distribution are attacked by more species than are local plants (Strong 1974). Plants which have been numerically abundant throughout recent geological time are fed upon by more herbivores than rarer plants (Southwood 1961, Birks 1980). Plants with complex three-dimensional structure (architecture) support more herbivores because they offer more scope for specialized feeding sites (Lawton & Price 1979). Long-lived plants have more enemies than short-lived species (Rhoades & Cates 1976). Plants with many close relatives in the

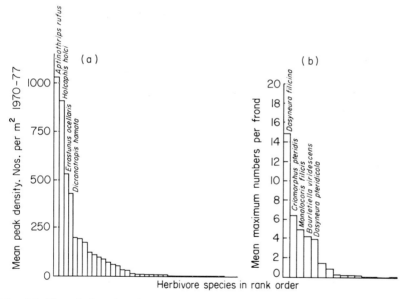

Fig. 5.2 The relative abundance of species in guilds of invertebrate herbivores. The guild typically comprises a small number of abundant species and a large number of less common species. Monophages tend to occur at the more abundant end of the spectrum; increasing sampling effort or increased patch size will tend to increase the species richness of the guild by adding generalist rather than specialist herbivores. (a) The insects of the grass *Holcus mollis* in southern England; (b) the insects of the fern *Pteridium aquilinum* in Yorkshire, both in ranked order of abundance (Lawton & McNeill 1979).

local flora support more herbivores than plants which are 'taxonomically isolated' (Lawton & Schroeder 1977). The guild of invertebrate herbivores on a plant consists of a small number of abundant species and a much larger number of scarce species (Fig. 5.2).

5.5.1 Species–area effects

More herbivore species are recorded from widespread plants than from more local species (Fig. 5.3). This is partly because widespread plants sample a greater number of local herbivore faunas, and partly because the probability of a plant being discovered by a mobile herbivore is proportional to the plant's geographic range. It does not necessarily follow that a widespread plant species will support more herbivores *per individual* than a similar-sized but more local plant.

Species–area effects account for 32% of the variance in the number of agromyzid fly species on different British Umbelliferae. Lawton and Price (1979) suggest that the number of species of miners on many common and widespread umbellifers is less than the number that could evolve to exploit these plants, and that the guilds of miners are not 'equilibrium assemblages'. Area effects often account for less of the variance than this; for example, only 10% of the variance in the number of mesophyll-feeding leafhoppers on British trees is accounted for by their geographic distributions. Despite similar areas of distribution in Britain, oak, *Quercus robur*, supports 10 leafhopper species but ash, *Fraxinus excelsior*, has none at all (Claridge & Wilson 1981).

Artificially established patches of nettles, *Urtica dioica*, were rapidly colonized by herbivorous insects. Large areas, close to established nettle beds were colonized more rapidly than smaller or more isolated clumps. There were marked differences in the arrival rates of different herbivores; for example, the weevil *Phyllobius pomaceus* took three years to reach an isolated plot whereas the psyllid *Trioza urticae* arrived almost immediately (Davis 1975).

Data are available to show how insect herbivore communities change as host-plant range contracts (Ward & Lakhani 1977) or expands (Winter 1974).

5.5.2 Plant architecture

The importance of vegetation structure and plant architecture are shown in Cameron's (1972) study of two salt-marsh communities. The low sward of

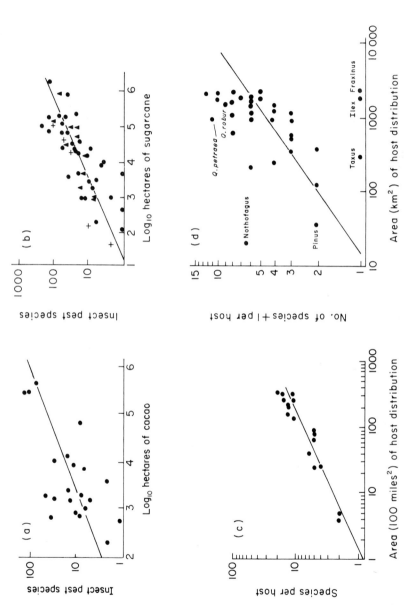

Fig. 5.3 Species–area curves. In tropical crops, the number of herbivorous pest insects is well correlated with the area of cultivation but not with the length of time that the crop has been cultivated: (a) cacao (Strong 1974); (b) sugarcane (Strong *et al.* 1977). Native insects of trees also show species area effects; more widespread plants tend to support more herbivorous arthopods; (c) leaf-mining insects of California oaks (Opler 1974) and (d) mesophyll-feeding leafhoppers of British trees (Claridge & Wilson 1981).

Salicornia contained 35 species of herbivorous insects, while the taller, *Spartina* community had 42. There were more species in each of the major insect orders, and more specialist feeders in all groups in the more structured *Spartina*.

Grassland Auchenorrhyncha are greatly influenced by the vertical structure of the vegetation they inhabit, and mowing (especially in July) can severely reduce the numbers of many species (Morris 1981).

Correlations between plant-species diversity and insect diversity are typically weak in old field communities (Murdoch *et al.* 1972, Southwood *et al.* 1979). For Homoptera, insect-species richness, evenness and diversity were all correlated with both plant diversity and vegetation structure (foliage height diversity) accounting for 79% of the variance in insect diversity when data from different fields were pooled. It was impossible, however, to separate the effects of plant-species diversity from those of plant architecture in determining the diversity of the insect community (Murdoch *et al.* 1972).

Herbivore diversity is sometimes negatively correlated with plant-species richness; for example, irrigation coupled with fertilizer application on short-grass prairie caused an increase in plant productivity but a reduction in plant-species richness. These changes led in turn to an increase in both the biomass and the diversity of the arthropod community (Kirchner 1977).

5.5.3   Plant longevity

The number of insect species found on weeds and other annual plants is less than the number on perennial plants of comparable size and geographic range (Lawton & Schroeder 1977). The proportion of the fauna found on annual plants made up of generalists is greater than on longer-lived but otherwise comparable plants (Lawton & Strong 1981; but see section 3.7.6).

The longer-lived the host plant, the less important is chance in structuring the species composition of the herbivore guild on a particular individual. Short-lived plants presumably have fewer herbivores because they are 'harder to find' in evolutionary time, and typically scarce or sporadic in their occurrence in ecological time (Feeny 1976, Rhoades & Cates 1976).

5.5.4   Time and the size of herbivore guilds

There is no reason to expect that the number of herbivore species on introduced plants should increase significantly over ecological time. In a

matter of years after their introduction, the plants will have accumulated all
the members of the local fauna that are pre-adapted to breeding on them.
Failure to pick up these adapted species will be due at this stage to species-
area effects. In the long term, a slow trickle of new species will establish on

Table 5.1 Time certain trees and shrubs have been continuously present in the
British Isles and estimated numbers of associated insects according to Southwood
(1961; Birks 1980).

	Time growing in Britain (years)	No. of associated insect species
Native trees and shrubs		
Quercus robur and *Q. petraea* (oak)	9000	284
Corylus avellana (hazel)	9500	73
Ulmus spp. (elm)	9000	82
Alnus glutinosa (alder)	7500	90
Betula spp. (birch)	13 000	229
Tilia spp. (lime or linden)	8000	31
Fraxinus excelsior (ash)	8000	41
Salix spp. (willows)	13 000	266
Crataegus spp. (hawthorn)	7000	149
Ilex aquifolium (holly)	7500	7
Pinus sylvestris (pine)	12 000	91
Taxus baccata (yew)	8000	1
Prunus spinosa (sloe)	8000	109
Fagus sylvatica (beech)	6000	64
Acer campestre (field maple)	5000	26
Carpinus betulus (hornbeam)	5000	28
Juniperus communis (juniper)	13 000	20
Sorbus aucuparia (rowan)	12 000	28
Malus sylvestris (apple)	4000	93
Populus spp. (poplars)	12 000	97
Introduced trees and shrubs		
Picea abies (spruce)	500	37
Abies spp. (firs)	400	16
Castanea sativa (sweet chestnut)	1900	5
Juglans regia (walnut)	600	3
Quercus ilex (holm oak)	400	2
Larix decidua (larch)	350	17
Acer pseudoplatanus (sycamore)	650	15
Robinia pseudacacia (acacia)	400	1
Platanus orientalis (plane)	450	0
Aesculus hippocastanum (horse chestnut)	400	4

the plants due to genetic mutations of previously unadapted herbivores or immigration of adapted exotic species. In the interim, there will be a few additions due to behavioural changes amongst the native fauna (in habitat selection, oviposition behaviour, etc.) that will increase herbivore species slightly. Even so, there is unlikely to be a strong correlation between time and species richness per plant in the range 10 to 10 000 years.

Birks (1980) presents data to show the time that different species of tree have been present in Britain since the completely treeless period which followed the last Ice Age (Table 5.1). Using all trees, there is a significant correlation between the number of herbivorous insect species that a tree supports and the length of time it has been in Britain. Using only native trees, there is no correlation. This suggests that for native trees, differences in herbivore number are best explained by species–area and other effects, while for recent introductions, time is more important.

The rapid build-up of species of herbivorous pest insects on exotic crop plants has been documented by Strong (1974) for cocoa and by Strong *et al.* (1977) for sugar-cane (Fig. 5.3); in both cases, the number of pests is related to the area of the plantings, to the fraction of the native fauna that can exploit the crop, and to the rate of importation of exotic pests. In neither case is the size of the pest guild correlated with the time that the crop has been cultured.

That time is perhaps the least important factor affecting guild structure, is demonstrated nicely by the fact that in Britain the alien but widespread sycamore *Acer pseudoplatanus* supports more species of mesophyll-feeding leafhoppers than the long-established but more local native *Acer campestre* (Claridge & Wilson 1981).

5.5.5   Taxonomic isolation

The more closely related plant species in the local flora, the greater the number of herbivores any one of them is likely to support. This is because, as a group, the plants are likely to have been more abundant and to have had a wider geographic distribution, offering greater scope for herbivore speciation.

For the British tree-feeding leaf-miners, only 23% of the variance in guild size is accounted for by the area of distribution of the tree, whereas 36% is accounted for by the taxonomic isolation of the host plant (measured as the number of genera per plant order; Godfray 1982). The factors that control the number of species found on an individual plant of a particular architecture, abundance and geographic distribution are almost completely unknown. It is difficult to find plant species that allow comparison but, for example, oak, *Quercus robur*, and ash, *Fraxinus excelsior*, have similar distributions in Britain (although ash is probably less numerous and is shorter-lived). Oak has 284 herbivorous insect species but ash only 41 (Birks 1980).

Oak is certainly less taxonomically isolated; it belongs, with most British trees, to the Fagales, whereas ash is the sole British tree in the order Contortae. Many other factors confound simple analysis, however. Ash may escape herbivores by leafing later than oak, or its leaves may contain a particularly effective cocktail of chemical defences (they are more toxic to cattle than oak leaves; Forsyth 1968). Forests where ash is the dominant tree are much less extensive than oak woods, and are largely confined to upland limestone areas.

Even after geographic distribution, architecture and taxonomic isolation have been considered, there is often great residual variation in the number of herbivore species per plant; for example, in Britain *Quercus robur* is host to almost the entire guild of gall-forming cynipid wasps, whereas *Q. petraea* has virtually none (Darlington 1974), and yet their ranges are of similar extent, they are of similar size and architecture, and are so similar taxonomically that it can take a hand lens to separate them in the field.

5.6   INTERSPECIFIC DYNAMICS OF HERBIVORES

Interspecific relationships between two herbivore species may be entirely neutral or they may be more or less closely interactive. They may be mutually beneficial (symbiotic) or beneficial to one species and neutral to the other (commensal). They may be mutually detrimental (competitive) or detrimental to one species and neutral to the other (amensal). The weight of the evidence suggests that asymmetric interactions (commensalism and amensalism) are more common amongst herbivores than symbiosis or competition; for example, the relationship between native insect herbivores tends to be weak and non-reciprocal, and to involve only a few of the species in the community (Lawton & Strong 1981).

5.6.1   Competition

There are examples, however, where the fitness of each of a pair of herbivore species is reduced by an increase in the abundance of the other; for instance, the two leafhoppers *Eupteryx urticae* and *E. cyclops* showed reduced fecundity in mixed populations on nettle, *Urtica dioica*. At a given density, fecundity depression per individual was greater for hoppers of the same species, showing that intraspecific competition was more intense than interspecific competition. Shortage of food and of oviposition sites are thought to be the factors that promote competition (Stiling 1980).

Of nine species of stem-boring insects which showed more than 70% overlap in their host-plant use, only three showed any evidence of competition, and in all these cases competition was due to aggressive interference between the larvae, rather than to exploitation of limiting food resources (Rathcke 1976).

The two tephritid flies *Urophora affinis* and *U. quadrifasciata* were introduced to Canada for the control of knapweeds *Centaurea diffusa* and *C. maculosa*. Despite the fact that the numbers of both species were reduced in flowerheads attacked by both flies, the total fly density was not significantly depressed, because *U. quadrifasciata* can attack some of the buds formed between the two generations of *U. affinis*, and can attack buds forming late in the season (Myers & Harris 1980).

Striking evidence of competition between root-feeding pasture scarabs and domestic livestock is provided by Roberts and Ridsdill–Smith (1979). For sheep alone, there was no statistical relationship between stocking rate and yield per animal ($r^2 = 0.01$). When scarab numbers were considered, the fit was greatly improved ($r^2 = 0.92$). At low stocking rates, pasture scarabs were so abundant (150 grubs m^{-2}) that grass growth was reduced to the extent that sheep growth was no better than at higher stocking rates.

Excluding hippopotamus from part of a Ugandan National Park by intensive shooting over an 11-year period, led to an increase of 167% in the numbers of other mammal species including buffalo, elephant and water-buck (Eltringham 1974). Clearly hippo reduced the fitness of these species under normal conditions but there is no evidence on whether an increase in the abundance of, say, buffalo would lead to a reduction in hippo numbers (but see also Sinclair 1977).

5.6.2   Amensalism

Most documented cases of 'competition' are actually non-reciprocal and one species has a markedly greater effect on the fitness of the other (Lawton & Hassell 1981). Considering the possibility of competition between the different bark lice feeding on the epiphytes of larch trees, Broadhead (1958) wrote that the 'early summer populations must certainly remove a considerable amount of food which would otherwise be available for the later summer species. The reverse influence is probably negligible, since the virtual absence of grazing during the winter months provides a respite for the food organisms'.

Manipulation of the densities of two limpets, *Acmaea scabra* and *A. digitalis*, which graze on encrusting microalgae on Central California rocky shores, showed that limpet grazing strongly limited the algal crop for most of the year, and that competition due to exploitation of the common food supply limited the growth rate and maximum size of the smaller of the two species, *A. scabra*. The reverse effect was much less pronounced and removal of the smaller limpet increased the growth of *A. digitalis* in only one case out of three (Haven 1973).

Removal experiments with voles have shown that the distribution and abundance of *Peromyscus maniculatus* can be reduced by the presence of *Microtus townsendii*. *Peromyscus* numbers increased almost as soon as the *Microtus* were removed, and crashed when *Microtus* were later reintroduced. The cause of the exclusion is aggressive behaviour by the *Microtus* (Redfield *et al.* 1977; see also Morris & Grant 1972).

High rainfall during the dry season when food is usually most scarce, leads to an increase in the population densities of the large Serengeti ungulates like wildebeest and buffalo but not of the smaller species like topi, kongoni or impala. Sinclair (1979) interprets this as evidence of interspecific competition and suggests that the larger species dominate the structure of this herbivore guild.

Introduced domestic livestock in the arid lands of Australia led to the extinction of several small and medium sized native herbivores. It is not clear, however, whether this was due to competition for food or to a reduction in cover leading to increased predation (Newsome 1971). The impact of domestic livestock on the native fauna depends upon the degree of dietary overlap they demonstrate. In north-western Australia, for example, sheep led to displacement of the red kangaroo, *Megaleia rufa*, because the sheep and the kangaroo had similar diets, whereas the euro, *Osphranter robustus*, actually increased in abundance. The euro could survive on the spinifex *Triodia pungens* which established itself after overgrazing, while the sheep themselves could not subsist on a diet so low in protein (Ealey & Main 1967).

Winter grazing by barnacle geese, *Branta leucopsis*, and greylag geese, *Anser anser*, significantly depressed the summer yield of Scottish grasslands for cattle, removing forage equivalent to 80 and 90 cow-grazing days ha^{-1}, respectively. In both places, the birds showed a preference for sown *Lolium* over indigenous grass species (Patton & Frame 1981).

Invertebrate herbivores sometimes depress the fitness of vertebrate herbivores. The gall wasp, *Andricus quercus-calicis*, causes extremely high losses of acorns on *Quercus robur* in southern England (see section 2.4.3). The wasp is likely to have a significant impact on the autumn and winter feeding of vertebrates like grey squirrel, *Sciurus carolinensis*, and jay, *Garrulus glandarius*, for which acorns normally form an important part of the diet.

A further class of amensal effects operates via habitat destruction; for example, rooting by feral populations of the European wild boar, *Sus scrofa*, reduced the cover of the understorey from about 95% to about 10% in *Fagus grandifolia* forests in the Great Smokey Mountains (Bratton 1975). Introduced red deer, *Cervus elaphus*, in New Zealand are thought to compete with the gallinule, *Notornis mantelli*, for high-quality food, and to reduce the winter survival of the birds by destroying the beech forest understorey which they require as cover (Mills & Mark 1977).

Ants increased in abundance at the highest sheep stocking levels studied by Hutchinson and King (1980), but all other insect groups were significantly reduced in abundance. The reduction is due to the combined effects of reduced vegetation height and structural complexity, as well as reduced food availability for litter-feeding invertebrates. Herbivorous insects (mainly pasture scarabs) peaked in abundance at intermediate stocking densities, but there was no indication that their feeding led to reduced wool production by the sheep (Hutchinson & King 1980; but see also section 5.6.1). More grasshoppers were found on heavily grazed than on lightly grazed fescue rangelands. The species composition of the grasshopper guild was also affected by grazing practice. Three species were more common in the short, heavily grazed community and five were more common in the taller vegetation; eight other less common species were unaffected by grazing (Holmes *et al.* 1979).

5.6.3   Commensalism

I do not know of any obviously mutualistic (symbiotic) relationships between herbivore species other than those involving mimetic rings. There are, however, a number of examples of commensal behaviour. Perhaps the best known is the relationship between zebra, wildebeest and Thomson's gazelle in the Serengeti (McNaughton 1976). The ungulates graze a patch of grassland in a characteristic sequence. The zebra feed first, removing the taller stems. The wildebeest graze down the leafy grasses and the gazelle then feed on the short sward of herbs and regrowth grass leaves which remain after wildebeest grazing. Thus wildebeest benefit from zebra grazing because the grass leaves are more accessible, and the Thomson's gazelle benefit from wildebeest grazing because the protein-rich herbs are made available and there is a flush of nutritious regrowth which they can exploit (McNaughton 1979a).

There is no suggestion, however, that wildebeest numbers or patterns of dispersal are *dependent* upon zebra. There may be a benefit to wildebeest in some places, but wildebeest doubled in number following the extinction of rinderpest without a concomitant increase in zebra numbers (see section 4.2.2).

Most commensal interactions are likely to be a good deal more fortuitous, like the relationship between gophers and checkerspot butterflies. Many larvae of the butterfly *Euphydryas editha* starve because drought causes the premature senescence of their food plants (see section 3.2.4). Where digging by gophers, *Thomomys bottae*, thins out the *Plantago erecta* plants during the winter, the remaining plants grow larger because of reduced competition and because the soil is tilled. The plants therefore survive two or three weeks longer into the dry season, and allow the larvae to complete their development before plant senescence (Singer & Ehrlich 1979).

5.7 COEXISTENCE AND COMPETITIVE EXCLUSION AMONGST HERBIVORES

It is of great interest to know the importance of competition between the members of a herbivore guild, and to understand the extent to which membership of a guild is prevented by interspecific aggression or resource depletion or pre-emption (competitive exclusion).

Closely related species of similar body size tend to avoid competition by living in different places (habitat separation) or eating foods of different kinds or in different ways (resource partitioning). The most frequent explanation of coexistence is that the species demonstrate habitat (or microhabitat) selection and feed in different kinds of places (e.g. Rosenzweig 1973). When species live in different habitats they encounter foods of different kinds and in different relative abundances; thus the diets of species showing habitat separation are almost bound to be different.

Where herbivores coexist within a more or less uniform habitat it is common to find that they differ in body size or in the timing of their feeding so that competition for food is less intense than might at first appear.

It is becoming clear both from theory (Skellam 1951, Levins & Culver 1971, Slatkin 1974) and experiment (Armstrong 1976, De Jong 1979) that spatial heterogeneity is of great importance in the coexistence of herbivore species. Theory suggests that on divided or ephemeral resources, for example, coexistence between two species can be promoted by aggregation of the superior competitor (allowing competitor-free space for the other species) or by dividing the resource into smaller breeding units (Atkinson & Shorrocks 1981).

Perhaps the most common cause of coexistence, however, is that the herbivore guild does not represent an equilibrium assemblage, and population growth does not continue long enough for the species ever to become sufficiently abundant to compete. Random catastrophes due to weather or landslip, changes in the plant community due to succession, and population regulation by natural enemies, are all factors tending to increase herbivore diversity by preventing the development of an equilibrium herbivore guild structured by interspecific competition.

An example of a moderately complex guild is given by Waloff (1968) who studied the herbivores feeding on broom, *Cytisus scoparius*. The group comprises 9 Lepidoptera, 5 Diptera, 1 Hymenopteran, 7 Coleoptera and 13 Hemiptera which regularly feed on that plant in Silwood Park. The species show a range of microhabitat and feeding preferences within the plant; 2 species feed on seeds, 3 on tissues inside the pod, 6 produce galls, 5 eat the leaves, 9 suck plant juices, 1 feeds on root nodules, 3 are stem-miners, 1 is a leaf-miner and 2 mine bark.

It is by no means clear how these species interact with one another and whether or not they compete. The five closely related mirid bugs, for instance, exhibit peak population densities at different times of year. *Orthotylus adenocarpi* and *Asciodema obsoletum* do show temporal overlap but *A. obsoletum* has an alternative food plant in gorse *Ulex europaeus* which often grows close to broom. These bugs are predatory as well as herbivorous and the larger mirids feed on the smaller individuals, even of their own species. Thus the later hatching species are disadvantaged compared to the early hatching kinds. The latest to emerge in England is *O. concolor* and it is the rarest of the five species. It is interesting that *O. concolor* is extremely abundant in California where it was introduced on broom plants, and thrives in the absence of the other species. Interrelations are further complicated by the fact that certain of the mirids share parasites (Waloff 1968).

Broadhead's (1958) study of bark lice demonstrates the typical kind of result from an investigation into coexistence among apparently (or potentially) competing herbivore species. Two food resources, epiphytic alga *Pleurococcus* sp. and lichen *Lecanora conizaeoides*, were specialized upon by groups of psocids. Within each group of potentially competing species, however, the most closely related species almost always showed microhabitat separation. Thus, for example, *Amphigerontia contaminata* feeds on the epiphytes of living twigs, *A. bifasciata* on dead ones; *Elipsocus westwoodi* lives on living branches, *E. mclachani* on dead ones while *E. hyalinus* is intermediate.

What appear to be good examples of competitive exclusion may turn out to be due to quite different causes; for example, the contraction of the range of Arctic hare *Lepus arcticus* in Newfoundland following the introduction of snowshoe hare, *Lepus americanus*, about 100 years ago has been described as a classic case of competitive exclusion (MacArthur 1968). A more plausible explanation now appears to be that the exclusion of *L. arcticus* from woodland (and also from parts of its alpine habitat which have not been invaded by *L. americanus*) is due to predation by *Lynx canadensis*. Predator numbers increased following the introduction of snowshoe hare, and the increase in predation rate on arctic hare was sufficient to account for the restriction of its range to habitats where cover from predation is available (Grant 1972).

5.7.1   Coexistence and resource partitioning

Evidence of competition is often sought from species that show different feeding behaviour in regions where the species live together (sympatry) and where they live separately (allopatry). Sympatric and allopatric populations of arctic ptarmigan have been studied by Bryant and Kuropat (1980). Where willow ptarmigan and rock ptarmigan live together in Alaska and Fennoscandia, the willow ptarmigan feeds preferentially on willow, taking birch only when willow is not available. The rock ptarmigan feeds preferentially on birch staminate catkins and, when these are depleted, on birch leaf buds. When these two birds live separately, the feeding behaviour of willow ptarmigan is unaltered but the behaviour of rock ptarmigan changes radically. In Iceland where rock ptarmigan is allopatric it feeds preferentially on willow throughout the winter, only taking birch catkins after the willows is buried by snow. Once the catkins are depleted it takes birch leaf buds and twigs.

A guild of herbivores inhabits the flowerheads of nodding thistle, *Carduus nutans*; more than 70% of thistle populations supported two or more coexisting insect species (average 3.4), and more than half the individual flowerheads showed evidence of competition between different species (Zwolfer 1979). The most important members of the guild are the picture wing fly, *Urophora solstitialis*, and the weevils *Larinus sternus* and *Rhinocyllus conicus*. Interactions between the different species within a single flowerhead are various. *Urophora solstitialis* is a gregarious species whose larvae inhabit large thistle heads which they turn into many-celled galls with lignified exteriors that other species are unable to penetrate. *Larinus sternus* can only outcompete *Urophora* when it attacks the flowerhead before the gall has hardened. More commonly, this weevil inhabits thistle heads which are too small for the fly gall and *Larinus* females show a significant preference for ovipositing in smaller heads. *Rhinocyllus conicus* loses out to both species in competition within an individual flowerhead. It survives because of its high fecundity and its habit of spreading its eggs over as many thistle heads as possible; larvae survive in those heads where there are no competitors (Zwolfer 1979). It is interesting that of the three species, it is *Rhinocyllus* that has been successful in the biological control of nodding thistle in Virginia (Kok & Surles 1975; see also section 4.1).

The guild of seed-feeding rodents in the sand dunes of desert Nevada was studied by Brown and Lieberman (1973). Up to five species can coexist in the most favoured sites where rainfall is locally higher. Each species is at least 1.5 times heavier than the next in the hierarchy of sizes and the larger animals feed preferentially on larger seeds. The species also show microhabitat separation, foraging in different areas relative to the cover of perennial plants. Diet overlap is higher (up to 80%) in areas where seed production is high, and diversity is low (species are excluded) where seed production is low and (by implication) competition is more severe. It was not tested whether supplementing food would increase species diversity. Differences in body size and seed-husking ability are also thought to be the basis of coexistence in other heteromyid rodents (Rosenzweig & Sterner 1970).

Individuals of different body sizes do not necessarily feed on foods of different sizes. Smigel and Rosenzweig (1974) found that *Dipodomys merriami* and *Perognathus penicillatus* selected the same set of seed sizes. Both species were more selective at high seed densities than at low. Despite considerable differences in bill sizes, five sympatric finch species showed almost 100% overlap in food size utilization. However, bill size was related to the proportion of large seeds in the diet (Pulliam & Enders 1971).

For a guild of seven sap-feeding insects on salt-marsh grasses, Denno (1980) suggests that the smaller species exploit a greater range of resources than larger ones (though they are not necessarily more catholic taxonomically; see section 3.4.1).

5.7.2   Coexistence and habitat separation

African ungulates show patterns of habitat separation between the species that prefer open grassland (Grant's gazelle, wildebeest and zebra) and those preferring wooded areas (dikdik, lesser kudu, etc.). There are differences in feeding height between many of the coexisting species, most obviously between animals like giraffe, rhinoceros and dikdik (Sinclair & Norton-Griffiths 1979).

Chew (1981) found that two pierid butterflies showing a pattern of distribution suggestive of competitive exclusion, actually practiced habitat selection based on the autecological requirements of their preferred host plants. They demonstrated neither behavioural interference nor exploitation of a common food resource.

Only one of the guild of seven sap-feeding insects on the salt marsh grass *Spartina patens* has retained the ability to fly. This leafhopper *Amplicephalus simplex* can, therefore, exploit competitor-free patches of grass created by the catastrophic effects of spring tides (Denno 1980).

In fact, many cases of coexistence are likely to be due to spatial heterogeneity of one kind or another (Southwood 1977) but documented cases are scarce. Two eastern Australian herbivorous gastropods live together on the same rocky surfaces of the intertidal zone. Caging experiments have shown that the superior competitor *Nerita atromentosa* can depress the population density of the second species *Cellana tramoserica* by reducing the availability of microalgal food. Coexistence of the molluscs is allowed because the inferior competitor *Cellana* has a spatial refuge in the subtidal zone where *Nerita* does not live. Also, the distribution of establishment of young *Nerita* is patchy and *Cellana* may persist temporarily in the *Nerita*-free places (Underwood 1978).

Interspecific competition between seven specis of *Drosophila* in a fruit and vegetable market was suggested by the negative correlation between the density of *D. melanogaster* and the body size of several other species. The species coexist by partitioning breeding sites and because the guild may never reach equilibrium (Atkinson 1979).

Species that feed on the same tissue may show microhabitat selection. Different species of aphid chose different regions of the underside of broad bean leaves to settle on; *Aphis fabae* chose veins near the centre, *Myzus persicae* settled near the edge of the middle of the leaf and *Acyrthosiphon pisum* occupied intermediate positions on the lamina. That the choice of microhabitat affects the success of the animals was shown by caging the different species on different parts of the leaf. Not surprisingly, they fed most rapidly and grew most quickly on the parts of the leaf they normally chose in the field (Lowe 1967). Leaf-mining species on oaks in California tend to be found on leaves of different ages or at different times of year. Where species do occur simultaneously, however, one tends to mine the leaf base near the petiole, another mines in mid leaf and a third attacks the mesophyll cells in the tip of the oak leaf (Opler 1974). The guild of insects feeding on the leaves of beech, *Fagus sylvatica*, shows separation within the canopy (most species prefer the lower crown in mature woodlands) and within individual leaves (Nielsen 1978).

The members of the guild of eight species of *Eurythroneura* (Homoptera), on *Platanus occidentalis* studied by McClure and Price (1975, 1976), compete for food in those years when numbers are high due to good weather early in the season. They could find insufficient differences between the species, however, to account for their continued coexistence in an equilibrium community. They suggest that competitive exclusion does not occur because the species peak in abundance in different parts of the geographic range of the guild and because intraspecific competition is more intense than interspecific. Lawton and Strong (1981) conclude that to account for the patterns exhibited by leaf-feeding insect communities we need a theory that stresses 'the autecological responses to weather, phenology, and host chemistry of individual species of plants and insects, the effects of isolation, migration and habitat heterogeneity, geographical factors of interspecies associations, and the action of predators and parasitoids'. While it is certainly true that competition between herbivorous insects has rarely been demonstrated unequivocally, and that other factors are bound to be important in structuring herbivore guilds, it seems unwise to write off competition as a potent force. We need carefully controlled, manipulative field experiments in which the densities of several members of a guild are varied in factorial combinations. Only then will we be able to weigh the relative importance of competition and other processes.

5.8   PREDATORS, PARASITES AND DISEASES IN COMMUNITY DYNAMICS

The role of predators in some plant–herbivore systems is straightforward. They keep herbivore numbers so low that competition between the herbivores is negligible and the plant suffers little, if any, reduction in fitness due to herbivore feeding. It is when predators are absent (as with alien herbivores) or reduced by some disturbance (often at the hand of man) that herbivores become spectacularly abundant.

The importance of natural enemies in maintaining low equilibrium levels of many herbivorous insects has been shown where application of broad-spectrum insecticide has 'created pests'. De Bach (1974) catalogues numerous examples where insect numbers increased dramatically when their parasitoids and predators were inadvertently killed.

Annecke *et al.* (1969) turned this syndrome to good effect in their work on the biological control of prickly pear in Cape Province, South Africa. The cochineal insect *Dactylopius opuntiae* imported to control the *Opuntia* was ineffective because it fell prey to coccinellid predators (cochineals have no known parasitoids). Annecke sprayed the cactus with the same low dosages of DDT that had caused pest outbreaks in the USA; because the ladybirds were active, they accumulated DDT over a long period and the desired effect was achieved. The coccinellids died in sufficient numbers that the cochineal insects were able to increase in density and destroy the cactus.

The role of natural enemies can also be gauged from the different degrees of success achieved in various attempts at biological weed control. Weed control by introduced insects has been reduced in effectiveness by native predators as varied as spiders, birds, ants, mites, bugs, coccinellids, mecoptera, lizards and mice (Goeden & Louda 1976). Control of prickly pear in South Africa was hindered by baboons feeding on *Cactoblastis*-infested cactus cladodes (Pettey 1947). Accidental predation by noctuid larvae (*Heliothis* sp.) of the larva and pupae of the fly *Procecidocharea utilis* by tunnelling through its galls destroyed 40–50% of the galls on Crofton weed, *Eupatorium adenophorum*, in some parts of Australia (Dodd 1961). Parasitic insects have had rather less effect in disrupting attempts at weed control. The polyphagous egg parasite *Trichogramma minutum* which had been introduced into Hawaii to control pest insects, turned against the moth *Bactra truculenta* which was established to control purple nutsedge, *Cyperus rotundus* (Goeden & Louda 1976).

Predation effects have varied from outright prevention of the establish-

ment of potential control agents (e.g. cinnabar moth, *Tyria jacobaeae*, on
ragwort, *Senecio jacobaea*, in Australia and New Zealand) to a reduction in
the effectiveness of established populations (e.g. the cochineal insect
Dactylopius opuntiae on South African populations of prickly pear) or to no
appreciable reduction in the destruction of plants (see the successful cases of
weed control in section 4.1). The effects of parasitism are generally less severe
than those of predation. No failure to establish a newly colonized species of
phytophagous insect has been ascribed solely to the activities of indigenous
parasitoids. Where weed control has been less successful, it is because
parasites reduced the initially high population densities of their adopted
hosts, or slowed their rate of increase. In most cases, however, parasites have
had little influence in reducing the effectiveness of biological weed control
(Goeden & Louda 1976).

Predation by the sea otter, *Enhydra lutris*, off islands in the Aleutian
chain limits epibenthic invertebrates, especially sea-urchins, *Strongylo-
centrotus polyacanthus*, which, in turn, allows a luxuriant development of the
macroalgal canopy. Estes *et al.* (1978) show that where sea otters are
common, sea-urchins are small and scarce, and the macroalgae are
predominately limited by competition. On islands where sea otters are
absent, the urchins are large and abundant and the fleshy microalgae have
been eliminated (Simenstad *et al.* 1978).

The impact of natural enemies on vegetation is seen in the changes that
took place in the wake of the myxomatosis epidemic of the 1950s in Britain.
Thomas (1960) writes that 'After the death of the rabbits, the turf increased
in height and there was a spectacular increase in the abundance of flowers,
especially of orchids in some places. But already there is an indication that in
the absence of rabbit grazing the turf may lose its density, and possibly its
botanical interest, as some plants—even the grasses—show a tendency to
decrease in quantity. There has been an increase of woody plants. Brambles
have grown greatly, and so have gorse bushes; their seedlings and those of
many other woody plants are becoming much more common in grasslands
and in woods'. The short-term effect of reduced herbivore density was an
increase in plant richness as previously suppressed plants expressed their
presence by flowering. The longer-term effect of the virus was a reduction in
plant-species richness, increased dominance by a small number of rank
grasses, and hastened succession towards woodland.

Primary production in rangelands is divided between the domestic
livestock and their vertebrate and invertebrate competitors. The yields from
certain heavily grazed pastures in Texas, for example, were increased by 44%

following the introduction of a parasite *Neodusmetia sangwani* to control the rhodesgrass scale insect *Antonina graminis* (Schuster *et al.* 1971).

Herbivore populations may be more stable in the presence of a sympatric species when they share a polyphagous natural enemy. Whittaker (1973) suggests that numbers of the hopper *Philaenus spumarius* are more stable when it occurs with *Neophilaenus lineatus*, because each host species acts as a reservoir for the parasitoid *Verrallia aucta* in years when the other host is less abundant.

The structure of the plant community is an important determinant of the predation rate. Structurally complex habitats may harbour more enemies than simple ones (see section 5.4). With cinnabar moth larvae, for instance, Dempster (1971) noted over 80% predation in the tall, dense vegetation at Monks Wood but only 30 to 60% in the heavily grazed, short community at Weeting Heath.

Even the structure of the surface of the plant itself can alter the impact of natural enemies. The parasitoid *Encarsia formosa* is much less effective in its control of the greenhouse whitefly on hairy-leaved varieties of cucumber than on glabrous forms. It walks three times faster on the glabrous plants and becomes less fouled by whitefly honeydew. Thus less time is spent grooming and more in hunting and parasitizing whiteflies (Price *et al.* 1980).

As we saw in section 3.1.2 the concentration of allylisothiocyanate in Brussels sprouts affects their susceptibility to attack by cabbage aphid, *Brevicoryne brassicae*; the compound acts as a feeding stimulant for this specialist herbivore and 'resistant' varieties like 'Early Half Tall' have low concentrations of the chemical. The parasitic wasp *Diaratiella rapae*, however, is attracted to cruciferous plants rather than to aggregations of the aphids upon them, and the rate of attraction is greater to plants rich in allylisothiocyanate. Thus the aphid population on the 'susceptible' variety 'Winter Harvest' suffered a higher rate of parasitism, and the pest population on the 'resistant' crop eventually outstripped that on the so-called 'susceptible' crop (van Emden 1978)!

CHAPTER 6
CONCLUSIONS

This chapter summarizes the principal features of plant–herbivore dynamics and highlights those avenues of research which are most likely to provide important new insights. It is inevitable in a brief summary, unfettered by the usual ifs and buts, that a number of the assertions will appear highly contentious. While it would be lamentable if this were the cause of confusion or irritation, it would be admirable if the contentions were the catalyst of really critical experiments. It is partly in the hope of being proved wrong that the general patterns emerging from this study of herbivory are listed below. For convenience, they are grouped loosely under a series of topics.

6.1   PLANT–HERBIVORE INTERACTIONS

6.1.1 There is a fundamental asymmetry in plant–herbivore interactions. Plants have much more impact on the dynamics of herbivores than herbivores have on the dynamics of plants. The effects of plant factors (abundance, quality, spatial distribution) on herbivore population change are usually a great deal more pronounced than the impact of herbivore feeding on plant numbers.

6.1.2 Mammalian and insect herbivore populations tend to be regulated in different ways. A relatively high proportion of mammalian herbivore populations is food-limited and a comparatively high proportion of insect herbivore populations is regulated by predators, parasites and diseases. This has important consequences for the impact of these two groups of herbivores on plants.

6.1.3 Invertebrate herbivores *can* reduce equilibrium plant populations well below their carrying capacities, as witnessed by the biological control of alien weeds such as prickly pear and St. John's wort by insects which were introduced without their natural enemies. The degree to which plant numbers are depressed below K depends upon the searching efficiency and aggregative responses of the herbivores, and upon the powers of compensation demonstrated by the plants.

6.1.4 Vertebrate herbivore populations are often limited by the availability of high quality food. The strong positive correlation between the biomass of vertebrate herbivores in African grasslands and the rate of plant productivity is strongly suggestive that the herbivores are food-limited. If the herbivores were predator-limited, increases in plant productivity would simply lead to increased predator abundance, leaving herbivore equilibrium unaltered.

6.1.5 Monophagous invertebrate herbivores in their native lands tend to have little impact on equilibrium plant abundance. Even when the insects are food-limited, the plants are rarely herbivore regulated. Such asymmetrical interaction is exemplified by the ragwort/cinnabar moth system, where the caterpillars are food-limited, and regularly strip the plants of all their leaves and flowers. The insects, however, are univolitine and pupate before the plant growing season is over. This provides the plant with a period of herbivore-free regrowth that is long enough for it to restore its competitive ability. Ragwort numbers are regulated by the availability of germination microsites (in turn, related to summer rainfall) and by competition with other plant species.

6.1.6 Invertebrate herbivores that are regulated at low densities by natural enemies have virtually no impact on plant dynamics when regrowth or other forms of compensation occur.

6.1.7 It is sometimes *assumed* that high rates of parasitism or predation mean that the herbivore population is predator-regulated. This could fruitfully be tested using various experimental techniques of predator exclusion; for example, spraying prickly pear with DDT improved weed control because cochineal insects had previously been regulated at low densities by ladybirds, and insecticide killed the predators but not the cochineals. Also, low rates of disease incidence or parasitism cannot be taken as evidence that these factors do *not* regulate herbivore density, particularly if the parasites or disease agents have rapid turnover rates.

6.1.8 Measures of energy flow are poor indicators of which population interactions are important in regulation. Energy flow is an especially poor estimate of feeding impact when there is wastage (as with herbivores that clip off, or induce premature abscission of plant parts), when the damage is highly localized (as with meristem feeders, stem borers or ring barkers) or is caused by viral, bacterial or fungal diseases transmitted by the herbivore.

6.1.9 The minimum information necessary to understand the dynamics of field populations of one plant and a single species of monophagous herbivore consists of:

(a) the intrinsic rate of increase of the plant;
(b) the carrying capacity of the plant in the absence of herbivores;
(c) the functional response of the herbivore;
(d) the intrinsic rate of increase of the herbivore;
(e) the per capita impact of herbivore feeding on plant performance; and,
(f) the nature of density dependence in the herbivore's numerical response.

A full understanding will only be obtained when the details of plant compensation are known, and when the importance of plant refuges, aggregative herbivore behaviour, and the role of natural enemies are elucidated. We do not have even the minimum data set for a single plant–herbivore system.

6.1.10 To understand the dynamics of a polyphagous vertebrate herbivore feeding on a multispecies plant community demands that a whole new set of interactions is considered, including:

(a) the relative competitive abilities of the plants in the absence of herbivores;
(b) the effect of defoliation on competitive ability;
(c) the preferences of the herbivores for different plant species;
(d) frequency-dependent and plant-density-dependent changes in food selection (switching);
(e) the effect of herbivore feeding on food quality; and,
(f) the effect of diet composition on herbivore numerical responses.

Needless to say, estimation of these parameters presents almost insuperable difficulties for the field ecologist. Nonetheless, unravelling the dynamics of these multispecies systems presents one of the most intriguing challenges in modern biology.

6.2   PLANT POPULATION DYNAMICS

6.2.1 It is impossible to obtain estimates of a plant's intrinsic rate of increase without equally precise information on both seed production *and* survivorship. In most population studies, plant fecundity receives considerably more attention than mortality, and we know a great deal more about how many seeds are produced per plant than about what fraction of the seeds survives to produce seed themselves. There is great scope for field studies which attempt to estimate both the intrinsic rate of increase of plants, and the actual rates of increase under various levels of herbivore attack.

6.2.2 Attempts should be made to produce 'damage functions' in which plant performance is plotted against herbivore feeding. Ideally, plant performance should be measured in terms of the number of offspring surviving to produce seed (fitness) but seed production *and* plant survivorship may be all that it is practical to measure. Experimental increases and reductions in herbivore numbers may be necessary to obtain a sufficient range of herbivore feeding rates.

6.2.3 Seed production is extremely sensitive to herbivore attack and reduced seed production is typically the plant's first response to defoliation. Whether or not reduced seed production leads to reduced plant fitness depends upon the factors regulating plant recruitment, and only when recruitment is seed limited will herbivores that reduce seed production regulate the plant population.

6.2.4 Seed production and seed predation are closely interrelated in their impact on population dynamics. If the seed predator population is regulated by factors other than seed availability, or is only capable of a delayed numerical response to increased seed availability, then high rates of seed production may lead to predator satiation, and to the establishment of a seedling population. Here, a defoliating insect that caused only a small reduction in seed output might completely prevent seedling establishment by reducing seed fall below the threshold necessary for predator satiation.

6.2.5 Herbivores influence the timing of sexual maturation in plants (e.g. defoliation delays flower initiation). If herbivore feeding causes plant mortality, and thereby reduces plant density, development time may be reduced (in perennials) or increased (in annuals that show precocious flowering at high densities). Whether or not delayed flowering affects plant fitness should be established by comparing the fates of seeds produced at different times.

6.2.6 Plant populations in which recruitment is limited by microsite availability may be influenced by herbivore numbers either when the herbivores create germination sites directly (by their digs, scrapes or hoof prints) or indirectly (by reducing the spread, shade, or litter production of neighbouring plants).

6.2.7 Plant survivorship must be measured from *seed* to seeding plant rather than from seedling to seeding plant. Many of the most important demographic processes operate between dispersal of the seed from the parent plant and the emergence of a recognizable seedling. Work on this aspect of plant ecology is exceptionally laborious but is critical to a full understanding of population dynamics.

6.2.8 In assessing the impact of herbivore attack on the death rate of herbaceous plants it is essential to determine whether or not the genet actually dies; for example, casual observation suggests that ragwort is biennial, and that flowering always kills the plant. In fact, neither severe defoliation nor flowering kills more than a small proportion of the genets and most plants grow back from buds on the root stock or from root fragments. Herbivore feeding can even lead to an *increase* in plant numbers when several shoots develop from the rootstock of one defoliated plant.

6.2.9 Plant death rates are determined largely by competition with other plants. In plant monocultures, increases in death rate due to herbivore feeding will typically be compensated by reductions in the natural rate of self-thinning, so that herbivore-caused mortality has little impact on the equilibrium population density of mature plants. In mixed plant communities, selective feeding by herbivores that increases plant death rate will almost certainly bring about a reduction in plant population density, because other plant species benefit from competitor-release.

6.2.10 Plant productivity is affected by the rate of defoliation. In monocultures, or where widely spaced plants suffer little competition, light to moderate levels of defoliation will not depress the rate of plant production (and may even increase it). Plants in dense, mixed communities, however, will normally not be able to compensate even for quite low levels of selective defoliation, because neighbouring plants which have not been attacked will pre-empt the unoccupied canopy space. In grasslands, the productivity of the plant community as a whole will almost always be greater under moderate grazing pressure than in herbivore-free exclosures, because self-shading is minimized and the age structure of the leaves is altered in favour of younger, more productive tissues.

6.2.11 There is limitless scope for field experimentation in which plant populations in different parts of their range are subjected to different regimes of defoliation and soil disturbance, and from which herbivores are variously excluded or increased in abundance. Every effort should be made to estimate the impact of these treatments on plant fitness, and not just on certain components of fitness such as shoot growth or seed production.

6.3   HERBIVORE POPULATION DYNAMICS

6.3.1 Herbivore fecundity is generally density dependent. With invertebrate herbivores this occurs because reduced feeding success in the larval stage leads to the production of smaller, less fecund adults. In mammals, high density leads to a delay in the onset of sexual maturity and/or to a reduction in the length of the breeding season. Density dependence in fecundity can occur more or less linearly for invertebrates, but many mammalian herbivores show strongly non-linear density dependence in fecundity.

6.3.2 Mortality may be density dependent in some stages of the life cycle but not in others. In many mammal populations, for example, juvenile survival is density dependent, but adult survival is density independent. With insects, generation survival may be density independent in certain species, while others suffer density-dependent mortality from predators, parasitoids or disease. Not all natural enemy attack is density dependent, and, where enemy numbers are regulated by factors other than prey abundance, it may be *inversely* density dependent.

6.3.3 Emigration is almost universally density dependent, and is often the most immediate response by herbivores to declining food quantity or quality.

6.3.4 Patterns of herbivore dispersal are intimately linked to the availability of high quality food. Herbivorous insects show a range of dispersal capabilities associated with the predictability of appearance of their host plants. Insects of permanent habitats or long-lived plants frequently have wingless forms, while insects exploiting ephemeral plants or habitats are almost invariably winged. Serengeti wildebeest migrate thousands of kilometres per year in search of widely spaced patches of green vegetation. Other vertebrates may migrate to avoid natural enemies during the breeding season.

6.3.5 Intrinsic rate of increase is inversely related to body size so that high rates of population increase are associated with low individual rates of feeding. The rate of feeding per gram of body weight is greater for small herbivores than for large, and for warm blooded compared with poikilothermic animals of the same size.

6.4   SPATIAL DISTRIBUTIONS

6.4.1 The density and spatial distribution of food plants interact with the herbivore's powers of dispersal and its plant-finding ability in determining herbivore mortality, especially for insects which must move from one plant to another in the course of development.

6.4.2 Exploitation patterns are determined by the spatial pattern of the plant population and by the searching and aggregative behaviour of the herbivores. Herbivore populations that exploit their food plants in a coarse-grained way tend to be more stable and to occur at lower equilibrium densities than otherwise similar, fine-grained species, because their feeding behaviour creates a refuge for the plant.

6.4.3 The existence of refuges in which a species can escape over-exploitation by its enemies or exclusion by its competitors tends to increase population stability and to enhance species richness. Identifying, counting, and plotting seasonal changes in the spatial distribution of these refuges offers a tremendous challenge to field ecologists.

6.4.4 Population densities tend to be lower, and population equilibria more stable, in heterogeneous than in homogeneous environments. Spatial heterogeneity of plant distribution tends to increase the grain response of the herbivores, and to provide refuges for the plant from over-exploitation.

6.5   SEASONAL FACTORS

6.5.1 Temporal heterogeneity (due to fluctuating weather conditions, plant successional change or soil disturbance) is the key factor affecting population change in most herbivores. Species with high intrinsic rates of increase tend to track temporal changes, while species with low rates of increase tend to show reduced equilibrium densities in highly variable environments. Temporal heterogeneity has one of its most profound effects on plant–herbivore dynamics when the phenologies of the plant and the herbivore are altered in different ways or to different degrees; for example, the synchrony between peak food quality and peak herbivore feeding is often critical in determining whether herbivore abundance increases or declines.

6.5.2   In seasonal environments, pulses of food production are followed by prolonged periods in which there is virtually no high quality food. Most vertebrates build up body reserves during the pulse of high quality growth, then gradually lose body weight over the rest of the year, feeding on inferior, often low-protein diets. Invertebrates spend the period of low food quality in diapause (or other resting state) and rely on external cues (temperature, moisture or day length) to trigger the emergence of the feeding stages so that they coincide with the pulse of food availability.

6.5.3   Most outbreaks of native herbivores and crop pests are caused by changes in patterns of rainfall. Increased rainfall may increase plant productivity, allowing herbivores to escape the control of natural enemies. Reduced rainfall may bring about drought-induced changes in food plant quality which favour herbivore increase. Associated temperature changes may lead to a reduction in the control exerted by invertebrate predators and parasites because of differential changes in predator and prey development rates.

6.6   CYCLES

6.6.1   There is strong evidence that herbivore population cycles are caused by plant–herbivore interactions (rather than by predator–prey, disease–host or plant–environment interactions). In the ten-year cycles of snowshoe hare and larch budmoth, heavy exploitation during a period of high food-quality is followed by an extended period when the regrowth foliage is of low quality. The herbivores' rate of increase is negative throughout this post-defoliation period, and only becomes positive once the quality of the food plant recovers its initial, higher levels. Peak numbers of arctic lemmings occur following winter breeding beneath the snow. Four-year cycles in abundance are probably the result of impaired plant regrowth after these peaks of lemming feeding, when the carbohydrate reserves of the plants are severely depleted and numerous meristems are destroyed.

6.6.2 Herbivores can alter the rate of nutrient cycling. When large, and normally long-lived nutrient stores (like tree trunks) are released because the plant tissues are killed prematurely, a pulse of nutrients is added to the soil by decomposers. This may be lost to the ecosystem, if a rapid burst of plant growth does not recapture the liberated nutrients, and an irreversible change in plant community structure and productivity may result. This, in turn, may lead to a permanent reduction in equilibrium herbivore abundance.

6.7 HARVESTING

6.7.1 The response of a herbivore population to harvesting depends upon the degree of non-linearity in its density dependence. If density dependence only becomes intense at high densities (close to K), then maximum productivity is obtained at densities only little below the carrying capacity (not at $K/2$ as predicted by the logistic on the assumption of linear density dependence). Thus for the many ungulates which exhibit pronounced non-linear density dependence, management aimed at maintaining the population at $K/2$ will produce lower and more unstable yields than could be harvested at a higher equilibrium density.

6.7.2 Criteria for maximum sustained yield, estimated from deterministic models, are likely to lead to over-exploitation when applied in stochastic environments.

6.8 MONOPHAGY AND POLYPHAGY

6.8.1 Strict monophagy is common only amongst invertebrate herbivores. For insects, the fraction of monophages in the herbivore guild tends to be lower on annuals than on perennials. Amongst perennial plants, however, trees support a smaller proportion of monophagous insects than do herbaceous perennials. Monophagy is more common amongst mining and galling insects than amongst leaf-chewing and other surface-feeding groups. This suggests that generalist predators pose more of a threat to the evolution of monophagy than do parasites, since gall formers and leaf miners often suffer high rates of attack from generalist parasitoids. Where monophagous insects *do* feed on the plant surface, they tend to be distasteful to predators, and to show aposematic colouration.

6.8.2 In vertebrates, increased body size is correlated with increased polyphagy (and with a reduction in average dietary protein levels). Polyphagous vertebrate herbivores exhibit distinct preferences for plants of different species and tissues of different kinds. Many species also demonstrate diet supplementation and take novel foods to rectify protein, mineral or vitamin deficiencies.

6.9   DIVERSITY OF PLANTS AND HERBIVORES

6.9.1 Selective grazing by vertebrate herbivores alters the quality of food available to them subsequently. Short-term exploitation leads to reduced relative abundance, and thus to reduced rates of feeding from preferred food types. The plant's facultative defences may further reduce food quality. Long-term selective feeding modifies the competitive abilities of grazed and ungrazed plant species, and may lead to the development of a distinctly different plant community. The grazing-produced community may support a higher or a lower equilibrium density of herbivores than the vegetation it replaces (depending upon the effect of feeding on primary productivity and plant species composition).

6.9.2 Selective feeding by herbivores may affect plant species richness. When herbivores feed preferentially on what would otherwise be the dominant plant species, richness is normally increased. Selective feeding on already uncompetitive plants inevitably leads to reduced plant diversity. Note that mowing machines are *selective* herbivores. They defoliate to a uniform height, but they take a greater proportion of the shoot biomass of upright plants than of rosette-forming species. Since mowing is selective it can, and often does, lead to increases in plant species richness.

6.9.3 The number of species of herbivorous insects attacking a given plant species is determined by a complex of ecological, evolutionary and biogeographic factors. Large plants support more species than small plants because they provide more feeding niches and support larger populations. Long-lived species support more herbivores than short-lived plants of the same size, because they represent a more predictable food resource. Plants with wide geographic distributions have more herbivorous insects than local plants because they recruit herbivores from a variety of geographically more restricted insects. Plants that grow in many habitats are exposed to a larger herbivorous insect fauna than plants growing in a single habitat. Plants with more closely related species in the local flora will support more herbivore species than plants which are taxonomically isolated because there is more scope for adaptive radiation amongst the insects. Abundant plants will support more herbivores than rare plants because the extinction rate of species populations will be greater on rare or very local plants.

6.9.4 The importance of competition and mutualism in structuring herbivore guilds remains to be clarified. Great scope exists for manipulative experiments in which the abundances of putative competitor or mutualist species are increased and reduced in various factorial combinations. Monitoring subsequent changes in herbivore numbers and calculating the impact on plant performance will provide invaluable information on community dynamics.

6.9.5 The occurrence of competition and mutualism may be highly patchy in space and sporadic in time. Failure to obtain evidence of competition or mutualism from small, random samples of the habitat cannot be taken to indicate that the processes are unimportant in population dynamics.

6.9.6 Insect numbers per plant and per square metre are higher in monocultures than in plant species mixtures (polycultures). This 'resource concentration' effect appears to be due mainly to the non-host plants simply hiding the hosts, or, more subtly, to their masking the host plant's chemical attractants. When the insects spend time investigating the non-host plants, or show an increased tendency for re-take-off and flight following contact with a non-host plant, their foraging time may be reduced to such an extent that the host plants experience a lower rate of attack. These effects are termed 'associational resistance'. Higher rates of predation and lower rates of insect reproduction in the polyculture may enhance resource concentration in some cases.

6.9.7 The concept of plant apparency has stimulated much productive research and debate but contributes little to an understanding of plant–herbivore dynamics, largely because apparency is unmeasureable. It may be revealing to compare the dynamics of long-lived and short-lived plants, or to contrast the biology of plants from ephemeral and more permanent habitats; apparency itself, however, is little more than *r*- and *K*-selection in another guise.

6.9.8 Generalizations about the ecological role of plant chemicals are few. Even some of the 'classic' examples do not stand detailed scrutiny; for example, the close correlation between oak leaf tannins and impaired caterpillar performance was quite logically taken to indicate that tannins interfere with digestion. However, recent experiments have failed to detect any influence of tannins on the rate of digestion of plant material by either generalist or specialist insect herbivores. If tannins do affect insect fitness, it is probably via some other mechanism.

6.10 INTRASPECIFIC VARIABILITY

6.10.1 Ecologists have been as remiss in ignoring genetics as geneticists have been in ignoring population dynamics. Plant populations are typically polymorphic in their quality as food for herbivores. This may be because they vary in the kind or in the concentration of secondary compounds they contain, or because they differ in their concentrations of nutrients, attractants or feeding stimulants. Genetic variability in plants may affect the size and stability of the herbivore population that develops on them. Insects may establish more readily on polymorphic populations but will tend to maintain relatively low average densities because so many of the plants are unsuitable. Sessile insects on long-lived host plants may evolve local races that perform well only on the plants upon which they developed.

6.10.2 Any model that attempts to predict herbivore abundance from plant numbers must take account of the proportions of plants of different phenotypes, and the performance of insect phenotypes on plants of different kinds. There is great scope for transfer experiments in which a comparison is made between the fecundity and survivorship of insects that are moved to other plants or to different parts of the same plant. In this way, it may be possible to distinguish between plants that are avoided because they are unsuitable as food (i.e. unavailable) and those that are simply not exploited (due to the aggregative behaviour or functional responses of the herbivores).

6.11 HERBIVORES AND PLANT SUCCESSION

6.11.1 Vertebrate herbivores have more impact than invertebrates on plant secondary succession and plant community structure. This is readily seen by

Chapter 6

comparing the dramatic effects of fencing to exclude vertebrate herbivores, with the virtually imperceptible effects of excluding insects through the application of broad-spectrum insecticides. Whether the difference in impact is due to the fact that vertebrates tend more often to be food-limited, or to the fact that they are larger, more starvation-tolerant and have higher per-capita damage rates than invertebrates remains to be determined. It is clear, however, that warm blooded herbivores of a given size and standing crop biomass would have a greater impact on the plant community than poikilotherms because their feeding rates are higher.

6.11.2 The hypothesis that plants from different successional stages are differentially palatable to generalist herbivores is untestable. Generalist herbivores are themselves *characteristic* of particular successional communities, so that experiments tend to confirm such unremarkable facts as that early successional herbivores prefer early successional plants!

EPILOGUE

The study of plant–herbivore dynamics is in its infancy. For all the wealth of natural history in the preceding chapters, and despite a vast amount of pure and applied ecological research in botany and in zoology, the studies in which populations of animals and plants have been given equal, detailed attention can be counted on one hand. We do know enough, however, to appreciate that generalizations are going to be few and far between. We can return to the 'food limitation hypothesis' with which we began, confident in the knowledge that it is impossible to divide up the world's herbivorous species into those that are food limited and those that are not. Food will be of importance to *all* herbivores, some of the time. It will be important *all* the time to those species where low levels of disease, scarce natural enemies, plentiful cover and breeding sites and clement weather ensure an otherwise constantly benign environment.

Other herbivore populations may not be regulated at all; numbers increase exponentially whenever conditions permit, but local extinctions are frequent when conditions deteriorate. The population may be re-established after a shorter or longer period by immigration of animals from other areas. On the other hand, populations followed at a small spatial scale may appear to be unregulated, despite the fact that regulation occurs at a higher spatial scale; for example, the immigrants that re-establish local populations after extinction, may be the individuals driven out by competition from high-density patches elsewhere. Further, a failure to detect density dependence in small-scale populations may reflect nothing more than the fact that a sample of low-density patches has been taken from a heterogeneous mosaic, and that the scarce high-density patches (in which density dependence is intense) have simply not been sampled.

It is important to understand that long runs of population density data on their own are worthless in attempting to understand the mechanisms of population dynamics. We should not expect to be able to explain the behaviour of plant–herbivore systems simply by observing them; for example, we know a great deal about what happens to animals during population cycles but we do not know what *causes* the cycles. There is no shortage of plausible hypotheses, and models of disease–host, predator–prey and plant–herbivore interactions all produce cycles which look realistic. Only by carefully designed, critical manipulative experiments can we hope to determine the causes of population change.

This, of course, is a great deal easier said than done. There are formidable problems in making such experiments realistic; for instance, it is hard to make them sufficiently large scale so that one treatment does not influence its neighbours. In enrichment studies, animals emigrating from the poorer to the richer patches will tend to exaggerate any differences there might be in within-patch dynamics. Again, it is extremely difficult to replicate ecological field experiments, such is the variation in weather from year to year, and in microhabitat from place to place.

The role of herbivory in the regulation of plant populations is even less well understood. We can be reasonably certain that plant carrying capacity (K, the maximum biomass per unit area) is determined by soil nutrient and water availability and by the abundance and competitive ability of other plant species. It is affected by herbivores only to the extent that they influence these other factors. Plant numbers may be depressed below K by herbivore feeding. The degree of depression depends upon the rate of increase of the herbivore, its growth efficiency, the importance of *its* natural enemies, the pattern in which it attacks the plant population (whether random or aggregated), and the kind and degree of density dependence in the herbivore's demographic parameters. Some plant populations have been reduced to a tiny fraction of their carrying capacity by insect herbivores during biological weed control programmes. It is not clear, however, whether the low-density equilibrium plant population is maintained by herbivore feeding. It may be, for example, that improved grazing management following the initial reduction in weed populations leads to greater vigour in the pasture species so that low levels of weed abundance are regulated by plant competition rather than by insect feeding.

Plant abundance need not necessarily be depressed below K by the attentions of herbivores however, if (a) animal numbers are not food limited (their populations are determined by natural enemies or by habitat requisites), and (b) the plant can compensate for herbivore feeding. Again, plants do not submit passively to herbivore attack, but respond to feeding by changing their shape, their chemistry, their photosynthetic rate and their physical defences. The rate of herbivore feeding in one generation can, therefore, affect the food quality and the microclimate experienced by subsequent generations.

Herbivore feeding may allow an increase in the abundance of non-food or less-preferred plant species (competitor release) or lead to increased plant-species richness when the herbivore feeds preferentially on the most competitive plant species (herbivore-mediated coexistence).

Plant populations are buffered against over-exploitation by herbivores by the existence of various kinds of refuges including long-lived seed banks, seed immigration, ungrazable reserves of tissue and by spatial distributions which cause the herbivore to exploit the plant populations in a coarse-grained way.

With native herbivores and native host plants, there are very few convincing data to show that plant dynamics are driven by herbivore feeding. For our best-documented case, ragwort abundance appears to determine cinnabar moth numbers, but plant recruitment is regulated by abiotic factors. Despite the fact that caterpillars regularly strip the plants to bare, flowerless stalks, the regenerative powers of the plant are so great that some seed is set from regrowth shoots on almost all the damaged plants, and some rootstocks survive to produce numerous shoots in the following year. Ragwort recruitment is limited by the availability of establishment microsites which, on dry sandy soils is determined by the amount of early summer rainfall, and in mesic pastures is affected by the way in which grazing or cutting practice provide competition-free gaps in the sward.

In alien communities, cinnabar moth *can* reduce ragwort numbers to low levels provided (a) that the growing season is short and (b) that ground cover by other plant species is high (Harris *et al.* 1978). When the growing season is substantially longer than the feeding period of the moth, plant regrowth can compensate for loss of seed and replenish carbohydrate reserves before conditions deteriorate. When the plant grows on bare, open soils, competition-free sites are present in excess and the probability of a seed producing an established plant is so high that herbivore regulation is virtually impossible. This example emphasizes again the paramount importance of phenology in plant–herbivore dynamics. Dissynchrony between plant development and the herbivores' demand for high-quality food will often be the key factor determining herbivore numbers in the following generation.

Finally, a word must be said about the bizarre notion that herbivory is good for plants. In its extreme form the view manifests itself in statements to the effect that grasses 'encourage high levels of consumption' (Owen 1980). This, of course, is absurd. Certain grasses can tolerate grazing, but only at a cost; they are unable to compete with rank species in the absence of defoliation by herbivores. Thus while the persistence of a grass species may depend on continued defoliation of the community by herbivores, the genetic fitness of individual plants is most unlikely to be enhanced by defoliation. Of course, the fitness of individual grass plants may be exceptionally difficult to assess. The typical effect of defoliation on pasture grasses is to delay flowering and to increase longevity. Thus in a grassland dominated by vigorous vegetative clones, and where establishment of seedlings is unikely due to intense plant competition, fitness measured in terms of seed production may be almost irrelevant. A particular genetic individual may cover several hectares and survive for hundreds of years without leaving a single progeny derived from seed.

While we aim to assess the impact of herbivore feeding on plant fitness, we must realize that the concept of plant fitness is an elusive spectre. The problem is that herbivore feeding changes the rules; a plant genotype that is favoured when it is scarce may be selected against if it becomes abundant. By selective feeding and by switching, herbivores may ensure that fitness is both frequency dependent and density dependent.

Our knowledge of the theory of plant–herbivore dynamics now outstrips empirical understanding to an almost embarrassing extent. Theoretical studies have established the role and importance of density-dependence, spatial and temporal heterogeneity, time-lags and age-structure effects that are almost universally accepted. I do not know of a single study where these processes have been elucidated for both the plant and the herbivore involved. Plant–herbivore dynamics will come of age only when we have amassed a wide variety of carefully designed, long-term, manipulative field experiments.

REFERENCES

Abbott H.G. & Quink T.F. (1970) Ecology of eastern white pine seed caches made by small forest mammals. *Ecology* **51**, 271–8.

Abrahamson W.G. (1975) Reproductive strategies in dewberries. *Ecology* **56**, 721–6.

A'Brook J. (1968) The effect of plant spacing on the numbers of aphids trapped over the groundnut crop. *Annals of Applied Biology* **61**, 289–94.

Adams M.W. (1967) Basis of yield component compensation in crop plants with special reference to the field bean, *Phaseolus vulgaris*. *Crop Science* **7**, 505–10.

Adams S.N. (1975) Sheep and cattle grazing in forests: a review. *Journal of Applied Ecology* **12**, 143–52.

Alberta Th. (1962) Actual and potential production of agricultural crops. *Netherlands Journal of Agricultural Science* **10**, 325–32.

Alcock M.B. (1964) The physiological significance of defoliation on the subsequent regrowth of grass–clover mixtures and cereals. In *Grazing in Terrestrial and Marine Environments* (Ed. D.J. Crisp) pp. 25–41. Blackwell Scientific Publications, Oxford.

Aleksiuk M. (1970) The seasonal food regime of arctic beavers. *Ecology* **51**, 264–70.

Alexandre D.Y. (1978) Le role disseminateur des elephants en Foret de Tai, Cote D'Ivoire. *Terre et Vie* **32**, 47–72.

Allden W.G. & Whittaker I.A. McD. (1970) The determinants of herbage intake by grazing sheep: the interrelationship of factors influencing herbage intake and availability. *Australian Journal of Agricultural Research* **21**, 755–66.

Altieri M.A., Schoonhoven A. van & Doll J. (1977) The ecological role of weeds in insect pest management systems: a review illustrated by bean (*Phaseolus vulgaris*) cropping systems. *PANS* **23**, 195–205.

Ames D.R., & Brink D.R. (1977) Effect of temperature on lamb performance and protein efficiency ratio. *Journal of Animal Science* **44**, 136–40.

Anderson R.C. & Loucks O.L. (1979) White-tail deer (*Odocoileus virginineus*) influence on structure and composition of *Tsuga canadensis* forests. *Journal of Applied Ecology* **16**, 855–61.

Anderson R.M., Gordon D.M., Crawley M.J. & Hassell M.P. (1982) Variability in the abundance of animal and plant species. *Nature* **296**, 245–8.

Anderson R.M. & May R.M. (1978) Regulation and stability of host–parasite population interactions: I Regulatory processes. *Journal of Animal Ecology* **47**, 219–47.

Anderson R.M. & May R.M. (1980) Infectious diseases and population cycles in forest insects. *Science* **210**, 658–61.

Anderson R.M. & May R.M. (1981) The population dynamics of microparasites and their invertebrate hosts. *Philosophical Transactions of the Royal Society London B* **291**, 451–524.

Andres L.A., Davis C.J., Harris P. & Wapshere A.J. (1976) Biological control of weeds. In *Theory and Practice of Biological Control* (Eds C.B. Huffaker & P.S. Messenger) pp. 481–99. Academic Press, New York.

Andres L.A. & Goeden R.D. (1971) The biological control of weeds by introduced natural enemies. In *Biological Control* (Ed. C.B. Huffaker) pp. 143–64. Plenum Press, London.

Andrewartha H.G. & Birch L.C. (1954) *The Distribution and Abundance of Animals.* University of Chicago Press, Chicago.

Andrzejewska L. (1967) Estimation of the effects of feeding of the sucking insect *Cicadella viridis* L. (Homoptera: Auchenorrhyncha) on plants. In *Secondary Productivity in Terrestrial Ecosystems* (Ed. K. Petrusewicz) vol. II, pp. 791–805.

Andrzejewska L. & Gyllenberg G. (1980) Small herbivore subsystem. In *Grasslands, Systems Analysis and Man* (Eds A.I. Breymeyer & G.M. Van Dyne) pp. 201–67. Cambridge University Press, Cambridge.

Andrzejewska L. & Wojcik Z. (1970) The influence of Acridoidea on the primary production of a meadow (field experiment). *Ekologia Polska* **28**, 89–109.

Annecke D.P., Karny M. & Burger W.A. (1969) Improved biological control of the prickly pear, *Opuntia megacantha* Salm-Dyck, in South Africa through the use of an insecticide. *Phytophylactica* **1**, 9–13.

Anon (1981) *Farmers Diary.* Collins. London.

Antonovics J. (1972) Population dynamics of the grass *Anthoxanthum odoratum* on a zinc mine. *Journal of Ecology* **60**, 351–66.

ARC (1965) *The nutritional requirements of farm livestock. No. 2 Ruminants* pp. 1–264. Agricultural Research Council. London.

Archer M. (1973) The species preferences of grazing horses. *Journal of the British Grassland Society* **28**, 123–38.

Armstrong R.A. (1976) Fugitive species: experiments with fungi and some theoretical considerations. *Ecology* **57**, 953–63.

Arnold G.W. (1962) Effects of pasture maturity on the diet of sheep. *Australian Journal of Agricultural Research* **13**, 701–6.

Arnold G.W. (1964) Factors within plant associations affecting the behaviour and performance of grazing animals. In *Grazing in Terrestrial and Marine Environments* (Ed. D.J. Crisp) pp. 133–54. Blackwell Scientific Publications, Oxford.

Arnold G.W., de Boer E.S. & Boundy C.A.P. (1980) The influence of odour and taste on the food preferences and food intake of sheep. *Australian Journal of Agricultural Research* **31**, 571–87.

Arnold G.W. & Dudzinski M.L. (1967) Studies on the diet of the grazing animal III. The effect of pasture species and pasture structure on the herbage intake of sheep. *Australian Journal of Agricultural Research* **18**, 657–66.

Ashton D.H. (1979) Seed harvesting by ants in forests of *Eucalyptus regnans* F Muell. in central Victoria. *Australian Journal of Ecology* **4**, 265–77.

Atkinson W.D. (1979) A field investigation of larval competition in domestic *Drosophila*. *Journal of Animal Ecology* **48**, 91–102.

Atkinson W.D. & Shorrocks B. (1981) Competition on a divided and ephemeral resource: a simulation model. *Journal of Animal Ecology* **50**, 461–71.

Atsatt P.R. & O'Dowd D.J. (1976) Plant defense guilds. *Science* **193**, 24–9.

Auer C. (1961) Ergebnisse zwölfjähriger quantitativer Untersuchungen der

Populationsbewegung des Grauen Larchenwicklers *Zeiraphera griseana* Hubner (= *diniana* Guenee), in Oberengadin (1940/60). *Mitt. Eidg. Anst. förstl. Vers wes* **37**, 173–263.

Auerbach M.J. & Strong D.R. (1981) Nutritional ecology of *Heliconia* herbivores: experiments with plant fertilization and alternative hosts. *Ecological Monographs* **51**, 63–83.

Augspurger C.K. (1981) Reproductive synchrony of a tropical shrub: experimental studies on effects of pollinators and seed predators on *Hybanthus prunifolius* (Violaceae). *Ecology* **62**, 775–88.

Bach C.E. (1980a) Effects of plant diversity and time of colonization on an herbivore–plant interaction. *Oecologia* **44**, 319–26.

Bach C.E. (1980b) Effects of plant density and diversity on the population dynamics of a specialist herbivore, the striped cucumber beetle, *Acalymma vittata* (Fab.). *Ecology* **61**, 1515–30.

Baile C.A. (1975) Control of food intake in ruminants. In *Digestion and Metabolism in the Ruminant* (Eds I.W. McDonald & A.C.I. Warner). University of New England Publishing Unit, Armidale.

Bailey C.G. & Riegert P.W. (1973) Energy dynamics of *Encoptolophus sordidus costalis* (Scudder) (Orthoptera: Acrididae) in a grassland ecosystem. *Canadian Journal of Zoology* **51**, 91–100.

Baker F.S. (1950) *Principles of Silviculture.* McGraw Hill, New York.

Balch R.E., Clark R.C. & Brown N.R. (1958) *Adelges piceae* (Ratz.) in Canada with reference to biological control. *Proceedings of the 10th International Congress of Entomology* **4**, 807–17.

Baltensweiler W. (1964) *Zeiraphera griseana* Hubner (Lepidoptera: Tortricidae) in the European Alps. A contribution to the problem of cycles. *Canadian Entomologist* **96**, 792–800.

Baltensweiler W. (1977) Colour-polymorphism and dynamics of larch budmoth populations (*Zeiraphera diniana* Gn., Lep. Tortricidae). *Mitteilungen der Schweizerischen Entomologischen Gesellschaft* **50**, 15–23.

Baltensweiler W., Benz G., Bovey P. & Delucchi V. (1977) Dynamics of larch budmoth populations. *Annual Review of Ecology and Systematics* **22**, 79–100.

Baltensweiler W. & Fischlin A. (1979) The role of migration for the population dynamics of the larch budmoth, *Zeiraphera diniana* Gn. *Mitteilungen der Schweizerischen Entomologischen Gesellschaft* **52**, 259–71.

Banks C.J. & Macaulay E.D.M. (1967) Effects of *Aphis fabae* and its attendant ants and insect predators on yields of field beans (*Vicia faba* L.). *Annals of Applied Biology* **60**, 445–53.

Bardner R. (1968) Wheat bulb fly *Leptohylemyia coarctata* Fall., and its effect on the growth and yield of wheat. *Annals of Applied Biology* **61**, 1–11.

Bardner R. & Fletcher K.E. (1974) Insect infestations and their effects on the growth and yield of field crops: a review. *Bulletin of Entomological Research* **64**, 141–60.

Barnes D.L. (1979) Cattle ranching in east and southern Africa. In *Management of Semi-arid Ecosystems* (Ed. B.H. Walker) pp. 9–54. Elsevier, Amsterdam.

Barnett N.M. & Naylor A.W. (1966) Amino acid and protein metabolism in bermuda grass during water stress. *Plant Physiology* **41**, 1222–30.

Batzli G.O. (1981) Populations and energetics of small mammals in the tundra ecosystem. In *Tundra Ecosystems: A Comparative Analysis* (Eds L.C. Bliss, O.W. Heal & J.J. Moore) pp. 377–96. Cambridge University Press, Cambridge.

Batzli G.O., Jung H.G. & Guntenspergen G. (1981a) Nutritional ecology of microtine rodents: linear forage-rate curves for brown lemmings. *Oikos* **37**, 112–16.

Batzli G.O. & Pitelka F.A. (1970) Influence of meadow mouse populations on California grassland. *Ecology* **51**, 1027–39.

Batzli G.O. & Pitelka F.A. (1975) Vole cycles: test of another hypothesis. *American Naturalist* **109**, 482–7.

Beacham T.D. (1979) Size and growth characteristics of dispersing voles, *Microtus townsendii*. *Oecologia* **42**, 1–10.

Beatley J.C. (1969) Dependence of desert rodent on winter annuals and precipitation. *Ecology* **50**, 721–4.

Beatley J.C. (1976) Rainfall and fluctuating plant populaions in relation to distributions and numbers of desert rodents in southern Nevada. *Oecologia* **24**, 21–42.

Beck S.D. (1965) Resistance of plants to insects. *Annual Review of Entomology* **10**, 207–32.

Beck S.D. & Schoonhoven L.M. (1979) Insect behaviour and plant resistance. In *Breeding Plants Resistant to Insects* (Eds F.G. Maxwell & P.R. Jennings) pp. 115–35. John Wiley & Sons, New York.

Beddington J.R. (1974) Age structure, sex ratio and population density in the harvesting of natural animal populations. *Journal of Applied Ecology* **11**, 915–24.

Beddington J.R., Free C.A. & Lawton J.H. (1978) Characteristics of successful natural enemies in models of biological control of insect pests. *Nature* **273**, 513–19.

Beddington J.R. & May R.M. (1977) Harvesting natural populations in a randomly fluctuating environment. *Science* **197**, 463–5.

Beddington J.R. & Taylor D.B. (1973) Optimum age specific harvesting of a population. *Biometrics* **29**, 801–9.

Bell R.H.V. (1970) The use of the herb layer by grazing ungulates in the Serengeti. In *Animal Populations in Relation to their Food Resources* (Ed. A. Watson) pp. 111–24. Blackwell Scientific Publications, Oxford.

Belovsky G.E. (1978) Diet optimization in a generalist herbivore: the moose. *Theoretical Population Biology* **14**, 105–34.

Belovsky G.E. (1981) Food plant selection by a generalist herbivore: the moose. *Ecology* **62**, 1020–30.

Belyea R.M. (1952) Death and deterioration of balsam fir weakened by spruce budworm defoliation in Ontario. *Journal of Forestry* **50**, 729–38.

Bennett F.D. (1970) Recent investigations on the biological control of some tropical and subtropical weeds. *Proceedings of the 10th British Weed Control Conference*, pp. 660–8.

Bennett F.D. & Hughes I.W. (1959) Biological control of insect pests in Bermuda. *Bulletin of Entomological Research* **50**, 423–36.

Benson W.W. (1978) Resource partitioning in passion vine butterflies. *Evolution* **32**, 493–518.

Bentley S. & Whittaker J.B. (1979) Effects of grazing by a chrysomelid beetle, *Gastrophysa viridula*, on competition between *Rumex obtusifolius* and *Rumex crispus*. *Journal of Ecology* **67**, 79–90.

Bentley S., Whittaker J.B. & Malloch A.J.C. (1980) Field experiments on the effects of grazing by a chrysomelid beetle (*Gastrophysa viridula*) on seed production and quality in *Rumex obtusifolius* and *Rumex crispus*. *Journal of Ecology* **68**, 671–4.

Benz G. (1974) Negative feed-back by competition for food and space, and by cyclic induced changes in the nutritional base as regulatory principles in the population dynamics of the larch budmoth, *Zeiraphera diniana* (Guenee) (Lep., Tortricidae). *Zeitschrift für Angewandte Entomologie* **76**, 196–228.

Berenbaum M. (1981) Effects of linear furanocoumarins on an adapted specialist insect (*Papilio polyxenes*). *Ecological Entomology* **6**, 345–51.

Berger P.J., Sanders E.H., Gardner P.D. & Negus N.C. (1977) Phenolic plant compounds functioning as reproductive inhibitors in *Microtus montanus*. *Science* **195**, 575–7.

Bergerud A.T. (1971) The population dynamics of Newfoundland caribou. *Wildlife Monographs* **25**, 1–55.

Bernays E.A. (1978) Tannins: an alternative viewpoint. *Entomologia Experimentalis et Applicata* **24**, 44–53.

Bernays E.A. (1981) Plant tannins and insect herbivores: an appraisal. *Ecological Entomology* **6**, 353–60.

Bernays E.A. & Chapman R.F. (1976) Antifeedant properties of seedling grasses. *Symp. Biol. Hungary* **16**, 41–6.

Bernays E.A. & Chapman R.F. (1978) Plant chemistry and acridoid feeding behaviour. In *Biochemical Aspects of Plant and Animal Coevolution* (Ed. J.B. Harborne) pp. 99–141. Academic Press, London.

Berryman A.A. (1976) Theoretical explanation of mountain pine beetle dynamics in lodgepole pine forests. *Environmental Entomology* **5**, 1225–33.

Bickoff E.M. (1968) Oestrogenic constituents of forage plants. *Pastures and Field Crops*, pp. 1–39. Review Series 1. Commonwealth Agricultural Bureau. Hurley, Berkshire.

Bines J.A. (1976) Factors influencing the voluntary food intake in cattle. In *Principles of Cattle Production* (Eds H. Swan & W.H. Broster). Butterworth, London.

Binnie R.C., Chestnut D.M.B. & Murdoch J.C. (1980) The effect of time of initial defoliation and height of defoliation on the productivity of perennial ryegrass swards. *Grass and Forage Science* **35**, 267–73.

Birks H.J.B. (1980) British trees and insects: a test of the time hypothesis over the last 13,000 years. *American Naturalist* **115**, 600–5.

Birney E.C., Grant W.E. & Baird D.D. (1976) Importance of vegetative cover to cycles of *Microtus* populations. *Ecology* **57**, 1043–51.

Black J.N. (1957) Seed size as a factor in the growth of subterranean clover (*T. subterraneum* L.) under spaced and sward conditions. *Australian Journal of Agricultural Research* **8**, 335–51.

Black J.N. (1964) An analysis of the potential production of swards of subterranean clover (*Trifolium subterraneum* L.) at Adelaide, South Australia. *Journal of Applied Ecology* **1**, 3–18.

Blakley N. (1981) Life history significance of size-triggered metamorphosis in milkweed bugs (*Oncopeltus*). *Ecology* **62**, 57–64.

Blakley N.R. & Dingle H. (1978) Competition: butterflies eliminate milkweed bugs from a Caribbean island. *Oecologia* **37**, 133–6.

Blaxter K.L. (1962) *The Energy Metabolism of Ruminants*. Hutchinson, London.

Blaxter K.L., Wainman F.W. & Wilson R.S. (1961) The regulation of food intake by sheep. *Animal Production* **3**, 51–61.

Bobek B. (1977) Summer food as the factor limiting roe deer population size. *Nature* **268**, 47–9.

Boggs C.L. & Gilbert L.E. (1979) Male contribution to egg production in butterflies: evidence for transfer of nutrients at mating. *Science* **206**, 83–4.

Bossema I. (1968) Recovery of acorns in the European jay (*Garrulus g. glandarius* L.). *Verhandelingen van der Koninklijke Vlaamse Academie voor Wetenschappen* **71**, 1–5.

Bottrell D.G. & Adkisson P.L. (1977) Cotton insect pest management. *Annual Review of Entomology* **22**, 451–81.

Boutton T.W., Cameron G.N. & Smith B.N. (1978) Insect herbivory on C_3 and C_4 grasses. *Oecologia* **36**, 21–32.

Brambell F.W.R. (1944) The reproduction of the wild rabbit, *Oryctolagus cuniculus* (L.). *Proceedings of the Zoological Society of London* **114**, 1–45.

Branson F.A. (1953) Two new factors affecting resistance of grasses to grazing. *Journal of Range Management* **6**, 165–71.

Bratton S.P. (1975) The effects of the European wild boar *Sus scrofa* on gray beech forest in the Great Smoky Mountains. *Ecology* **56**, 1356–66.

Brattsten L.B., Wilkinson C.F. & Eisner T. (1977) Herbivore–plant interactions: mixed function oxidases and secondary plant substances. *Science* **196**, 1349–52.

Bray J.R. (1964) Primary consumption in three forest canopies. *Ecology* **45**, 165–7.

Breedlove D.E. & Ehrlich P.R. (1968) Plant–herbivore coevolution: lupines and lycaenids. *Science* **162**, 671–2.

Breedlove D.E. & Ehrlich P.R. (1972) Coevolution: patterns of legume predation by a lycaenid butterfly. *Oecologia* **10**, 99–104.

Broadhead E. (1958) The psocid fauna of larch trees in northern England—an ecological study of mixed species populations exploiting a common resource. *Journal of Animal Ecology* **27**, 217–63.

Brody S. (1945) *Bioenergetics and Growth*. Hafner, New York.

Brougham R.W. (1958) Interception of light by the foliage of pure and mixed stands of pasture plants. *Australian Journal of Agricultural Research* **9**, 39–52.

Brouwer R. (1962) Distribution of dry matter in the plant. *Netherlands Journal of Agricultural Science* **10**, 361–76.

Brower L.P. (1969) Ecological chemistry. *Scientific American* **220**, 22–9.

Brown E.S., Betts. E. & Rainey R.C. (1969) Season changes in distribution of the African armyworm, *Spodoptera exempta* (Wlk.) (Lep. Noctuidae), with special reference to eastern Africa. *Bulletin of Entomological Research* **58**, 661–728.

Brown J.H. & Lieberman G.A. (1973) Resource utilization and coexistence of seed-eating desert rodents in sand dune habitats. *Ecology* **54**, 788–97.

Brown J.H., Lieberman G.A. & Dengler W.F. (1972) Woodrats and cholla: dependence of a small mammal population on the density of cacti. *Ecology* **53**, 310–13.

Brown R.H. & Blaser R.E. (1968) Leaf area index in pasture growth. *Herbage Abstracts* **38**, 1–9.

Brown R.H., Cooper R.B. & Blaser R.E. (1966) Effects of leaf age on efficiency. *Crop Science* **6**, 206–9.

Browne L.B. (1975) Regulatory mechanisms in insect feeding. *Recent Advances in Insect Physiology* **11**, 1–116.

Brues C.T. (1946) *Insect Dietary*. Harvard University Press, Cambridge, Mass.

Bryant J.P. (1980) The regulation of snowshoe hare feeding behaviour during winter by plant antiherbivore chemistry. In *Proceedings of the First International Lagomorph Conference* (Ed. K. Myers) pp. 69–98. Guelph University Press, Guelph.

Bryant J.P. & Kuropat P.J. (1980) Selection of winter forage by subarctic browsing vertebrates: the role of plant chemistry. *Annual Review of Ecology and Systematics* **11**, 261–85.

Buchner P. (1965) *Endosymbiosis of Animals with Plant Microorganisms*. John Wiley & Sons, New York.

Buckner C.H. (1966) The role of vertebrate predators in the biological control of forest insects. *Annual Review of Entomology* **11**, 449–70.

Buechner H.K. & Dawkins H.C. (1961) Vegetation change induced by elephants and fire in Murchison Falls National Park, Uganda. *Ecology* **42**, 752–66.

Buffington L.C. & Herbel C.H. (1965) Vegetational changes on a semidesert grassland range. *Ecological Monographs* **35**, 139–64.

Bukovskii V. (1936) *The population of invertebrates of the Crimean beech forest*. Translated by J.D. Jackson. Oxford University Translations 241 F 159 B, Bureau of Animal Population, Oxford.

Bullen F.T. (1970) A review of the assessment of crop losses caused by locusts and grasshoppers. *Proceedings of the International Study Conference on Current and Future Problems of Acridology*, London, pp. 163–71.

Bulmer M.G. (1977) Periodical insects. *American Naturalist* **111**, 1099–117.

Busch R.H. & Ergun Y. (1973) Yield adjustment for missing units in spring wheat. *Crop Science* **13**, 126–30.

Butterfield J. & Coulson J.C. (1975) Insect food of adult red grouse *Lagopus lagopus scoticus* (Lath.). *Journal of Animal Ecology* **44**, 601–8.

Cameron E. (1935) A study of the natural control of ragwort (*Senecio jacobaea* L.). *Journal of Ecology* **23**, 265–322.

Cameron G.N. (1972) Analysis of insect trophic diversity in two salt marsh communities. *Ecology* **53**, 58–73.

Campbell A.G. (1966) The dynamics of grazed mesophytic pastures. *Proceedings of the 10th International Grassland Congress*, Helsinki, pp. 458–63.

Campbell B.C. & Duffy S.S. (1979) Tomatine and parasitic wasps: potential incompatability of plant antibiosis with biological control. *Science* **205**, 700–2.

Cantlon J.E. (1969) The stability of natural populations and their sensitivity to technology. *Brookhaven Symposium in Biology* **22**, 197–205.

Capinera J.L. (1978) Studies of host plant preference and suitability exhibited by early instar range caterpillar larvae. *Environmental Entomology* **7**, 738–40.

Capinera J.L. & Roltsch W.J. (1980) Response of wheat seedlings to actual and simulated migratory grasshopper defoliation. *Journal of Economic Entomology* **73**, 258–61.

Carne P.B. (1965) Distribution of the eucalypt-defoliating sawfly *Perga affinis affinis* (Hymenoptera). *Australian Journal of Zoology* **13**, 593–612.

Carne P.B. (1966) Ecological characteristics of the eucalypt-defoliating chrysomelid *Paropsis atomaria* Ol. *Australian Journal of Zoology* **14**, 647–72.

Carne P.B. (1969) On the population dynamics of the eucalypt-defoliating sawfly *Perga affinis affinis* Kirby (Hymenoptera). *Australian Journal of Zoology* **17**, 113–41.

Caruso J.L. (1970) Early seedling survival of *Melilotus* in bluegrass sod. *Ecology* **51**, 553–4.

Caswell H., Reed F., Stephenson S.N. & Werner P.A. (1973) Photosynthetic pathways and selective herbivory: a hypothesis. *American Naturalist* **107**, 465–80.

Cates R.G. (1975) The interface between slugs and wild ginger: some evolutionary aspects. *Ecology* **56**, 391–400.

Cates R.G. & Orians G.H. (1975) Successional status and the palatability of plants to generalist herbivores. *Ecology* **56**, 410–18.

Caughley G. (1970) Eruption of ungulate populations, with emphasis on Himalayan thar in New Zealand. *Ecology* **51**, 53–72.

Caughley G. (1977) *Analysis of Vertebrate Populations.* John Wiley & Sons, New York.

Caughley G. & Lawton J.H. (1981) Plant–herbivore systems. In *Theoretical Ecology*, 2e (Ed. R.M. May) pp. 132–66. Blackwell Scientific Publications, Oxford.

Chadwick M.J. (1960) *Nardus stricta* L. *Journal of Ecology* **48**, 255–67.

Chapin F.S. (1977) Nutrient/carbon costs associated with tundra adaptations to a cold nutrient-poor environment. *Proceedings of the Circumpolar Conference on Northern Ecology*, pp. 1183–94. National Research Council of Canada, Ottawa.

Chapin F.S. & Slack M. (1979) Effect of defoliation upon root growth, phosphate absorbtion and respiration in nutrient-limited tundra graminoids. *Oecologia* **42**, 67–79.

Chapman R.F. (1974) The chemical inhibition of feeding by phytophagous insects: a review. *Bulletin of Entomological Research* **64**, 339–63.

Charman K. (1979) Feeding ecology and energetics of the dark-bellied brent goose (*Branta bernicla bernicla*) in Essex and Kent. In *Ecological Processes in Coastal Environments* (Eds R.L. Jefferies & A.J. Davy) pp. 451–65. Blackwell Scientific Publications, Oxford.

Charnov E.L. & Schaffer W.M. (1973) Life history consequences of natural selection: Cole's result revisited. *American Naturalist* **107**, 791–3.

Chater E.H. (1931) A contribution to the study of the natural control of gorse. *Bulletin of Entomological Research* **22**, 225–35.

Chatterjee S. & Chatterjee S.K. (1977) Influence of reproduction on the abscission process of cotton (*Gossypium barbadense* L.) leaves. *Annals of Botany* **41**, 517–25.

Chettleburgh M.R. (1952) Observations on the collection and burial of acorns by jays in Hainault Forest. *British Birds* **45**, 359–64.

Chew F.S. (1977) Coevolution of pierid butterflies and their cruciferous food plants. II The distribution of eggs on potential foodplants. *Evolution* **31**, 568–79.

Chew F.S. (1981) Coexistence and local extinction in two pierid butterflies. *American Naturalist* **118**, 655–72.

Chew F.S. & Rodman J.E. (1979) Plant resources for chemical defense. In *Herbivores: Their Interaction with Secondary Plant Metabolites* (Eds. G.A. Rosenthal & D.H. Janzen) pp. 271–307. Academic Press, London.

Chew R.M. (1974) Consumers as regulators of ecosystems: an alternative view to energetics. *Ohio Journal of Science* **74**, 359–70.

Chew R.M. & Chew A.E. (1970) Energy relationships of the mammals of a desert shrub (*Larrea tridentata*) community. *Ecological Monographs* **40**, 1–21.

Child G., Smith M.B.E. & von Richter W. (1970) Tsetse control hunting as a measure of large mammal population trends in the Okavango Delta, Botswana. *Mammalia* **34**, 34–75.

Chlodny J. (1967) The amount of food consumed and production output of larvae of the Colorado potato beetle (*Leptinotarsa decemlineata* Say). *Ekologia Polska A* **15**, 531–41.

Christian J.J., Lloyd J.A. & Davis D.E. (1965) The role of endocrines in the self-regulation of mammalian populations. *Recent Progress in Hormone Research* **21**, 501–78.

Churchill G.B., John H.H., Duncan D.P. & Hodson A.C. (1964) Long-term effects of defoliation of aspen by the forest tent caterpillar. *Ecology* **45**, 630–3.

Claridge M.F. & Wilson M.R. (1978) Oviposition behaviour as an ecological factor in woodland canopy leafhoppers. *Entomologia Experimentalis et Applicata* **24**, 101–9.

Claridge M.F. & Wilson M.R. (1981) Host plant associations, diversity and species–area relationships of mesophyll-feeding leafhoppers of trees and shrubs in Britain. *Ecological Entomology* **6**, 217–38.

Clark L.R. (1953) The ecology of *Chrysomela gemellata* Rossi and *C. hyperici* Forst., and their effect on St John's wort in the Bright District, Victoria. *Australian Journal of Zoology* **1**, 1–69.

Clark L.R. (1964) The population dynamics of *Cardiaspina albitextura* (Psyllidae). *Australian Journal of Zoology* **12**, 362–80.

Clark W.C. & Holling C.S. (1979) Process models, equilibrium structures, and population dynamics: on the formulation and testing of realistic theory in ecology. In *Population Ecology*. (Eds U. Halbach & J. Jacobs) Fortschritte der Zoologie vol. 25: (2/3), pp. 29–52. Gustav Fischer Verlag, Stuttgart.

Clark-Martin S. (1975) *Ecology and management of south-western semidesert grass-*

shrub ranges; the status of our knowledge. USDA Forest Service Research RM-156, Fort Collins, Colorado.

Clements F.E. (1910) *The life history of lodgepole burn forests.* US Forest Service Bulletin 79, Washington.

Clements R.O. (1978) The benefits and some long-term consequences of controlling invertebrates in a perennial ryegrass sward. *Scientific Proceedings of the Royal Dublin Society, Series A* **6**, 335–41.

Clements R.O. & Henderson I.F. (1979) Insects as a cause of botanical change in swards. *Journal of the British Grassland Society, Occasional Symposium* **10**, 157–60.

Clutton-Brock T.H. (Ed.) (1977) *Primate Ecology: Studies of Feeding and Ranging Behaviour in Lemurs, Monkeys and Apes.* Academic Press, London.

Cock M.J.W. (1978) The assessment of preference. *Journal of Animal Ecology* **47**, 805–16.

Coe M.J., Bourn D. & Swingland I.R. (1979) The biomass, production and carrying capacity of giant tortoises on Aldabra. *Philosophical Transactions of the Royal Society London, Series B* **286**, 163–76.

Coe M.J., Cumming D.H. & Phillipson J. (1976) Biomass and production of large African herbivores in relation to rainfall and primary production. *Oecologia* **22**, 341–54.

Cohen D. (1968) A general model of optimal reproduction in a randomly varying environment. *Journal of Ecology* **56**, 219–28.

Cole F.R. & Batzli G.O. (1979) Nutrition and population dynamics of the prairie vole, *Microtus ochrogaster*, in central Illinois. *Journal of Animal Ecology* **48**, 455–70.

Cole L.C. (1951) Population cycles and random oscillations. *Journal of Wildlife Management* **15**, 233–51.

Cole L.C. (1954) The population consequences of life history phenomena. *Quarterly Review of Biology* **29**, 103–37.

Collins S. (1961) Benefits to understory from canopy defoliation by gypsy moth larvae. *Ecology* **42**, 836–8.

Collins W.J. (1981) The effects of length of growing season, with and without defoliation, on seed yield and seed hardiness in swards of subterranean clover. *Australian Journal of Agricultural Research* **32**, 783–92.

Collins W.J. & Aitken Y. (1970) The effect of leaf removal on flowering time in subterranean clover. *Australian Journal of Agricultural Research* **21**, 893–903.

Colvill K.E. & Marshall C. (1981) The patterns of growth, assimilation of $^{14}CO_2$ and distribution of ^{14}C-assimilate within vegetative plants of *Lolium perenne* at low and high density. *Annals of Applied Biology* **99**, 179–90.

Connell J.H. (1971) On the role of natural enemies in preventing competitive exclusion in some marine animals and in rain forests. In *Dynamics of Population* (Eds P.J. den Boer & G.R. Gradwell) pp. 298–312. Pudoc, Wageningen.

Connell J.H. (1979) Tropical rain forests and coral reefs as open non-equilibrium systems. In *Population Dynamics* (Eds R.M. Anderson, B.D. Turner & L.R. Taylor) pp. 141–63. Blackwell Scientific Publications, Oxford.

Connell J.H. & Slatyer R.O. (1977) Mechanisms of succession in natural communities and their role in community stability and organization. *American Naturalist* **111**, 1119–44.

Connor E.F. & Simberloff D. (1979) The assembly of species communities: chance or competition? *Ecology* **60**, 1132–40.

Cook C.W., Stoddart L.A. & Kinsinger F. (1968) Responses of crested wheatgrass to various clipping treatments. *Ecological Monographs* **28**, 237–72.

Cooper-Driver G.A. & Swain T. (1976) Cyanogenic polymorphism in bracken in relation to herbivore predation. *Nature* **260**, 604.

Corbett D.C.M. (1972) The effect of *Pratylenchus fallax* on wheat, barley and sugar beet roots. *Nematologica* **18**, 303–6.

Cotton D. (1968) An investigation of the structure, morphology and ecology of some Pennine soil microtopographic features. Unpublished PhD Thesis, University of Lancaster.

Coulson R.N. (1979) Population dynamics of bark beetles. *Annual Review of Entomology* **24**, 417–47.

Coupland R.T. (Ed.) (1979) *Grassland Ecosystems of the World: Analysis of Grasslands and their Uses*. Cambridge University Press, Cambridge.

Coyne P.I. & Cook C.W. (1970) Seasonal carbohydrate reserve cycles in eight desert range species. *Journal of Range Management* **23**, 438–44.

Crawford-Sidebotham T.J. (1972) The role of slugs and snails in the maintenance of the cyanogenesis polymorphisms of *Lotus corniculatus* L. and *Trifolium repens* L. *Heredity* **28**, 405–11.

Crawley M.J. (1975) The numerical responses of insect predators to changes in prey density. *Journal of Animal Ecology* **44**, 877–92.

Cromartie W.J. (1975) The effect of stand size and vegetational background on the colonization of cruciferous plants by herbivorous insects. *Journal of Applied Ecology* **12**, 517–33.

Crosby T.K. & Pottinger R.P. (Eds) (1979) *Proceedings of the 2nd Australasian Conference on Grassland Invertebrate Ecology*. Government Printer, Wellington.

Crossett R.N., Campbell D.J. & Stewart H.E. (1975) Compensatory growth in cereal root systems. *Plant and Soil* **42**, 673–83.

Da Costa C.P. & Jones C.M. (1971) Cucumber beetle resistance and mite susceptibility controlled by the bitter gene in *Cucumis sativus* L. *Science* **172**, 1145–6.

Dadd R.H. & Krieger D.L. (1968) Dietary amino acid requirements of the aphid, *Myzus persicae*. *Journal of Insect Physiology* **14**, 741–64.

Darlington A. (1974) The galls on oak. In *The British Oak* (Eds M.G. Morris & F.H. Perring) pp. 298–311. Classey, Faringdon.

Darwin C. (1859) *The Origin of Species*. John Murray, London.

Davidson J. (1922) Biological studies of *Aphis rumicis* Linn. Reproduction on varieties of *Vicia faba*. *Annals of Applied Biology* **9**, 135–42.

Davidson J.L. & Donald C.M. (1958) The growth of swards of subterranean clover with particular reference to leaf area. *Australian Journal of Agricultural Research* **9**, 53–72.

Davidson R.L. (1979) Effects of root feeding on foliage yield. In *Proceedings of the*

2nd Australasian Conference on Grassland Invertebrate Ecology (Eds T.K. Crosby & R.P. Pottinger) pp. 117–20. Government Printer, Wellington.

Davis B.N.K. (1975) The colonization of isolated patches of nettles (*Urtica dioica*) by insects. *Journal of Applied Ecology* **12**, 1–14.

Davis L.D. (1957) Flowering and alternate bearing. *Proceedings of the American Society of Horticultural Science* **70**, 545–56.

Davy A.J. (1980) *Deschampsia caespitosa. Journal of Ecology* **68**, 1075–96.

DeAngelis D.L., Travis C.C. & Post W.M. (1979) Persistence and stability of seed-dispersed species in a patchy environment. *Theoretical Population Biology* **16**, 107–25.

De Bach P. (1974) *Biological Control by Natural Enemies.* Cambridge University Press, Cambridge.

Degabriele R. (1980) The physiology of the koala. *Scientific American* **243**, 94–9.

De Jong G. (1979) The influence of the distribution of juveniles over patches of food on the dynamics of a population. *Netherlands Journal of Zoology* **29**, 33–51.

Dement W.A. & Mooney H.A. (1974) Seasonal variation in the production of tannins and cyanogenic glucosides in the chaparral shrub, *Heteromeles arbutifolia. Oecologia* **15**, 65–76.

Dempster J.P. (1968) Intraspecific competition and dispersal: as exemplified by a psyllid and its anthocorid predator. In *Insect Abundance* (Ed. T.R.E. Southwood) pp. 8–17. Blackwell Scientific Publications, Oxford.

Dempster J.P. (1969) Some effects of weed control on the numbers of the small cabbage white (*Pieris rapae* L.) on brussels sprouts. *Journal of Applied Ecology* **6**, 339–45.

Dempster J.P. (1971) The population ecology of the cinnabar moth, *Tyria jacobaeae* L. (Lepidoptera, Arctiidae). *Oecologia* **7**, 26–67.

Dempster J.P. (1975) *Animal Population Ecology.* Academic Press, London.

Dempster J.P. & Hall M.L. (1980) An attempt at re-establishing the swallowtail butterfly at Wicken Fen. *Ecological Entomology* **5**, 327–34.

Dempster J.P. & Lakhani K.H. (1979) A population model for cinnabar moth and its food plant, ragwort. *Journal of Animal Ecology* **48**, 143–64.

Denno R.F. (1980) Ecotope differentiation in a guild of sap-feeding insects on the salt marsh grass, *Spartina patens. Ecology* **61**, 702–14.

De Steven D. (1981) Abundance and survival of a seed-infesting weevil, *Pseudanthonomus hamamelidis* (Coleoptera: Curculionidae), on its variable-fruiting host plant, witch-hazel (*Hamamelis virginiana*). *Ecological Entomology* **6**, 387–96.

Dethier V.G. (1959a) Egg laying habits of Lepidoptera in relation to available food. *Canadian Entomologist* **91**, 554–61.

Dethier V.G. (1959b) Food-plant distribution and density and larval dispersal as factors affecting insect populations. *Canadian Entomologist* **91**, 581–96.

Dethier V.G. & MacArthur R.H. (1964) A field's capacity to support a butterfly population. *Nature* **201**, 728–9.

Detling J.K. & Dyer M.I. (1981) Evidence for potential plant growth regulators in grasshoppers. *Ecology* **62**, 485–8.

Detling J.K., Winn D.T., Proctor-Gregg C. & Painter E.L. (1980) Effects of simulated grazing by below ground herbivores on growth, CO_2 exchange and carbon allocation patterns of *Bouteloua gracilis*. *Journal of Applied Ecology* **17**, 771–8.

De Wilde J. & Schoonhoven L.M. (Eds) (1969) Insect and host plant. *Entomologia Experimentalis et Applicata* **12**, 471–810.

De Wit C.T. (1960) On Competition. *Verh. Landbouwk. Onderz.* **66**, 1–82.

De Wit C.T., Brouwer R. & Penning de Vries F.W.T. (1970) The simulation of photosynthetic systems. In *Prediction and Measurement of Photosynthetic Productivity*, pp. 47–69. Proceedings of IBP/PP Technical Meeting, Trebon, 14–21 September 1969. Pudoc, Wageningen.

Diamond J.M. & May R.M. (1981) Island biogeography and the design of natural reserves. In *Theoretical Ecology*, 2e (Ed. R.M. May) pp. 228–52. Blackwell Scientific Publications, Oxford.

Dimock E.J., Silen R.R. & Allen V.E. (1976) Genetic resistance in Douglas-fir to damage by snowshoe hare and black-tailed deer. *Forest Science* **22**, 106–21.

Dingle H. (1972) Migration strategies of insects. *Science* **175**, 1327–34.

Dinus R.J. & Yates H.O. (1975) Protection of seed orchards. In *Seed Orchards*. (Ed. R. Faulkner) Forestry Commission Bulletin vol. 54, pp. 58–71. HMSO, London.

Dinesman L.G. (1967) Influence of vertebrates on primary production of terrestrial communities. In *Secondary Productivity in Terrestrial Ecosystems* (Ed. K. Petrusewicz) vol. I, pp. 261–6.

Dirzo R. & Harper J.L. (1980) Experimental studies on slug–plant interactions. II. The effect of grazing by slugs on high density monocultures of *Capsella bursa-pastoris* and *Poa annua*. *Journal of Ecology* **68**, 999–1011.

Dittmer H.J. (1973) Clipping effects on Bermuda grass biomass. *Ecology* **54**, 217–19.

Dixon A.F.G. (1971a) The role of aphids in wood formation. I. The effect of the sycamore aphid, *Drepanosiphum platanoidis* (Schr.) (Aphididae), on the growth of sycamore, *Acer pseudoplatanus* (L.). *Journal of Applied Ecology* **8**, 165–79.

Dixon A.F.G. (1971b) The role of aphids in wood formation. II. The effect of the lime aphid, *Eucallipterus tiliae* L. (Aphididae), on the growth of lime *Tilia* x *vulgaris* Hayne. *Journal of Applied Ecology* **8**, 393–9.

Dixon A.F.G. (1975) Effect of population density and food quality on autumnal reproductivity activity in the sycamore aphid, *Drepanosiphum platanoidis* (Schr.). *Journal of Animal Ecology* **44**, 297–304.

Dixon A.F.G. (1976) Timing of egg hatch and viability of the sycamore aphid, *Drepanosiphum platanoidis* (Schr.), at bud burst of sycamore, *Acer pseudo-platanus* L. *Journal of Animal Ecology* **45**, 593–603.

Dodd A.P. (1940) *The Biological Campaign Against Prickly Pear*. Commonwealth Prickly Pear Board, pp. 1–177. Government Printer, Brisbane.

Dodd A.P. (1961) Biological control of *Eupatorium adenophorum* in Queensland. *Australian Journal of Science* **23**, 356–65.

Dolinger P.M., Ehrlich P.R., Fitch W.L. & Breedlove D.E. (1973) Alkaloid and predation patterns in Colorado lupine populations. *Oecologia* **13**, 191–204.

Dowden P.B. (1961) The gypsy moth egg parasite *Ocencyrtus kuwanai*, in southern Connecticut in 1960. *Journal of Economic Entomology* **54**, 876–8.

Dowden P.B., Jaynes H.A. & Carolin V.M. (1953) The role of birds in a spruce budworm outbreak in Maine. *Journal of Economic Entomology* **46**, 307–12.

Dowding P., Chapin F.S., Wielgolaski F.E. & Kilfeather P. (1981) Nutrients in tundra ecosystems. In *Tundra Ecosystems: A Comparative Analysis* (Eds L.C. Bliss, O.W. Heal & J.J. Moore) pp. 647–83. Cambridge University Press, Cambridge.

Duffey E., Morris M.G., Sheail J., Ward L.K., Wells D.A. & Wells T.C.E. (1974) *Grassland Ecology and Wildlife Management.* Chapman & Hall, London.

Duncalf W.G. (1976) *The Guinness Book of Plant Facts and Feats.* Guinness Superlatives Ltd, London.

Duncan D.P. & Hodson A.C. (1958) Influence of the forest tent caterpillar upon the aspen forests of Minnesota. *Forest Science* **4**, 71–93.

Dyer M.I. & Bokhari U.G. (1976) Plant–animal interactions: studies of the effects of grasshopper grazing on blue grama grass. *Ecology* **57**, 762–72.

Dymond J.R. (1947) Fluctuations in animal populations with special reference to those of Canada. *Transactions of the Royal Society of Canada* **41(5)**, 1–34.

Eadie J. (1970) Sheep production and pastoral resources. In *Animal Populations in Relation to their Food Resources* (Ed. A. Watson) pp. 7–24. Blackwell Scientific Publications, Oxford.

Ealey E.H. & Main A.R. (1967) Ecology of the euro, *Macropus robustus* (Gould), in north-western Australia. III. Seasonal changes in nutrition. *CSIRO Wildlife Research* **12**, 53–65.

East R. (1974) Predation on the soil-dwelling stages of the winter moth at Wytham Woods, Berkshire. *Journal of Animal Ecology* **43**, 611–26.

Ebell L.F. (1967) Cone production induced by drought in potted Douglas fir. *Canadian Forest Service Bimonthly Research Notes* **23**, 26–7.

Ebell L.F. (1971) Girdling: its effect on carbohydrate status and on reproductive bud and cone development of Douglas fir. *Canadian Journal of Botany* **49**, 453–66.

Edgar J.A., Culvenor C.C.J. & Pliske T.E. (1974) Coevolution of danaid butterflies with their host plants. *Nature* **250**, 646–8.

Edmunds G.F. & Alstad D.N. (1978) Coevolution in insect herbivores and conifers. *Science* **199**, 941–5.

Ehrlich P.R. (1965) The population biology of the butterfly *Euphydryas editha* II. The structure of the Jasper Ridge colony. *Evolution* **19**, 327–36.

Ehrlich P.R. & Birch L.C. (1967) The 'balance of nature' and 'population control'. *American Naturalist* **101**, 97–107.

Ehrlich P.R., Breedlove D.E., Brussard P.F. & Sharp M.A. (1972) Weather and the "regulation" of subalpine populations. *Ecology* **53**, 243–7.

Ehrlich P.R. & Gilbert L.E. (1973) Population structure and dynamics of the tropical butterfly *Heliconius ethilla*. *Biotropica* **5**, 69–82.

Ehrlich P.R. & Raven P.H. (1965) Butterflies and plants: a study in coevolution. *Evolution* **18**, 586–608.

Eisenberg R.M. (1970) The role of food in the regulation of the pond snail, *Lymnaea elodes*. *Ecology* **51**, 680–4.

Ellison L. (1960) Influence of grazing on plant succession of rangelands. *Botanical Review* **26**, 1–78.

Elton C. (1942) *Voles, Mice and Lemmings: Problems in Population Dynamics.* Clarendon Press, Oxford.

Eltringham S.K. (1974) Changes in the large mammal community of Mweya Peninsula, Rwenzori National Park, Uganda, following removal of hippopotamus. *Journal of Applied Ecology* **11**, 855–65.

Embree D.G. (1967) Effects of winter moth on growth and mortality of red oak in Nova Scotia. *Forest Science* **13**, 295–9.

Erickson J.M. & Feeny P. (1974) Sinigrin: a chemical barrier to the black swallowtail butterfly, *Papilio polyxenes. Ecology* **55**, 103–11.

Ericsson A., Larsson S. & Tenow O. (1980) Effects of early and late season defoliation on growth and carbohydrate dynamics in Scots pine. *Journal of Applied Ecology* **17**, 747–69.

Estes J.A., Smith N.S. & Palmisano J.F. (1978) Sea otter predation and community organization in the Western Aleutian Islands, Alaska. *Ecology* **59**, 822–33.

Evans D.M. (1973) Seasonal variations in the body composition and nutrition of the vole *Microtus agrestis. Journal of Animal Ecology* **42**, 1–18.

Evans G.C. (1972) *The Quantitative Analysis of Plant Growth.* Blackwell Scientific Publications, Oxford.

Evans G.R. & Tisdale E.W. (1972) Ecological characteristics of *Aristida longiseta* and *Agropyron spicatum* in west-central Idaho. *Ecology* **53**, 137–42.

Evans P.S. (1972) Root growth of *Lolium perenne* L. III. Investigation of the mechanism of defoliation-induced suppression of elongation. *New Zealand Journal of Agricultural Research* **15**, 347–55.

Faeth S.H., Connor E.F. & Simberloff D. (1981) Early leaf abscission: a neglected source of mortality for folivores. *American Naturalist* **117**, 409–15.

Farrow E.P. (1925) *Plant Life on East Anglian Heaths.* Cambridge University Press, Cambridge.

Feeny P.P. (1969) Inhibitory effect of oak leaf tannins on the hydrolysis of proteins by trypsin. *Phytochemistry* **8**, 2119–26.

Feeny P. (1970) Seasonal changes in the oak leaf tannins and nutrients as a cause of spring feeding by winter moth caterpillars. *Ecology* **51**, 565–81.

Feeny P. (1976) Plant apparency and chemical defense. *Recent Advances in Phytochemistry* **10**, 1–40.

Fidler J.H. (1936) Some notes on the biology and economics of some British chafers. *Annals of Applied Biology* **23**, 409–27.

Field C.R. (1976) Palatability factors and nutritive values of the food of buffaloes (*Syncerus caffer*) in Uganda. *East African Wildlife Journal* **14**, 181–201.

Field C.R. & Laws R.M. (1970) The distribution of the larger herbivores in the Queen Elizabeth National Park, Uganda. *Journal of Applied Ecology* **7**, 273–94.

Finnegan R.J. (1965) The pine needle miner *Exoteleia pinifoliella* (Chamb.) (Lepidoptera: Gelechiidae), in Quebec. *Canadian Entomologist* **97**, 744–50.

Fischlin A. & Baltensweiler W. (1979) Systems analysis of the larch budmoth system. Part I. The larch–larch budmoth relationship. In *Dispersal of Forest Insects* (Eds V. Delucchi & W. Baltensweiler) pp. 273–89. IUFRO, Zuoz, Switzerland.

Fisher R.A. (1930) *The Genetical Theory of Natural Selection.* Clarendon Press, Oxford.

Flaherty D. (1969) Vineyard trophic complexity and densities of the Willamette mite, *Eotetranychus willamettei* Ewing (Acarina: Tetranychidae) *Ecology* **50**, 911–16.

Flowerdew J.R. (1972) The effect of supplementary food on a population of woodmice (*Apodemus sylvaticus*). *Journal of Animal Ecology* **41**, 553–66.

Ford E.B. (1945) *Butterflies*. Collins, London.

Ford E.D. (1975) Competition and stand structure in some even-aged plant monocultures. *Journal of Ecology* **63**, 311–33.

Fordham R.A. (1971) Field populations of deermice with supplemental food. *Ecology* **52**, 138–46.

Formosov A.N. (1933) The crop of cedar nuts, invasions into Europe of the Siberian nutcracker (*Nucifraga caryocatactes macrorhynchus* Brehm) and fluctuations in numbers of the squirrel (*Sciurus vulgaris* L.). *Journal of Animal Ecology* **2**, 70–81.

Forsyth A.A. (1968) *British Poisonous Plants*. Ministry of Agriculture, Fisheries and Food Bulletin 161, pp. 1–131. HMSO, London.

Foulds W. & Grime J.P. (1972) The response of cyanogenic and acyanogenic phenotypes of *Trifolium repens* to soil moisture supply. *Heredity* **28**, 181–7.

Fowden L., Lewis D. & Tristram H. (1967) Toxic amino acids: their action as antimetabolites. *Advances in Enzymology* **29**, 89–163.

Fowler C.W. (1981) Density dependence as related to life history strategy. *Ecology* **62**, 602–10.

Fox L.R. & Macauley B.J. (1977) Insect grazing on *Eucalyptus* in response to variation in leaf tannins and nitrogen. *Oecologia* **29**, 145–62.

Fraenkel G.S. (1959) The *raison d'être* of secondary plant substances. *Science* **129**, 1466–70.

Frame J. (1966) The evaluation of herbage production under cutting and grazing regimes. *Proceedings of the 10th International Grassland Congress*, Helsinki, pp. 291–7.

Franklin R.T. (1970) Insect influences in the forst canopy. In *Analysis of Temperate Forest Ecosystems* (Ed. D.E. Reichle) pp. 86–99. Springer Verlag, New York.

Free C.A., Beddington J.R. & Lawton J.H. (1977) On the inadequacy of simple models of mutual interference for parasitism and predation. *Journal of Animal Ecology* **46**, 543–54.

Freeland W.J. (1974) Vole cycles: another hypothesis. *American Naturalist* **108**, 238–45.

Freeland W.J. (1980) Mangabey (*Cercocebus albigena*) movement patterns in relation to food availability and fecal contamination. *Ecology* **61**, 1297–303.

Freeland W.J. & Janzen D.H. (1974) Strategies in herbivory by mammals: the role of plant secondary compounds. *American Naturalist* **108**, 269–89.

Fretwell S.D. & Lucas H.L. (1969) On territorial behaviour and other factors influencing habitat distribution in birds I. Theoretical development. *Acta Biotheoretica* **19**, 16–36.

Fuller W.A. (1967) Winter ecology of lemmings and fluctuations of their populations. *Terre et la Vie* **2**, 97–115.

Fuller W.A., Martell A.M., Smith R.F.C. & Speller S.W. (1975) High-arctic lemmings *Dicrostonyx groenlandicus*. II. Demography. *Canadian Journal of Zoology* **53**, 867–78.

Futuyma D.J. (1976) Food plant specialization and environmental predictability in

Lepidoptera. *American Naturalist* **110**, 285–92.

Futuyma D.J. & Gould F. (1979) Associations of plants and insects in a deciduous forest. *Ecological Monographs* **49**, 33–50.

Futuyma D.J. & Wasserman S.S. (1980) Resource concentration and herbivory in oak forests. *Science* **210**, 920–2.

Gardner G. (1977) The reproductive capacity of *Fraxinus excelsior* on the Derbyshire limestone. *Journal of Ecology* **65**, 107–18.

Garrison G.A. (1972) Carbohydrate reserves and response to use. In *Wildland Shrubs—Their Biology and Utilization*, pp. 271–8. USDA Forest Service General Technical Report INT 1. Utah State University, Logan.

Gashwiler J.S. (1970) Further study of conifer seed survival in a western Oregon clearcut. *Ecology* **51**, 849–54.

Gates C.T. (1968) Water deficits and growth of herbaceous plants. In *Water Deficits and Plant Growth* (Ed. T.T. Kozlowski) vol. II, pp. 135–90. Academic Press, London.

Gause G.F. (1934) *The Struggle for Existence*. Waverly Press, Baltimore.

Geier P.W. (1963) The life history of codling moth, *Cydia pomonella* (L.) (Lepidoptera: Tortricidae), in the Australian Capital Territory. *Australian Journal of Zoology* **11**, 323–67.

Geist V. (1971) *Mountain Sheep: A Study in Behaviour and Evolution*. University of Chicago Press, Chicago.

Ghent A.W. (1958) Studies of regeneration in forest stands devastated by the spruce budworm. II. Age, height, growth and related studies of balsam fir seedlings. *Forest Science* **4**, 135–46.

Ghent A.W. (1960) A study of group-feeding behaviour of larvae of the jack pine sawfly, *Neodiprion pratti banksianae* Roh. *Behaviour* **16**, 110–48.

Gibson R.M. & Guinness F.E. (1980) Differential reproduction among red deer (*Cervus elaphus*) stags on Rhum. *Journal of Animal Ecology* **49**, 199–208.

Giertych M. (1970) The influence of defoliation on flowering in pine, *Pinus sylvestris* L. *Arboretum Kornickie* **15**, 93–8.

Gifford R.M. & Marshall C. (1973) Photosynthesis and assimilate distribution in *Lolium multiflorum* Lans. following differential defoliation. *Australian Journal of Biological Science* **26**, 517–26.

Gilbert L.E. (1975) Ecological consequences of a coevolved mutualism between butterflies and plants. In *Coevolution of Animals and Plants* (Eds L.E. Gilbert & P.H. Raven) pp. 210–14. University of Texas Press, Austin.

Gilbert L.E. (1979) Development of theory in the analysis of insect–plant interactions. In *Analysis of Ecological Systems* (Eds D.J. Horn, G.R. Stairs & R.D. Mitchell) pp. 117–54. Ohio State University Press, Columbus.

Glover J. (1963) The elephant problem at Tsavo. *East African Wildlife Journal* **1**, 1–10.

Godfray H.C.J. (1982) Leaf miners and their parasites in relation to succession. Unpublished PhD Thesis, University of London.

Goeden R.D. & Louda S.M. (1976) Biotic interference with insects imported for weed control. *Annual Review of Entomology* **21**, 325–42.

Goldberg M., Tabroff N.R. & Tamarin R.H. (1980) Nutrient variation in beach grass in relation to beach vole feeding. *Ecology* **61**, 1029–33.

Golley F.B. & Buechner H.K. (1968) *A Practical Guide to the Study of the*

Productivity of Large Herbivores. IBP Handbook 7. Blackwell Scientific Publications, Oxford.

Golley F., Odum H.T. & Wilson R.F. (1962) The structure and metabolism of a Puerto Rican red mangrove forest in May. *Ecology* **43**, 9–19.

Goodall D.W. (1967) Computer simulation of changes in vegetation subject to grazing. *Journal of the Indian Botanical Society* **46**, 356–62.

Goodall J. (1963) Feeding behaviour of wild chimpanzees. *Symposia of the Zoological Society of London* **10**, 39–47.

Goodrum P.D., Reid V.H. & Boyd C.E. (1971) Acorn yields, characteristics and management criteria of oaks for wildlife. *Journal of Wildlife Management* **35**, 520–32.

Gosling L.M., Watt A.D. & Baker S.J. (1981) Continuous retrospective census of the East Anglian coypu population between 1970 and 1979. *Journal of Animal Ecology* **50**, 885–901.

Gossard T.W. & Jones R.E. (1977) The effects of age and weather on egg-laying in *Pieris rapae* L. *Journal of Applied Ecology* **14**, 65–71.

Gough H.C. (1946) Studies on the wheat bulb fly *Leptohylemia coarctata* Fall II. Numbers in relation to crop damage. *Bulletin of Entomological Research* **37**, 439–54.

Gradwell G. (1974) The effect of defoliators on tree growth. In *The British Oak* (Eds M.G. Morris & F.H. Perring) pp. 182–93. Classey, Faringdon.

Grant P.R. (1972) Interspecific competition between rodents. *Annual Review of Ecology and Systematics* **3**, 79–106.

Grant P.R. & Abbot I. (1980) Interspecific competition, island biogeography and null hypotheses. *Evolution* **34**, 332–41.

Grant S.A., Barthram G.T. & Torvell L. (1981) Components of regrowth in grazed and cut *Lolium perenne* swards. *Grass and Forage Science* **36**, 155–68.

Gray A.J. & Scott R. (1977) *Puccinellia maritima* (Huds.) Parl. *Journal of Ecology* **65**, 699–716.

Green R. (1978) Factors affecting the diet of farmland skylarks, *Alauda arvensis*. *Journal of Animal Ecology* **47**, 913–28.

Green T.R. & Ryan C.A. (1972) Wound-induced proteinase inhibitor in plant leaves: a possible defense mechanism against insects. *Science* **175**, 776–7.

Green T.W. & Palmblad I.G. (1975) Effects of insect seed predators on *Astragalus cibarius* and *Astragalus utahensis* (Leguminosae). *Ecology* **56**, 1435–40.

Grime J.P. & Jeffrey D.W. (1965) Seedling establishment in vertical gradients of sunlight. *Journal of Ecology* **53**, 612–42.

Grodzinski W., Gorecki A., Janas K. & Migula P. (1966) Effect of rodents on the primary productivity of alpine meadows in Bierzezady Mountains. *Acta Theriologica* **11**, 419–31.

Gross J.E. (1969) Optimum yield in deer and elk populations. 34th North American Wildlife Conference, pp. 372–86.

Gruys P. (1970) Growth in *Bupalus piniarius* (Lepidoptera: Geometridae) in relation to larval population density. *Verhandelingen Rijksinstituut voor Natuurbeheer* **1**, 1–127. Pudoc, Wageningen.

Guinness F.E., Clutton-Brock T.H. & Albon S.D. (1978) Factors affecting mortality in red deer (*Cervus elaphus*). *Journal of Animal Ecology* **47**, 817–32.

Gutierrez A.P., Wang Y. & Regev U. (1979) A optimization model for *Lygus hesperus* (Heteroptera: Miridae) damage in cotton: the economic threshold revisited. *Canadian Entomologist* **111**, 41–54.

Gwynne M.D. & Bell R.H.V. (1968) Selection of vegetation components by grazing ungulates in the Serengeti National Park. *Nature* **220**, 390–3.

Hafez E.S.E. (Ed.) (1962) *The Behaviour of Domestic Animals.* Baillière, Tindall & Cassell, London.

Hairston N.G., Smith F.E. & Slobodkin L.B. (1960) Community structure, population control and competition. *American Naturalist* **94**, 421–5.

Hamilton B.A., Hutchinson K.J., Annis P.C. & Donnelly J.B. (1973) Relationships between the diet selected by grazing sheep and the herbage on offer. *Australian Journal of Agricultural Research* **24**, 271–7.

Hanover J.W. (1966) Genetics of terpenes. I. Gene control of monoterpene levels in *Pinus monticola* Dougl. *Heredity* **21**, 73–84.

Hanson W.C. & Eberhardt L.L. (1971) A Columbia River Canada geese population 1950–1970. *Wildlife Monographs* **28**, 15–56.

Hansson L. (1971) Small rodent food, feeding and population dynamics. *Oikos* **22**, 183–98.

Harberd D.J. (1961) Observations on the population structure and longevity of *Festuca rubra. New Phytologist* **60**, 184–206.

Harberd D.J. (1962) Some observations on natural clones in *Festuca ovina. New Phytologist* **61**, 85–100.

Harborne J.B. (Ed.) (1972) *Phytochemical Ecology.* Academic Press, London.

Hare R.C. (1966) Physiology of resistance to fungal diseases in plants. *Botanical Review* **32**, 95–137.

Harley K.L.S. & Thorsteinson A.J. (1967) The influence of plant chemicals on the feeding behaviour, development and survival of the two-striped-grasshopper, *Melanoplus bivittatus* (Say), Acrididae: Orthoptera. *Canadian Journal of Zoology* **45**, 305–19.

Harper J.L. (1959) The ecological significance of dormancy. *Proceedings of the 4th International Congress of Crop Protection* (Hamburg 1957), pp. 415–20.

Harper J.L. (1961) Approaches to the study of plant competition. In *Mechanisms in Biological Competiton.* (Ed. F.L. Milthorpe) Symposia of the Society for Experimental Biology, vol. **15**, pp. 1–39.

Harper J.L. (1967) A Darwinian approach to plant ecology. *Journal of Ecology* **55**, 247–70.

Harper J.L. (1969) The role of predation in vegetational diversity. *Brookhaven Symposia in Biology* **22**, 48–62.

Harper J.L. (1977) *Population Biology of Plants.* Academic Press, London.

Harper J.L. & Lovell P.H. & Moore K.G. (1970) The shapes and sizes of seeds. *Annual Review of Ecology and Systematics* **1**, 327–56.

Harper J.L. & White J. (1974) The demography of plants. *Annual Review of Ecology and Systematics* **5**, 419–63.

Harris P. (1960) Natural mortality of the pine shoot moth *Rhyacionia buoliana* (Schiff) in England. *Canadian Journal of Zoology* **38**, 755–68.

Harris P. (1974a) A possible explanation of plant yield increases following insect damage. *Agro-Ecosystems* **1**, 219–25.

Harris P. (1974b) The impact of cinnabar moth on ragwort in eastern and western Canada and its implication for biological control strategy. *Commonwealth Institute of Biological Control, Trinidad. Miscellaneous Publication* **105**, 119–23.

Harris P. & Peschken D.P. (1971) *Hypericum perforatum* L., St. John's wort (Hypericaceae). *Commonwealth Institute of Biological Control, Trinidad. Technical Communication* **4**, 89–94.

Harris P., Wilkinson A.T.S., Thompson L.S. & Neary M. (1978) Interaction between the cinnabar moth, *Tyria jacobaeae* L. (Lep.: Arctiidae) and ragwort, *Senecio jacobaea* L. (Compositae) in Canada. *Proceedings of the 4th International Symposium on the Biological Control of Weeds*, pp. 174–80.

Hassell M.P. (1975) Density dependence in single-species populations. *Journal of Animal Ecology* **44**, 282–95.

Hassell M.P. (1978) *The Dynamics of Arthropod Predator–Prey Systems*. Princeton University Press, Princeton.

Hassell M.P. & Comins H.N. (1978) Sigmoid functional responses and population stability. *Theoretical Population Biology* **9**, 202–21.

Hassell M.P., Lawton J.H. & Beddington J.R. (1977) Sigmoid functional responses by invertebrate predators and parasitoids. *Journal of Animal Ecology* **46**, 249–62.

Hassell M.P., Lawton J.H. & May R.M. (1976) Patterns of dynamical behaviour in single-species populations. *Journal of Animal Ecology* **45**, 471–86.

Hassell M.P. & May R.M. (1974) Aggregation of predators and insect parasites and its effect on stability. *Journal of Animal Ecology* **43**, 567–94.

Hassell M.P. & Varley G.C. (1969) New inductive population model for insect parasites and its bearing on biological control. *Nature* **223**, 1133–6.

Haukioja E. (1980) On the role of plant defences in the fluctuation of herbivore populations. *Oikos* **35**, 202–13.

Haukioja E. (1981) Invertebrate herbivory at tundra sites. In *Tundra Ecosystems: A Comparative Analysis* (Eds L.C. Bliss, O.W. Heal & J.J. Moore) pp. 547–55. Cambridge University Press, Cambridge.

Haukioja E. & Niemela P. (1977) Retarded growth of a geometrid larva after mechanical damage of its host tree. *Annales Zoologici Fennici* **14**, 48–52.

Haven S.B. (1973) Competition for food between the intertidal gastropods *Acmaea scabra* and *A. digitalis*. *Ecology* **54**, 143–51.

Hay M.E. (1981) Herbivory, algal distribution, and the maintenance of between-habitat diversity on a tropical fringing reef. *American Naturalist* **118**, 520–40.

Hay M.E. & Fuller P.J. (1981) Seed escape from heteromyid rodents: the importance of microhabitat and seed preference. *Ecology* **62**, 1395–9.

Heady H.F. (1975) *Rangeland Management*. McGraw-Hill, New York.

Heagle A.S. (1973) Interactions between air pollutants and plant parasites. *Annual Review of Phytopathology* **11**, 365–88.

Heichel G.H. & Turner N.C. (1976) Phenology and leaf growth of defoliated hardwood trees. In *Perspectives in Forest Entomology* (Eds J.F. Anderson & H.K. Kaya) pp. 31–40. Academic Press, New York.

Henderson I.F. & Clements R.O. (1979) Differential susceptibility to pest damage in agricultural grasses. *Journal of Agricultural Science, Cambridge* **93**, 465–72.

Hendrix S.D. (1979) Compensatory reproduction in a biennial herb following insect defloration. *Oecologia* **42**, 107–18.

Hercus J.M. (1961) What do sheep eat on tussock grassland? *New Zealand Journal of Agriculture* **103**, 73–6.

Hett J.M. & Loucks O.L. (1976) Age structure models of balsam fir and eastern hemlock. *Journal of Ecology* **64**, 1029–44.

Hewson R. (1977) Food selection by brown hares (*Lepus capensis*) on cereal and turnip crops in north-east Scotland. *Journal of Applied Ecology* **14**, 779–85.

Hill M.O. & Stevens P.A. (1981) The density of viable seed in soils of forest plantations in upland Britain. *Journal of Ecology* **69**, 693–709.

Hinton H.E. (1981) *The Biology of Insect Eggs*. 3 vols. Pergamon Press, Oxford.

Hirose Y., Suzuki Y., Takagi M., Hiehata K., Yamasaki M., Kimoto H., Yamanaka M., Iga M. & Yamaguchi K. (1980) Population dynamics of the citrus swallowtail, *Papilio xuthus* Linne (Lepidoptera: Papilionidae): mechanisms stabilizing its numbers. *Researches in Population Ecology* **21**, 260–85.

Hirst J.M., Hide G.A., Stedman O.J. & Griffith R.L. (1973) Yield compensation in gappy potato crops and methods to measure effects of fungi pathogenic on seed tubers. *Annals of Applied Biology* **73**, 143–50.

Hodgkinson K.C. (1974) Influence of partial defoliation on photosynthesis, photo-respiration and transpiration by lucerne leaves of different ages. *Australian Journal of Plant Physiology* **1**, 561–78.

Hodgson J. (1966) The frequency of defoliation of individual tillers in a set stocked sward. *Journal of the British Grassland Society* **21**, 258–63.

Hodkinson I.D. (1973) The population dynamics and host plant interactions of *Strophingia ericae* (Curt.) (Homoptera: Psylloidea). *Journal of Animal Ecology* **42**, 565–83.

Hoffmann R.S. (1958) The role of reproduction and mortality in population fluctuations of voles (*Microtus*). *Ecological Monographs* **28**, 79–109.

Holling C.S. (1965) The functional response of predators to prey density and its role in mimicry and population regulation. *Memoirs of the Entomological Society of Canada* **45**, 1–60.

Holmes N.D., Smith D.S. & Johnston A. (1979) Effect of grazing by cattle on the abundance of grasshoppers. *Journal of Range Management* **23**, 310–11.

Holmes W. (Ed.) (1980) *Grass: Its Production and Utilization*. Blackwell Scientific Publications, Oxford.

Holt B.R. (1972) Effect of arrival time on recruitment, mortality and reproduction in successional plant populations. *Ecology* **53**, 669–73.

Hoppensteadt F.C. & Keller J.B. (1976) Synchronization of periodical cicada emergences. *Science* **194**, 335–7.

Horn H. (1981) Succession. In *Theoretical Ecology*, 2e (Ed. R.M. May) pp. 253–71. Blackwell Scientific Publications, Oxford.

Horsfield D. (1977) Relationship between feeding of *Philaenus spumarius* (L.) and the amino acid concentration in the xylem sap. *Ecological Entomology* **2**, 259–66.

Horton K. (1964) Deer prefer jack pine. *Journal of Forestry* **62**, 497–9.

Houpt T.R. (1959) Utilization of blood urea in ruminants. *American Journal of Physiology* **197**, 115–20.

House H.L. (1965) Effects of low levels of the nutrient content of a food and of a nutrient imbalance on the feeding and nutrition of a phytophagous larva, *Celerio euphorbiae* (Linnaeus) (Lepidoptera: Sphingidae) *Canadian Entomologist* **97**, 62–8.

House H.L. (1972) Insect nutrition. In *Biology of Nutrition* (Ed. R.N.T-W-Fiennes) pp. 513–73. Pergamon Press, Oxford.

Howe H.F. (1977) Bird activity and seed dispersal of a tropical wet forest tree. *Ecology* **58**, 539–50.

Hubbell S.P. (1980) Seed predation and the coexistence of tree species in tropical forests. *Oikos* **35**, 214–29.

Hubbell S.P. & Werner P.A. (1979) On measuring the intrinsic rate of increase of populations with heterogeneous life histories. *American Naturalist* **113**, 277–93.

Huffaker C.B. (1953) Quantitative studies on the biological control of St John's wort (klamath weed) in California. *Proceedings of the Seventh Pacific Science Congress* **4**, 303–13.

Huffaker C.B. (Ed.) (1980) *New Technology of Pest Control.* John Wiley & Sons, New York.

Huffaker C.B. & Kennett C.E. (1959) A ten year study of vegetational changes associated with biological control of Klamath weed. *Journal of Range Management* **12**, 69–82.

Hughes R.D. (1975) Introduced dung beetles and Australian pasture ecosystems. *Journal of Applied Ecology* **12**, 819–37.

Hughes R.E. & Dale J. (1970) Trends in montane grasslands in Snowdonia, expressed in terms of 'relative entropy'. *Nature* **225**, 756–8.

Humphreys L.R. (1966) Pasture defoliation practice: a review. *Journal of the Australian Institute of Agricultural Science* **32**, 93–105.

Hungate R.E. (1975) The rumen microbial ecosystem. *Annual Review of Ecology and Systematics* **6**, 39–66.

Hunter R.F. (1964) Home range behaviour in hill sheep. In *Grazing in Terrestrial and Marine Environments* (Ed. D.J. Crisp) pp. 155–71 Blackwell Scientific Publications, Oxford.

Hussey N.W. & Parr W.J. (1963) The effect of glasshouse red spider mite (*Tetranychus urticae* Koch) on the yield of cucumbers. *Journal of Horticultural Science* **38**, 255–63.

Hutchinson K.J. & King K.L. (1980) The effects of sheep stocking level on invertebrate abundance, biomass and energy utilization in a temperate, sown grassland. *Journal of Applied Ecology* **17**, 369–87.

Inouye R.S., Byers G.S. & Brown J.H. (1980) Effects of predation and competition on survivorship, fecundity and community structure of desert annuals. *Ecology* **61**, 1344–51.

Ishii R., Tsunoda K. & Machida H. (1972) Studies on the effect of non-uniformity of planting on growth and yield of crops. II. Growth compensation and competition between hills in non-uniform rice populations consisting of the hills of different number of seedlings. *Proceedings of the Crop Science Society of Japan* **41**, 57–62.

Ito Y. (1980) *Comparative Ecology.* Cambridge University Press, Cambridge.

Ivlev V.S. (1961) *Experimental Ecology of the Feeding of Fishes.* Translated from

Russian (1955) by D Scott. Yale University Press, New Haven.

Iyer J.G. & Dosen R.C. (1974) Compensatory trends in forest growth. *Ecology* **55**, 211–12.

Jackson M.V. & Williams T.E. (1979) Response of grass swards to fertilizer N under cutting or grazing. *Journal of Agricultural Science, Cambridge* **92**, 549–62.

Jacobson J. & Crosby D.G. (Eds) (1971) *Naturally Occurring Insecticides.* Marcel Dekker Inc., New York.

Jaksic F.M. & Fuentes E.R. (1980) Why are native herbs in the Chilean matorral more abundant beneath bushes: microclimate or grazing? *Journal of Ecology* **68**, 665–9.

Jameson D.A. (1963) Responses of individual plants to harvesting. *Botanical Review* **29**, 532–94.

Janzen D.H. (1969) Seed-eaters versus seed size, number, toxicity and dispersal. *Evolution* **23**, 1–27.

Janzen D.H. (1970) Herbivores and the number of tree species in tropical forests. *American Naturalist* **104**, 501–28.

Janzen D.H. (1971a) Escape of *Cassia grandis* L. beans from predators in time and space. *Ecology* **52**, 964–79.

Janzen D.H. (1971b) Escape of juvenile *Dioclea megacarpa* (Leguminosae) vines from predators in a deciduous tropical forest. *American Naturalist* **105**, 97–112.

Janzen D.H. (1972) Association of a rain forest palm and seed-eating beetles in Puerto Rico. *Ecology* **53**, 258–61.

Janzen D.H. (1975) Intra- and inter-habitat variations in *Guazuma ulmifolia* (Sterculiaceae) seed predation by *Amblycerus cistelinus* (Bruchidae) in Costa Rica. *Ecology* **56**, 1009–13.

Janzen D.H. (1976) Why bamboos wait so long to flower. *Annual Review of Ecology and Systematics* **7**, 347–91.

Janzen D.H. (1979) New horizons in the biology of plant defenses. In *Herbivores: Their Interaction with Secondary Plant Metabolites* (Eds G.A. Rosenthal & D.H. Janzen) pp. 331–50. Academic Press, New York.

Jarman P.J. & Sinclair A.R.E. (1979) Feeding strategy and the pattern of resource partitioning in ungulates. In *Serengeti: Dynamics of an Ecosystem* (Eds A.R.E. Sinclair & M. Norton-Griffiths) pp. 130–63. University of Chicago Press, Chicago.

Jenkins S.H. (1980) A size-distance relation in food selection by beavers. *Ecology* **61**, 740–6.

Jewell P.A., Milner C. & Boyd J.M. (1974) *Island Survivors: The Ecology of the Soay Sheep of St Kilda.* Athlone Press, London.

Jewiss O.R. & Woledge J. (1967) The effect of age on the rate of apparent photosynthesis in leaves of tall fescue (*Festuca arundinacea* Shreb.). *Annals of Botany* **31**, 661–71.

Johnson C.G. (1969) *Migration and Dispersal of Insects by Flight.* Methuen, London.

Jones D.A. (1966) On the polymorphism of cyanogenesis in *Lotus corniculatus* L.I. Selection by animals. *Canadian Journal of Genetics and Cytology* **8**, 556–67.

Jones D.A. (1972) Cyanogenic glycosides and their function. In *Phytochemical Ecology* (Ed. J.B. Harborne) pp. 103–24. Academic Press, London.

Jones D.A. (1973) Coevolution and cyanogenesis. In *Taxonomy and Ecology* (Ed.

V.H. Heywood) pp. 213–42. Academic Press, London.

Jones E.W. (1959) *Quercus* L. *Journal of Ecology* **47**, 169–222.

Jones F.G.W., Dunning R.A. & Humphries K.P. (1955) The effects of defoliation and loss of stand upon yield of sugar beet. *Annals of Applied Biology* **43**, 63–70.

Jones F.G.W. & Jones M.G. (1974) *Pests of Field Crops*, 2e. Edward Arnold, London.

Jones M.G. (1933a) Grassland management and its influence on the sward. *Journal of the Royal Agricultural Society* **94**, 21–41.

Jones M.G. (1933b) Grassland management and its influence on the sward. *Empire Journal of Experimental Agriculture* **1**, 43–57; 122–8; 223–34; 361–7.

Jones R.E., Gilbert N., Guppy M. & Nealis V. (1980) Long-distance movement in *Pieris rapae*. *Journal of Animal Ecology* **49**, 629–42.

Jordan A.M. (1962) *Coleophora alticolella* Zell. (Lepidoptera) and its food plant *Juncus squarrosus* L. *Journal of Animal Ecology* **31**, 293–304.

Jordan P.A., Botkin D.B. & Wolfe M.L. (1971) Biomass dynamics in a moose population. *Ecology* **52**, 147–52.

Kaczmarek W. (1967) Elements of organization in the energy flow of forest ecosystems. In *Secondary Productivity of Terrestrial Ecosystems* (Ed. K. Petrusewicz) vol. II, pp. 663–78.

Kaczmarski F. (1966) Bioenergetics of pregnancy and lactation in the bank vole. *Acta Theriologica* **11**, 409–17.

Kamm J.A. (1979) Plant bugs: effects of feeding on grass seed development; and cultural control. *Environmental Entomology* **8**, 73–6.

Karban R. (1980) Periodical cicada nymphs impose periodical oak tree wood accumulation. *Nature* **287**, 326–7.

Kays S. & Harper J.L. (1974) The regulation of plant and tiller density in a grass sward. *Journal of Ecology* **62**, 97–106.

Keeler B. (1961) Damage to young plantations by the bank vole at Bernwood forest, 1958–60. *Journal of the Forestry Commission* **30**, 55–9.

Keith L.B. (1963) *Wildlife's Ten-Year Cycle*. University of Wisconsin Press, Madison.

Keith L.B. & Windberg L.A. (1978) A demographic analysis of the snowshoe hare cycle. *Wildlife Monographs* **58**, 1–70.

Kemp G.A. & Keith L.B. (1970) Dynamics and regulation of red squirrel (*Tamiasciurus hudsonicus*) populations. *Ecology* **51**, 763–79.

Kendall W.A. & Sherwood R.T. (1975) Palatability of leaves of tall fescue and reed canarygrass and some of their alkaloids to meadow voles. *Agronomy Journal* **67**, 667–71.

Kennedy J.S. (1951) Benefits to aphids from feeding on galled and virus-infected leaves. *Nature* **168**, 825.

Kennedy J.S. (1958) Physiological condition of the host plant and susceptibility to aphid attack. *Entomologia Experimentalis et Applicata* **1**, 50–65.

Kennedy J.S., Booth C.O. & Kershaw W.J.S. (1959) Host finding by aphids in the field. II. *Aphis fabae* Scop. (Gynoparae) and *Brevicoryne brassicae* L.; with a re-appraisal of the role of host finding behaviour in virus spread. *Annals of Applied Biology* **47**, 424–44.

Khanna S.K., Viswanathan P.N., Tewari C.P., Krishnan P.S. & Sanwal G.G. (1968) Biochemical aspects of parasitism by the angiosperm parasites: phenolics in parasites and hosts. *Physiologia Plantarum* **21**, 949–59.

Kigel J. (1980) Analysis of regrowth patterns and carbohydrate levels in *Lolium multiflorum* Lam. *Annals of Botany* **45**, 91–101.

Kimmins J.P. (1971) Variations in the foliar amino acid composition of flowering and non-flowering balsam fir (*Abies balsamea* (L.) Mill.) and white spruce (*Picea glauca* (Moench) Voss) in relation to outbreaks of the spruce budworm (*Choristoneura fumiferana* (Clem.)). *Canadian Journal of Zoology* **49**, 1005–11.

King R.W., Wardlaw I.F. & Evans L.T. (1967) Effects of assimilate utilization on photosynthetic rate in wheat. *Planta* **77**, 261–76.

Kirby E.J.M. (1968) Plant density and yield in cereals. *NAAS Quarterly Review* **80**, 139–45.

Kirchner T.B. (1977) The effects of resource enrichment on the diversity of plants and arthropods in a shortgrass prairie. *Ecology* **58**, 1334–44.

Kitching R.L. (1981) Egg clustering and the southern hemisphere lycaenids: comments on a paper by N.E. Stamp. *American Naturalist* **118**, 423–5.

Kitting C.L. (1980) Herbivore–plant interactions of individual limpets maintaining a mixed diet of intertidal marine algae. *Ecological Monographs* **50**, 527–50.

Kleiber M. (1961) *The Fire of Life: An Introduction to Animal Energetics.* John Wiley & Sons, New York.

Klein D.R. (1968) The introduction, increase, and crash of reindeer on St Matthew Island. *Journal of Wildlife Management* **32**, 350–67.

Klein D.R. (1970) Food selection by North American deer and their response to over-utilization of preferred plant species. In *Animal Populations in Relation to their Food Resources* (Ed. A. Watson) pp. 25–46. Blackwell Scientific Publications, Oxford.

Kloft W. (1957) Further investigations concerning the interrelationship between bark condition of *Abies alba* and infestation by *Adelges piceae typica* and *A. nusslini schneideri*. *Zeitschrift für Angewandte Entomologie* **41**, 438–42.

Kloft W. & Ehrhardt P. (1959) Untersuchungen über Saugtätigkeit und Schadwirkung der Sitkafichtenlaus, *Liosomaphis abietina* (Walk.). *Phytopathology* **35**, 401–10.

Klomp H. (1966) The dynamics of a field population of the pine looper, *Bupalus piniarius* L. (Lep., Geom.). *Advances in Ecological Research* **3**, 207–305.

Kogan M. (1977) The role of chemical factors in insect/plant relationships. *Proceedings of the 15th International Congress of Entomology*, pp. 211–27. Washington DC.

Kok L.T. & Surles W.W. (1975) Successful biocontrol of musk thistle by an introduced weevil, *Rhinocyllus conicus*. *Environmental Entomology* **4**, 1025–7.

Kozlowski T.T. (1971) *Growth and Development of Trees.* Academic Press, New York.

Krebs C.J. (1966) Demographic changes in fluctuating populations of *Microtus californicus*. *Ecological Monographs* **36**, 239–73.

Krebs C.J. (1978) *Ecology: The Experimental Analysis of Distribution and Abundance*, 2e. Harper & Row, New York.

Krebs C.J. (1979) Dispersal, spacing behaviour and genetics in relation to population fluctuations in the vole *Microtus townsendii*. In *Population Ecology* (Eds U. Halbach & J. Jacobs) *Fortschritte der Zoologie* **25(2/3)**, 61–77. Gustav Fischer Verlag, Stuttgart.

Krebs C.J. & DeLong K.T. (1965) A *Microtus* population with supplemental food. *Journal of Mammalogy* **46**, 566–73.

Krebs C.J., Gaines M.S., Keller B.L., Myers J.H. & Tamarin R.H. (1973) Population cycles in small rodents. *Science* **179**, 35–41.

Krebs C.J., Keller B.L. & Tamarin R.H. (1969) *Microtus* population biology: demographic changes in fluctuating populations of *M. ochrogaster* and *M. pennsylvanicus* in southern Indiana. *Ecology* **50**, 587–607.

Krebs C.J. & Myers J.H. (1974) Population cycles in small mammals. *Advances in Ecological Research* **8**, 267–399.

Krebs J.R., Houston A.I. & Charnov E.L. (1981) Some recent developments in optimal foraging. In *Foraging Behaviour* (Eds A.C. Kamil & T.D. Sargent) pp. 3–18. Garland STPM Press, New York.

Krefting L.W. & Roe E.I. (1949) The role of some birds and mammals in seed germination. *Ecological Monographs* **19**, 269–86.

Krieger R.I., Feeny P.P. & Wilkinson C.F. (1971) Detoxification enzymes in the guts of caterpillars: an evolutionary answer to plant defense? *Science* **172**, 579–81.

Krusberg L.R. (1963) Host responses to nematode infection. *Annual Review of Phytopathology* **1**, 219–40.

Kruuk H. (1972) *The Spotted Hyena*. University of Chicago Press, Chicago.

Kuc J. (1972) Phytoalexins. *Annual Review of Phytopathology* **10**, 207–32.

Kulman H.M. (1971) Effects of insect defoliation on growth and mortality of trees. *Annual Review of Entomology* **16**, 289–324.

Lack D. (1954) *The Natural Regulation of Animal Numbers*. Clarendon Press, Oxford.

Laing J.E. & Hamai J. (1976) Biological control of insect pests and weeds by imported parasites, predators, and pathogens. In *Theory and Practice of Biological Control* (Eds C.B. Huffaker & P.S. Messenger) pp. 685–743. Academic Press, New York.

Langer R.H.M. (1956) Growth and nutrition of timothy (*Phleum pratense*). I. The life history of individual tillers. *Annals of Applied Biology* **44**, 166–87.

Langlands J.P. & Bennett I.L. (1973) Stocking intensity and pastoral production. II. Herbage intake by Merino sheep grazed at different stocking rates. *Journal of Agricultural Science* **81**, 205–9.

Lanigan G.W. & Smith L.W. (1970) Metabolism of pyrrolizidine alkaloids in the ovine rumen. *Australian Journal of Agricultural Research* **21**, 493–500.

Larkin P.A. (1977) An epitaph for the concept of maximum sustained yield. *Transactions of the American Fisheries Society* **196**, 1–11.

Law R. (1981) The dynamics of a colonizing population of *Poa annua*. *Ecology* **62**, 1267–77.

Lawrence W.H. & Rediske J.H. (1962) Fate of sown Douglas-fir seed. *Forest Science* **8**, 211–18.

Laws R.M., Parker I.S.C. & Johnstone R.C.B. (1975) *Elephants and their Habitat.* Clarendon Press, Oxford.

Lawton J.H. (1976) The structure of the arthropod community on bracken. *Botanical Journal of the Linnean Society, London* **73**, 187–216.

Lawton J.H. (1978) Host-plant influences on insect diversity: the effects of space and time. In *Diversity of Insect Faunas* (Eds L.A. Mound & N. Waloff) pp. 105–25. Blackwell Scientific Publications, Oxford.

Lawton J.H. & Hassell M.P. (1981) Asymmetrical competition in insects. *Nature* **289**, 793–5.

Lawton J.H. & McNeill S. (1979) Between the devil and the deep blue sea: on the problem of being a herbivore. In *Population Dynamics* (Eds R.M. Anderson, B.D.Turner & L.R. Taylor) pp. 223–44. Blackwell Scientific Publications, Oxford.

Lawton J.H. & Price P.W. (1979) Species richness of parasites on hosts: agromyzid flies on the British Umbelliferae. *Journal of Animal Ecology* **48**, 619–37.

Lawton J.H. & Schroeder D. (1977) Effects of plant type, size of geographical range and taxonomic isolation on number of insect species associated with British plants. *Nature* **265**, 137–40.

Lawton J.H. & Strong D.R. (1981) Community patterns and competition in folivorous insects. *American Naturalist* **118**, 317–38.

Leader-Williams N. (1980) Population dynamics and mortality of reindeer introduced into South Georgia. *Journal of Wildlife Management* **44**, 640–57.

Le Baron A., Bond K.K., Aitken P. & Michaelson L. (1979) An explanation of the Bolivian highlands grazing–erosion syndrome. *Journal of Range Management* **32**, 201–8.

Le Cren E.D. & Holgate M.W. (Eds) (1962) *The Exploitation of Natural Animal Populations.* Blackwell Scientific Publications, Oxford.

Lee T.D. & Bazzaz F.A. (1980) Effects of defoliation and competition on growth and reproduction in the annual plant *Abutilon theophrasti. Journal of Ecology* **68**, 813–21.

Lemen C. (1981) Elm trees and elm leaf beetles: patterns of herbivory. *Oikos* **36**, 65–7.

Leng R. A. (1970) Formation and production of volatile fatty acids in the rumen. In *Physiology of Digestion and Metabolism in the Ruminant* (Ed. A.T. Phillipson) pp. 406–21. Oriel Press, Newcastle-upon-Tyne.

Leslie P.H. & Gower J.C. (1960) The properties of a stochastic model for the predator–prey type of interaction between two species. *Biometrica* **47**, 219–34.

Leverich W.J. & Levin D.A. (1979) Age-specific survivorship and reproduction in *Phlox drummondii. American Naturalist* **113**, 881–903.

Levin D.A. (1971) Plant phenolics: an ecological perspective. *American Naturalist* **105**, 157–81.

Levin D.A. (1973) The role of trichomes in plant defense. *Quarterly Review of Biology* **48**, 3–15.

Levin D.A. (1976) The chemical defenses of plants to pathogens and herbivores. *Annual Review of Ecology and Systematics* **7**, 121–59.

Levin S.A. (1974) Dispersion and population interactions. *American Naturalist* **108**, 207–28.

Levins R. & Culver D. (1971) Regional co-existence of species and competition between rare species. *Proceedings of the National Academy of Sciences of the United States of America* **68**, 1246–8.

Lewis A.C. (1979) Feeding preference for diseased and wilted sunflower in the grasshopper, *Melanoplus differentialis*. *Entomologia Experiments et Applicata* **26**, 202–7.

Lewis A.R. (1980) Patch use by gray squirrels and optimal foraging. *Ecology* **61**, 1371–9.

Lewis C.T. & Waloff N. (1964) The use of radioactive tracers in the study of dispersion of *Orthotylus virescens* (Douglas and Scott) (Miridae: Heteroptera). *Entomologia Experiments et Applicata* **7**, 15–24.

Lewis T. (1965) The effects of shelter on the distribution of insect pests. *Scientific Horticulture* **17**, 74–84.

Lieberman D., Hall J.B., Swaine M.D. & Lieberman M. (1979) Seed dispersal by baboons in the Shai Hills, Ghana. *Ecology* **60**, 65–75.

Ligon J.D. (1978) Reproductive interdependence of pinon jays and pinon pines. *Ecological Monographs* **48**, 111–26.

Lindquist B. (1938) Dalby Soderskog. *Acta Phytogeographica Sueccia* **10**, 1–21.

Linhart Y.B. (1976) Density dependent seed germination strategies in colonizing versus non-colonizing plant species. *Journal of Ecology* **64**, 375–80.

Livingston R.B. (1972) Influence of birds, stones and soil on the establishment of pasture juniper *Juniperus communis* and red cedar *J. virginiana* in New England pastures. *Ecology* **53**, 1141–7.

Lloyd M. & Dybas H.S. (1966) The periodical cicada problem. *Evolution* **20**, 133–49.

Lock J.M. (1972) The effects of hippopotamus grazing on grasslands. *Journal of Ecology* **60**, 445–67.

Longman K.A. & Coutts M.P. (1974) Physiology of the oak tree. In *The British Oak* (Eds M.G. Morris & F.H. Perring) pp. 194–221. Classey, Faringdon.

Lotka A.J. (1925) *Elements of Physical Biology*. Williams & Wilkins, Baltimore.

Lowe H.J.B. (1967) Interspecific differences in the biology of aphids (Homoptera: Aphididae) on leaves of *Vicia faba*. *Entomologia Experiments et Applicata* **10**, 347–57; 413–20.

Lowe V.P.W. (1969) Population dynamics of the red deer (*Cervus elaphus* L.) on Rhum. *Journal of Animal Ecology* **38**, 425–57.

Lubchenco J. (1978) Plant species diversity in marine intertidal community: importance of herbivore food preference and algal competitive abilities. *American Naturalist* **112**, 23–39.

Ludlow M.M. & Wilson G.L. (1971) Photosynthesis of tropical pasture plants. II. Temperature and illuminance history. III. Leaf age. *Australian Journal of Biological Sciences* **24**, 1065–75; 1077–87.

Ludwig D., Jones D.D. & Holling C.S. (1978) Qualitative analysis of insect outbreak systems: the spruce budworm and forest. *Journal of Animal Ecology* **47**, 315–32.

Luginbill P. & McNeal F.H. (1958) Influence of seeding density and row spacings on the resistance of spring wheats to the wheat stem saw fly. *Journal of Economic Entomology* **51**, 804–8.

Lundgren L. (1975) Natural plant chemicals acting as oviposition deterrents on cabbage butterflies (*Pieris brassicae* (L.), *P. rapae* (L.) and *P. napi* (L.)). *Zoologica Scripta* **4**, 253–8.

MacArthur R.H. (1968) The theory of the niche. In *Population Biology and Evolution* (Ed. R.C. Lewontin) pp. 159–76. Syracuse University Press, New York.

MacArthur R.H. & Wilson E.O. (1967) *The Theory of Island Biogeography*. Princeton University Press, Princeton.

Madden J.L. (1977) Physiological reaction of *Pinus radiata* to attack by the woodwasp, *Sirex noctilio* F. (Hymenoptera: Siricidae). *Bulletin of Entomological Research* **67**, 405–26.

Maddock L. (1979) The 'migration' and grazing succession. In *Serengeti: Dynamics of an Ecosystem* (Eds A.R.E. Sinclair & M. Norton-Griffiths) pp. 104–29. University of Chicago Press, Chicago.

Maiorana V.C. (1978) What kinds of plants do herbivores really prefer? *American Naturalist* **112**, 631–5.

Maggs D.H. (1964) Growth rates in relation to assimilate supply and demand. I. Leaves and roots as limiting regions. *Journal of Experimental Botany* **15**, 574–83.

Majak W., Quinton D.A. & Broersma K. (1980) Cyanogenic glycoside levels in Saskatoon serviceberry. *Journal of Range Management* **33**, 197–9.

Marks P.L. (1974) The role of pin cherry (*Prunus pennsylvanica* L.) in the maintenance of stability in northern hardwood ecosystems. *Ecological Monographs* **44**, 73–88.

Marsh R. & Campling R.C. (1970) Fouling of pastures by dung. *Herbage Abstracts* **40**, 123–30.

Marshall C. & Sagar G.R. (1968) The interdependence of tillers in *Lolium multiflorum* Lam.: a quantitative assessment. *Journal of Experimental Botany* **19**, 785–94.

Mathews C.P. & Westlake D.F. (1969) Estimation of production by populations of higher plants subject to high mortality. *Oikos* **20**, 156–60.

Mattheis P.J., Tieszen L.L. & Lewis M.C. (1976) Response of *Dupontia fisheri* to lemming grazing in an Alaskan arctic tundra. *Annals of Botany* **40**, 179–97.

Matthews J.D. (1963) Factors affecting the production of seed by forest trees. *Forestry Abstracts* **24**, 1–13.

Mattson W.J. (1980) Herbivory in relation to plant nitrogen content. *Annual Review of Ecology and Systematics* **11**, 119–61.

Mattson W.J. & Addy N.D. (1975) Phytophagous insects as regulators of forest primary production. *Science* **190**, 515–22.

Maun M.A. & Cavers P.B. (1971) Seed production in *Rumex crispus*. I. The effects of removal of cauline leaves at anthesis. II. The effects of removal of various proportion of flowers at anthesis. *Canadian Journal of Botany* **49**, 1123–30; 1841–8.

Maxwell F.G., Jenkins J.N. & Parrott W.L. (1972) Resistance of plants to insects. *Advances in Agronomy* **24**, 187–265.

Maxwell F.G. & Jennings P.R. (Eds) (1980) *Breeding Plants Resistant to Insects.* John Wiley & Sons, New York.

May L.H. (1960) The utilization of carbohydrate reserves in pasture species after defoliation. *Herbage Abstracts* **30**, 239–45.

May R.M. (1973) *Stability and Complexity in Model Ecosystems.* Princeton University Press, Princeton.

May R.M. (1976) Simple mathematical models with very complicated dynamics. *Nature* **261**, 459–67.

May R.M. (1977) Predators that switch. *Nature* **269**, 103–4.

May R.M. (1979) Periodical cicadas. *Nature* **277**, 347–9.

May R.M. (1981a) Models for single populations. In *Theoretical Ecology*, 2e (Ed. R.M. May) pp. 5–29. Blackwell Scientific Publications, Oxford.

May R.M. (1981b) Models for two interacting populations. In *Theoretical Ecology*, 2e (Ed. R.M. May) pp. 78–104. Blackwell Scientific Publications, Oxford.

May R.M. & Anderson R.M. (1978) Regulation and stability of host–parasite population interactions II. Destabilizing processes. *Journal of Animal Ecology* **47**, 249–67.

Mayse M.A. (1978) Effects of spacing between rows on soybean arthropod populations. *Journal of Applied Ecology* **15**, 439–50.

McBee R.H. (1971) Significance of intestinal microflora in herbivory. *Annual Review of Ecology and Systematics* **2**, 165–76.

McCambridge W.F. & Knight F.B. (1972) Factors affecting spruce beetles during a small outbreak. *Ecology* **53**, 830–9.

McClure M.S. & Price P.W. (1975) Competition among sympatric *Erythroneura* leaf hoppers (Homoptera: Cicadellidae) on American sycamore. *Ecology* **56**, 1388–97.

McClure M.S. & Price P.W. (1976) Ecotope characteristics of coexisting *Erythroneura* leafhoppers (Homoptera: Cicadellidae) on sycamore. *Ecology* **57**, 928–40.

McDonald P., Edwards R.A. & Greenhalgh J.F.D. (1981) *Animal Nutrition*, 3e. Longman, London.

McGilchrist C.A. & Trenbath B.R. (1971) A revised analysis of competition experiments. *Biometrics* **27**, 659–71.

McKey D. (1979) The distribution of secondary compounds within plants. In *Herbivores: Their Interaction with Secondary Plant Metabolites* (Eds G.A. Rosenthal & D.H. Janzen) pp. 55–133. Academic Press, London.

M'Closkey R.T. (1980) Spatial patterns in sizes of seeds collected by four species of heteromyid rodents. *Ecology* **61**, 486–9.

McNaughton S.J. (1976) Serengeti migratory wildebeest: facilitation of energy flow by grazing. *Science* **191**, 92–4.

McNaughton S.J. (1979a) Grazing as an optimization process: grass–ungulate relationships in the Serengeti. *American Naturalist* **113**, 691–703.

McNaughton S.J. (1979b) Grassland–herbivore dynamics. In *Serengeti: Dynamics of an Ecosystem* (Eds A.R.E. Sinclair & M. Norton-Griffiths) pp. 46–81. University of Chicago Press, Chicago.

McNeill S. (1971) The energetics of a population of *Leptopterna dolobrata*

(Heteroptera: Miridae). *Journal of Animal Ecology* **40**, 127–40.

McNeill S. & Lawton J.H. (1970) Annual production and respiration in animal populations. *Nature* **225**, 472–4.

McNeill S. & Southwood T.R.E. (1978) The role of nitrogen in the development of insect/plant relationships. In *Biochemical Aspects of Plant and Animal Co-evolution* (Ed. J.B. Harborne) pp. 77–98. Academic Press, London.

Mech L.D. (1966) *The Wolves of Isle Royale*. Fauna of National Parks of the United States. Fauna Series vol. 7, pp. 1–210. Government Printing Office, Washington.

Mellanby K. (1968) The effects of some mammals and birds on regeneration of oak. *Journal of Applied Ecology* **5**, 359–66.

Merton L.F.H., Bourn D.M. & Hnatiuk R.J. (1976) Giant tortoise and vegetation interactions on Aldabra Atoll. I. Inland. *Biological Conservation* **9**, 293–304.

Metcalf C.L. & Flint W.P. (1951) *Destructive and Useful Insects: Their Habits and Control*. McGraw Hill, New York.

Migula P. (1969) Bioenergetics of pregnancy and lactation in the European common vole. *Acta Theriologica* **14**, 167–79.

Miles P.W. (1968) Insect secretions in plants. *Annual Review of Phytopathology* **6**, 137–64.

Millar J.S. (1978) Energetics of reproduction in *Peromyscus leucopus*: the cost of lactation. *Ecology* **59**, 1055–61.

Miller D. (1970) *Biological control of weeds in New Zealand 1927–48*. New Zealand Department of Scientific and Industrial Research Information Series vol. 74, pp. 1–104.

Miller M.R. (1975) Gut morphology of mallards in relation to diet quality. *Journal of Wildlife Management* **39**, 168–73.

Miller P.R. (1973) Oxidant-induced community change in a mixed coniferous forest. In *Air Pollution Damage to Vegetation* (Ed. J. Neagle) pp. 101–17. American Chemical Society, Washington DC.

Mills J.A. & Mark A.F. (1977) Food preferences of takahe in Fiordland National Park, New Zealand, and the effect of competition from introduced red deer. *Journal of Animal Ecology* **46**, 939–58.

Milthorpe F.L. & Davidson J.L. (1966) Physiological aspects of regrowth in grasses. In *The Growth of Cereals and Grasses* (Eds. F.L. Milthorpe & J.D. Ivins) pp. 241–55. Butterworths, London.

Milthorpe F.L. & Moorby J. (1974) *An Introduction to Crop Physiology*. Cambridge University Press, Cambridge.

Milton K. (1981) Food choice and digestive strategies of two sympatric primate species. *American Naturalist* **117**, 496–505.

Mispagel M.E. (1978) The ecology and bioenergetics of the acridid grasshopper *Bootettix punctatus* on creosotebush, *Larrea tridentata* in the northern Mojave desert. *Ecology* **59**, 779–88.

Mittler T.E. (1958) Studies on the feeding and nutrition of *Tuberolachnus salignus* (Gmelin) (Homoptera: Aphididae). III. The nitrogen economy. *Journal of Experimental Biology* **35**, 626–38.

Mittler T.E. & Sutherland O.R.W. (1969) Dietary influences on aphid polymorphism. *Entomologia Experimentalis et Applicata* **12**, 703–13.

Miyashita K. (1963) Outbreaks and population fluctuations of insects, with special reference to agricultural insect pests in Japan. *Bulletin of the National Institute of Agricultural Sciences Series C* **15**, 99–170.

Moen A.N. (1978) Seasonal changes in heart rates, activity metabolism, and forage intake of white-tailed deer. *Journal of Wildlife Management* **42**, 715–38.

Mohler C.L., Marks P.L. & Sprugel D.C. (1978) Stand structure and allometry of trees during self thinning of pure stands. *Journal of Ecology* **66**, 559–614.

Monro J. (1967) The exploitation and conservation of resources by populations of insects. *Journal of Animal Ecology* **36**, 531–47.

Montgomery G. (Ed.) (1978) *Arboreal Folivores.* Smithsonian Press, Washington DC.

Mook L.J. (1963) Birds and the spruce budworm. *Memoirs of the Entomological Society of Canada* **31**, 268–71.

Moore B.P. (1965) Pheromones and insect control. *Australian Journal of Science* **28**, 243–5.

Moore L.R. (1978) Seed predation in the legume *Crotalaria. Oecologia* **34**, 185–202; 203–23.

Moran N. & Hamilton W.D. (1980) Low nutritive quality as defense against herbivores. *Journal of Theoretical Biology* **86**, 247–54.

Morris M.G. (1981) Responses of grassland invertebrates to management by cutting. III. Adverse effects on Auchenorrhyncha. *Journal of Applied Ecology* **18**, 107–23.

Morris R.D. & Grant P.R. (1972) Experimental studies of competitive interaction in a two-species system. IV. *Microtus* and *Clethrionomys* species in a single enclosure. *Journal of Animal Ecology* **41**, 275–90.

Morris R.F. (Ed.) (1963) The dynamics of epidemic spruce budworm populations. *Memoirs of the Entomological Society of Canada* **31**, 1–332.

Morris R.E. (1967) Influence of parental food quality on the survival of *Hyphantria cunea. Canadian Entomologist* **99**, 24–33.

Morris R.F., Cheshire W.F., Miller C.A. & Mott D.G. (1958) The numerical response of avian and mammalian predators during a gradation of the spruce budworm. *Ecology* **39**, 487–94.

Morrow P.A. & LaMarche V.C. (1978) Tree ring evidence for chronic insect suppression of productivity in subalpine *Eucalyptus. Science* **201**, 1224–6.

Morse D.H. (1975) Ecological aspects of adaptive radiation in birds. *Biological Reviews* **50**, 167–214.

Morton E.S. (1973) On the evolutionary advantages and disadvantages of fruit eating in tropical birds. *American Naturalist* **107**, 8–22.

Moss R. (1972) Food selection by red grouse (*Lagopus lagopus scoticus* (Lath.)) in relation to chemical composition. *Journal of Animal Ecology* **41**, 411–28.

Moss R., Miller G.R. & Allen S.E. (1972) Selection of heather by captive red grouse in relation to the age of the plant. *Journal of Applied Ecology* **9**, 771–81.

Moss R., Welch D. & Rothery P. (1981) Effects of grazing by mountain hares and red deer on the production and chemical composition of heather. *Journal of Applied Ecology* **18**, 487–96.

Mowat F. (1975) *The Snow Walker.* Heinemann, London.

Mueller-Dombois D. (1972) Crown distortion and elephant distribution in the woody vegetations of Ruhuna National Park, Ceylon. *Ecology* **53**, 208–26.

Mukerji M.K. & Guppy J.C. (1970) A quantitative study of food consumption and growth in *Pseudaletia unipuncta* (Lepidoptera: Noctuidae). *Canadian Entomologist* **102**, 1179–88.

Mulkern G.B. (1967) Food selection by grasshoppers. *Annual Review of Entomology* **12**, 59–78.

Muller H.J. (1958) The behaviour of *Aphis fabae* in selecting its host plants, especially different varieties of *Vicia faba. Entomologia Experimentalis et Applicata* **1**, 66–72.

Murdoch W.W. (1966) Community structure, population control, and competition—a critique. *American Naturalist* **100**, 219–26.

Murdoch W.W. (1969) Switching in general predators: experiments on predator specificity and stability of prey populations. *Ecological Monographs* **39**, 335–54.

Murdoch W.W., Evans F.C. & Peterson C.H. (1972) Diversity and pattern in plants and insects. *Ecology* **53**, 819–29.

Murdoch W.W. & Oaten A. (1975) Predation and population stability. *Advances in Ecological Research* **9**, 1–131.

Murton R.K. (1971) The significance of a specific search image in the feeding behaviour of the wood pigeon. *Behaviour* **40**, 10–42.

Murton R.K., Isaacson A.J. & Westwood N.J. (1966) The relationships between wood pigeons and their clover food supply and the mechanism of population control. *Journal of Applied Ecology* **3**, 55–93.

Murton R.K., Westwood N.J. & Isaacson A.J. (1974) A study of wood-pigeon shooting: the exploitation of a natural animal population. *Journal of Applied Ecology* **11**, 61–81.

Muthukrishnan J. & Delvi M.R. (1974) Effect of ration levels on food utilization in the grasshopper, *Poecilocerus pictus. Oecologia* **16**, 227–36.

Myers J.H. (1981) Interactions between western tent caterpillars and wild rose: a test of some general plant herbivore hypotheses. *Journal of Animal Ecology* **50**, 11–25.

Myers J.H. & Harris P. (1980) Distribution of *Urophora* galls in flower heads of diffuse and spotted knapweed in British Columbia. *Journal of Applied Ecology* **17**, 359–67.

Myers K., Hale C.S., Mykytowycz R. & Hughes R.C. (1971) The effects of varying density and space on sociality and health in animals. In *Behaviour and Environment*, pp. 148–87. Plenum Press, London.

Myers K. & Poole W.E. (1963) A study of the biology of the wild rabbit, *Orytolagus cuniculus* (L.), in confined populations. IV. The effects of rabbit grazing on sown pastures. *Journal of Ecology* **51**, 435–51.

Neales T.F. & Incoll L.D. (1968) The control of leaf photosynthesis rate by the level of assimilate concentration in the leaf: a review of the hypothesis. *Botanical Review* **34**, 107–25.

Neuenschwander P., Michalakis S., Mikros L. & Mathioudis M. (1980) Compensation for early fruit drop caused by *Dacus oleae* (Gmel.) (Diptera: Tephritidae) due to an increase in weight and oil content of the remaining olives. *Zeitschrift für Angewandte Entomologie* **89**, 514–25.

Newsom L.D., Kogan M., Miner F.D., Rabb R.L., Turnipseed S.G. & Whitcomb W.H. (1980) General accomplishments towards better pest control in soybean. In *New Technology of Pest Control* (Ed. C.B. Huffaker) pp. 51–98. John Wiley & Sons, New York.

Newsome A.E. (1970) An experimental attempt to produce a mouse plague. *Journal of Animal Ecology* **39**, 299–311.

Newsome A.E. (1971) Competition between wildlife and domestic livestock. *Australian Veterinary Journal* **47**, 577–86.

Newson R.M. (1966) Reproduction in the feral coypu (*Myocaster coypus*). In *Comparative Biology of Reproduction in Mammals* (Ed. I.W. Rowlands) pp. 323–34. Symposium of the Zoological Society of London, vol. 15.

Newton I. (1970) Irruptions of crossbills in Europe. In *Animal Populations in Relation to their Food Resources* (Ed. A. Watson) pp. 337–57. Blackwell Scientific Publications, Oxford.

Newton I. & Kerbes R.H. (1974) Breeding of greylag geese (*Anser anser*) on the Outer Hebrides, Scotland. *Journal of Animal Ecology* **43**, 771–83.

Ng F.S.P. (1978) Strategies of establishment in Malayan forest trees. In *Tropical Trees as Living Systems* (Eds P.B. Tomlinson & M.H. Zimmermann) pp. 129–62. Cambridge University Press, Cambridge.

Nichols J.O. (1968) Oak mortality in Pennsylvania—a ten year study. *Journal of Forestry* **66**, 681–94.

Nicholson I.A., Paterson I.S. & Currie A. (1970) A study of vegetational dynamics: selection by sheep and cattle in *Nardus* pasture. In *Animal Populations in Relation to their Food Resources* (Ed. A. Watson) pp. 129–43. British Ecological Society Symposium, 10. Blackwell Scientific Publications, Oxford.

Nielsen B.O. (1978) Food resource partition in the beech leaf-feeding guild. *Ecological Entomology* **3**, 193–201.

Nielsen B.O. & Ejlersen A. (1977) The distribution pattern of herbivory in a beech canopy. *Ecological Entomology* **2**, 293–9.

Nisbet R.M. & Gurney W.S.C. (1982) *Modelling Fluctuation Populations*. John Wiley & Sons, New York.

Nixon C.M. & McClain M.W. (1969) Squirrel population decline following a late spring frost. *Journal of Wildlife Management* **33**, 353–7.

Noy-Meir I. (1975) Stability of grazing systems: an application of predator–prey graphs. *Journal of Ecology* **63**, 459–81.

Oates J.F., Waterman P.G. & Choo G.M. (1980) Food selection by the south Indian leaf-monkey, *Presbytis johnii*, in relation to leaf chemistry. *Oecologia* **45**, 45–56.

O'Dowd D.J. & Hay M.E. (1980) Mutualism between harvester ants and a desert ephemeral: seed escape from rodents. *Ecology* **61**, 531–40.

Odum E.P. Connell C.E. & Davenport L.B. (1962) Populations energy flow of three primary consumer components of old field ecosystems. *Ecology* **43**, 88–96.

Ong C.K., Marshall C. & Sagar G.R. (1978) The physiology of tiller death in grasses.

2. Causes of tiller death in a grass sward. *Journal of the British Grassland Society* **33**, 205–11.

Onuf C.P., Teal J.M. & Valiela I. (1977) Interactions of nutrients: plant growth and herbivory in a mangrove ecosystem. *Ecology* **58**, 514–26.

Opler P.A. (1974) Oaks as evolutionary islands for leaf-mining insects. *American Scientist* **62**, 67–73.

Oswalt D.L., Bertrand A.R. & Teal M.R. (1959) Influences of nitrogen fertilization and clipping on grass roots. *Proceedings of the Soil Science Society of America* **23**, 228–30.

Otte D. (1975) Plant preference and plant succession: a consideration of evolution of plant preference in *Schistocerca*. *Oecologia* **18**, 129–44.

Otte D. & Joern A. (1975) Insect territoriality and its evolution: population studies of desert grasshoppers on creosote bushes. *Journal of Animal Ecology* **44**, 29–54.

Overhulser D., Gara R.I. & Johnsey R. (1972) Emergence of *Pissodes strobi* (Coleoptera: Curculionidae) from previously attacked Sitka spruce. *Annals of the Entomological Society of America* **65**, 1423–4.

Owen D.F. (1980) How plants may benefit from the animals that eat them. *Oikos* **35**, 230–5.

Owen M. (1971) The selection of feeding site by white-fronted geese in winter. *Journal of Applied Ecology* **8**, 905–19.

Owen M. (1972) Some factors affecting food intake and selection in white fronted geese. *Journal of Animal Ecology* **41**, 79–92.

Paine R.T. & Levin S.A. (1981) Intertidal landscapes: disturbance and the dynamics of pattern. *Ecological Monographs* **51**, 145–78.

Painter E.L. & Detling J.K. (1981) Effects of defoliation on net photosynthesis and regrowth of western wheatgrass. *Journal of Range Management* **34**, 68–71.

Palmblad I.G. (1968) Competition in experimental populations of weeds with emphasis on the regulation of population size. *Ecology* **49**, 26–34.

Parker M.A. & Root R.B. (1981) Insect herbivores limit habitat distribution of a native composite, *Machaeranthera canescens*. *Ecology* **62**, 1390–2.

Parsons J. & Rothschild M. (1964) Rhodanese in the larva and pupa of the common blue butterfly, (*Polyommatus icarus* (Rott.)) (Lepidoptera). *Entomologist's Gazette* **15**, 58–9.

Parsons K.A. & de la Cruz A.A. (1980) Energy flow and grazing behaviour of conocephaline grasshoppers in a *Juncus roemerianus* marsh. *Ecology* **61**, 1045–50.

Patton D.L.H. & Frame J. (1981) The effect of grazing in winter by wild geese on improved grasslands in West Scotland. *Journal of Applied Ecology* **18**, 311–25.

Pearce R.B., Fissel G. & Carlson G.E. (1969) Carbon uptake and distribution before and after defoliation of alfalfa. *Crop Science* **9**, 756–9.

Pearson L.C. (1965) Primary production in grazed and ungrazed desert communities of eastern Idaho. *Ecology* **46**, 278–85.

Pearson O.P. (1966) The prey of carnivores during one cycle of mouse abundance. *Journal of Animal Ecology* **35**, 217–23.

Pearson R. (1981) *The Wisley Book of Gardening*. Collingridge Books, London.

Pease J.L., Vowles R.H. & Keith L.B. (1979) Interaction of snowshoe hares and woody vegetation. *Journal of Wildlife Management* **43**, 43–60.

Pemadasa M.A. (1976) Interference in populations of three weed species. *Journal of Applied Ecology* **13**, 899–913.

Perkins R.C.L. & Swezey O.H. (1924) The introduction into Hawaii of insects that attack *Lantana*. Experimental Station of the Hawaiian Sugar Planters' Association. *Entomological Service Bulletin* **16**, 1–83.

Peterken G.F. (1966) Mortality of holly (*Ilex aquifolium*) seedlings in relation to natural regeneration in the New Forest. *Journal of Ecology* **54**, 259–69.

Peterken G.F. & Tubbs C.R. (1965) Woodland regeneration in the New Forest, Hampshire, since 1650. *Journal of Applied Ecology* **2**, 159–70.

Peterson R.A. (1962) Factors affecting resistance to heavy grazing in needle-and-thread grass. *Journal of Range Management* **15**, 183–9.

Petrides G.A. & Swank W.G. (1966) Estimating the productivity and energy relations of an African elephant population. *Proceedings of the 9th International Grassland Congress*, pp. 832–42. Sao Paulo, Brazil.

Pettey F.W. (1947) The biological control of prickly pears in South Africa. *Union of South Africa Department of Agricultural Science Bulletin* **271**, 1–161.

Pielou E.C. (1969) *An Introduction to Mathematical Ecology*. John Wiley & Sons, New York.

Pillemer E.A. & Tingey W.M. (1976) Hooked trichomes: a physical plant barrier to a major agricultural pest. *Science* **193**, 482–4.

Pimentel D. (1961) Species diversity and insect population outbreaks. *Annals of the Entomological Society of America* **54**, 76–86.

Pimentel D. (1976) World food crisis: energy and pests. *Bulletin of the Entomological Society of America* **22**, 20–6.

Pitelka F.A. (1957) Some aspects of population structure in the short-term cycle of the brown lemming in northern Alaska. *Cold Spring Harbor Symposium in Quantitative Biology* **22**, 237–51.

Platt W.J. (1975) The colonization and formation of equilibrium plant species associations on badger disturbances in a tall grass prairie. *Ecological Monographs* **45**, 285–305.

Platt W.J. (1976) The natural history of a fugitive prairie plant (*Mirabilis hirsuta* (Pursh) MacM.). *Oecologia* **22**, 399–409.

Platt W.J., Hill G.R. & Clark S. (1974) Seed production in a prairie legume (*Astragalus canadensis* L.). *Oecologia* **17**, 55–63.

Pollard E. (1969) The effect of removal of arthropod predators on an infestation of *Brevicoryne brassicae* (Hemiptera: Aphididae) on brussels sprouts. *Entomologia Experimentalis et Applicata* **12**, 118–24.

Pollard E. (1971) Hedges. VI. Habitat diversity and crop pests: a study of *Brevicoryne brassicae* and its syrphid predators. *Journal of Applied Ecology* **8**, 751–80.

Port G.R. & Thompson J.R. (1980) Outbreaks of insect herbivores on plants along motorways in the United Kingdom. *Journal of Applied Ecology* **17**, 649–56.

Prestidge R.A. & McNeill S. (1983) The role of nitrogen in the ecology of grassland Auchenorryncha. In *Nitrogen as an Ecological Factor*, (Eds J.A. Lee, S. McNeill & I.H. Rorison). 22nd Symposium of the British Ecological Society. Blackwell Scientific Publications, Oxford. (In preparation.)

Price M.R.S. (1978) The nutritional ecology of Coke's hartebeest (*Alcelaphus buselaphus cokei*) in Kenya. *Journal of Applied Ecology* **15**, 33–49.

Price M.V. (1978) The role of microhabitat in structuring desert rodent communities. *Ecology* **59**, 910–21.

Price P.W., Bouton C.E., Gross P., McPheron B.A., Thompson J.N. & Weis A.E. (1980) Interactions among three trophic levels: influence of plants on interactions between insect herbivores and natural enemies. *Annual Review of Ecology and Systematics* **11**, 41–65.

Price P.W. & Waldbauer G.P. (1975) Ecological aspects of insect pest management. In *Introduction to Insect Pest Management* (Eds R.L. Metcalf & W.H. Luckman) pp. 36–73. John Wiley & Sons, New York.

Priestley C.A. (1962) Carbohydrate resources within the perennial plants. *Commonwealth Bureau of Horticulture and Plantation Crops Technical Communication* **27**, 1–16.

Priestley C.A. (1970) Some observations on the effect of cropping on the carbohydrate content in trunks of apple trees over a long period. *Report of the East Malling Research Station* 1969, 121–3.

Pritchard G. (1969) The ecology of a natural population of Queensland fruit fly. *Dacus tryoni*. II. The distribution of eggs and its relation to behaviour. *Australian Journal of Zoology* **17**, 293–311.

Proctor M. & Yeo P. (1973) *The Pollination of Flowers*. Collins, London.

Pulliam H.R. (1975) Diet optimization with nutrient constraints. *American Naturalist* **109**, 765–8.

Pulliam H.R. & Enders F. (1971) The feeding ecology of five sympatric finch species. *Ecology* **52**, 557–66.

Puritch G.S. (1972) Cone production in conifers. *Canadian Forestry Service Information Report* BC-X-65.

Pyke G.H., Pulliam H.R. & Charnov E.L. (1977) Optimal foraging: a selective review of theory and tests. *Quarterly Review of Biology* **52**, 137–54.

Rabinowitz D. (1979) Bimodal distributions of seedling weight in relation to density of *Festuca paradoxa* Desv. *Nature* **277**, 297–8.

Rabinowitz D. & Rapp J.K. (1980) Seed rain in a North American tall grass prairie. *Journal of Applied Ecology* **17**, 793–802.

Radvanyi A. (1970) Small mammals and regeneration of white spruce forests in western Alberta. *Ecology* **51**, 1102–5.

Rafes P.M. (1970) Estimation of the effects of phytophagous insects on forest production. In *Analysis of Temperate Forest Ecosystems* (Ed. D.E. Reichle) pp. 100–6. Springer Verlag, New York.

Randall R.E. (1981) Effects of grazing on the machair of the Monach Isles. In *Sand Dune Machair* (Ed. D.S. Ranwell) vol. 3, pp. 7–11. Institute of Terrestrial Ecology, Cambridge.

Randolph P.A., Randolph J.C., Mattingly K. & Foster M.M. (1977) Energy costs of reproduction in the cotton rat, *Sigmodon hispidus*. *Ecology* **58**, 31–45.

Rathcke B.J. (1976) Competition and coexistence within a guild of herbivorous insects. *Ecology* **57**, 76–87.

Rathcke B.J. & Poole R.W. (1975) Coevolutionary race continues: butterfly larval adaptation to plant trichomes. *Science* **187**, 175–6.

Rausher M.D. (1981a) Host plant selection by *Battus philenor* butterflies: the roles of predation, nutrition, and plant chemistry. *Ecological Monographs* **51**, 1–20.

Rausher M.D. (1981b) The effect of native vegetation on the susceptibility of *Aristolochia reticulata* (Aristolochiaceae) to herbivore attack. *Ecology* **62**, 1187–95.

Rausher M.D. & Feeny P. (1980) Herbivory, plant density, and plant reproductive success: the effect of *Battus philenor* on *Aristolochia reticulata*. *Ecology* **61**, 905–17.

Rawes M. (1981) Further results of excluding sheep from high level grasslands in the north Pennines. *Journal of Ecology* **69**, 651–69.

Raymond W.F. (1969) The nutritive value of forage crops. *Advances in Agronomy* **21**, 1–108.

Reardon P.O., Leinweber C.L. & Merrill L.B. (1972) The effect of bovine saliva on grasses. *Journal of Animal Science* **34**, 897–8.

Reardon P.O., Leinweber C.L. & Merrill L.B. (1974) Response of sideoats gramma to animal saliva and thiamine. *Journal of Range Management* **27**, 400–1.

Redfield J.A., Krebs C.J. & Taitt M.J. (1977) Competition between *Peromyscus maniculatus* and *Microtus townsendii* in grasslands of coastal British Columbia. *Journal of Animal Ecology* **46**, 607–16.

Reeks W.A. & Barter C.W. (1951) Growth reduction and mortality of spruce caused by the European spruce sawfly, *Gilpinia hercyniae* (Htg.) (Hymenoptera: Diprionidae). *Forestry Chronicle* **27**(2), 1–16.

Rehr S.S., Feeny P.P. & Janzen D.H. (1973) Chemical defence in Central American non-ant-acacias. *Journal of Animal Ecology* **42**, 405–16.

Reichle D.E. (1968) Relation of body size to food intake, oxygen consumption, and trace element metabolism in forest floor arthropods. *Ecology* **49**, 538–42.

Reichle D.E., Goldstein R.A., VanHook R.I. & Dodson G.J. (1973) Analysis of insect consumption in a forest canopy. *Ecology* **54**, 1076–84.

Reichman O.J. & Oberstein D. (1977) Selection of seed distribution types by *Dipodomys merriami* and *Perognathus amplus*. *Ecology* **58**, 636–43.

Rhoades D.F. (1979) Evolution of chemical defense against herbivores. In *Herbivores: Their Interaction with Secondary Plant Metabolites* (Eds G.A. Rosenthal & D.H. Janzen) pp. 3–54. Academic Press, New York.

Rhoades D.F. & Cates R.G. (1976) Toward a general theory of plant antiherbivore chemistry. *Recent Advances in Phytochemistry* **10**, 168–213.

Richards O.W. & Waloff N. (1954) Studies on the biology and population dynamics of British grasshoppers. *Anti-Locust Bulletin* **17**, 1–182.

Ricker W.E. (1958) Maximum sustained yields from fluctuating environments and mixed stocks. *Journal of the Fisheries Research Board of Canada* **15**, 991–1006.

Riddiford L.M. & Williams C.M. (1967) Volatile principle from oak leaves: role in the sex life of the polyphemus moth. *Science* **155**, 589–90.

Ridley H.N. (1930) *The Dispersal of Plants Throughout the World*. L. Reeve & Co., Ashford, England.

Ridsdill Smith T.J. (1977) Effects of root-feeding by scarabaeid larvae on growth of perennial ryegrass plants. *Journal of Applied Ecology* **14**, 73–80.

Riney T. (1964) The impact of introduction of large herbivores on the tropical environment. *IUCN Publications NS* **4**, 261–73.

Risch S. (1980) The population dynamics of several herbivorous beetles in a tropical agroecosystem: the effects of intercropping corn, beans and squash in Costa Rica. *Journal of Applied Ecology* **17**, 593–612.

Risser P.G. (Ed.) (1972) A preliminary compartment model of a tallgrass prairie, Osage site, 1970. *US IBP Grassland Biome Technical Report* **159**, 1–21.

Roberts H.A. & Feast P.M. (1973) Emergence and longevity of seeds of annual weeds in cultivated and undisturbed soil. *Journal of Applied Ecology* **10**, 133–43.

Roberts R.J. & Ridsdill-Smith T.J. (1979) Assessing pasture damage and losses in animal production caused by pasture insects. *Proceedings of the 2nd Australasian Conference on Grassland Invertebrate Ecology*, pp. 124–5. Government Printer, Wellington.

Rockwood L.L. (1973) The effect of defoliation on seed production of six Costa Rican tree species. *Ecology* **54**, 1363–9.

Rockwood L.L. (1974) Seasonal changes in the susceptibility of *Crescentia alata* leaves to the flea beetle, *Oedionychus* sp. *Ecology* **55**, 142–8.

Rodriguez J.G. (1960) Nutrition of the host and reaction to pests. *American Association for the Advancement of Science* **61**, 149–67.

Roff D.A. (1974) Spatial heterogeneity and the persistence of populations. *Oecologia* **15**, 245–58.

Rogers D. & Hassell M.P. (1974) General models for insect parasite and predator searching behaviour: interference. *Journal of Animal Ecology* **43**, 239–53.

Rohmeder E. (1967) Beziehungen zwischen Frucht-bzw. Samenerzeugung und Holzerzeugung der Waldbäume. *Allgemeine Forstzeitung* **22**, 33–9.

Room P.M., Harley K.L.S., Forno I.W. & Sands D.P.A. (1981) Successful biological control of the floating weed *Salvinia*. *Nature* **294**, 78–80.

Root R.B. (1973) Organization of plant-arthropod association in simple and diverse habitats: the fauna of collards (*Brassica oleracea*). *Ecological Monographs* **43**, 95–124.

Rosenthal G.A. & Bell E.A. (1979) Naturally occurring, toxic nonprotein amino acids. In *Herbivores: Their Interaction with Secondary Plant Metabolites* (Eds G.A. Rosenthal & D.H. Janzen) pp. 353–85. Academic Press, New York.

Rosenthal G.A., Dahlman D.L. & Janzen D.H. (1976) A novel means for dealing with L-canavanine, a toxic metabolite. *Science* **192**, 256–8.

Rosenthal G.A. & Janzen D.H. (Eds) (1979) *Herbivores: Their Interaction with Secondary Plant Metabolites*. Academic Press, New York.

Rosenzweig M.L. (1968) Net primary productivity of terrestrial communities: prediction from climatological data. *American Naturalist* **102**, 67–74.

Rosenzweig M.L. (1971) Paradox of enrichment: destabilization of exploitation ecosystems in ecological time. *Science* **171**, 385–7.

Rosenzweig M.L. (1973) Habitat selection experiments with a pair of coexisting heteromyid rodent species. *Ecology* **54**, 111–17.

Rosenzweig M.L. & Sterner P.W. (1970) Population ecology of desert rodent communities: body size and seed husking as bases for heteromyid coexistence. *Ecology* **51**, 217–24.

Ross B.A., Bray J.R. & Marshall W.H. (1970) Effects of long-term deer exclusion on a *Pinus resinosa* forest in north-central Minnesota. *Ecology* **51**, 1088–93.

Rothacher J.S., Blow F.E. & Potts S.M. (1954) Estimating the quantity of tree foliage in oak stands in the Tennessee Valley. *Journal of Forestry* **52**, 169–73.

Rottger U. von & Klingauf F. (1976) Physiological changes in sugar beet leaves caused by the beet fly, *Pegomya betae* Curt. (Muscidae: Anthomyidae). *Zeitschrift für Angewandte Entomologie* **82**, 220–7.

Rowan W. (1954) Reflections on the biology of animal cycles. *Journal of Wildlife Management* **18**, 52–60.

Rozin P. (1977) The significance of learning mechanisms in food selection: some biology, psychology and sociology of science. In *Mechanisms in Food Selection* (Eds L.M. Baker, M.R. Best & M. Domjan) pp. 557–89. Baylor Scientific Press, New York.

Russell F.C. (1948) Diet in relation to reproduction and the viability of the young. Part 1 Rats and other laboratory animals. *Commonwealth Bureau of Animal Nutrition Technical Communication* **16**, 1–99.

Rust R.W. & Clement S.L. (1977) Entomophilous pollination of the self-compatible species *Collinsia sparsiflora* Fisher and Meyer. *Journal of the Kansas Entomological Society* **50**, 37–48.

Ryle G.J.A. (1970) Partition of assimilates in an annual and a perennial grass. *Journal of Applied Ecology* **7**, 217–27.

Ryle G.J.A. & Powell C.E. (1975) Defoliation and regrowth in the graminaceous plant: the role of current assimilate. *Annals of Botany* **39**, 297–310.

Sadleir R.M.F.S. (1965) The relationship between agonistic behaviour and population changes in the deermouse, *Peromyscus maniculatus* (Wagner). *Journal of Animal Ecology* **34**, 331–52.

Sadleir R.M.F.S. (1969) *The Ecology of Reproduction in Wild and Domestic Mammals*. Methuen, London.

Salisbury E.J. (1942) *The Reproductive Capacity of Plants: Studies in Quantitative Biology*. G. Bell, London.

Salisbury E.J. (1961) *Weeds and Aliens*. Collins, London.

Sanders E.H., Gardner P.D., Berger P.J. & Negus N.C. (1981) 6-methoxybenzoxazolinone: a plant derivative that stimulates reproduction in *Microtus montanus*. *Science* **214**, 67–9.

Sarukhan J. (1974) Studies on plant demography: *Ranunculus repens* L., *R. bulbosus* L. and *R. acris* L. II. Reproductive strategies and seed population dynamics. *Journal of Ecology* **62**, 151–77.

Sarukhan J. (1978) Studies on the demography of tropical trees. In *Tropical Trees as Living Systems* (Eds P.B. Tomlinson & M.H. Zimmermann) pp. 163–84. Cambridge University Press, Cambridge.

Sarukhan J. & Harper J.L. (1973) Studies on plant demography: *Ranunculus repens* L., *R. bulbosus* L. and *R. acris* L. I. Population flux and survivorship. *Journal of Ecology* **61**, 675–716.

Savage Z. (1966) Citrus yield per tree by age. Florida Agricultural Experimental Station. Agricultural Extension Service Economic Series, vol. 66, pp. 1–9.

Savory C.J. (1978) Food consumption of red grouse in relation to the age and productivity of heather. *Journal of Animal Ecology* **47**, 269–82.

Saxena K.N. & Goyal S. (1978) Host–plant relations of the citrus butterfly *Papilio demoleus* L.: orientational and ovipositional responses. *Entomologia Experimentalis et Applicata* **24**, 1–10.

Schaller G.B. (1967) *The Deer and the Tiger: A Study of Wildlife in India.* University of Chicago Press, Chicago.

Schaller G.B. (1980) *Stones of Silence: Journeys in the Himalaya.* Andre Deutsch, London.

Schemske D.W. (1980) The evolutionary significance of extrafloral nectar production by *Costus woodsonii* (Zingiberaceae): an experimental analysis of ant protection. *Journal of Ecology* **68**, 959–67.

Schmidt-Nielsen K. (1979) *Animal Physiology: Adaptation and Environment*, 2e. Cambridge University Press, Cambridge.

Schoener T.W. (1971) Theory of feeding strategies. *Annual Review of Ecology and Systematics* **2**, 369–404.

Schoonhoven L.M. (1972) Some aspects of host selection and feeding in phytophagous insects. Mededeling No 218 van het Laboratorium voor Entomologie, pp. 557–66. Wageningen.

Schreiber M.M. (1967) A technique for studying weed competition in forage legume establishment. *Weeds* **15**, 1–4.

Schroeder L.A. (1976) Energy, matter and nitrogen utilization by the larvae of the monarch butterfly *Danaus plexippus*. *Oikos* **27**, 259–64.

Schroeder L.A. & Dunlap D.G. (1970) Respiration of cecropia moth (*Hyalophora cecropia* L.) larvae. *Comparative Biochemistry and Physiology* **35**, 953–7.

Schultz A.M. (1969) A study of an ecosystem: the arctic tundra. In *The Ecosystem Concept in Natural Resource Management* (Ed. G.M. van Dyne) pp. 77–93. Academic Press, New York.

Schuster M.F., Boling J.C. & Morony J.J. (1971) Biological control of rhodesgrass scale by airplane release of an introduced parasite of limited dispersing ability. In *Biological Control* (Ed. C.B. Huffaker) pp. 227–50. Plenum Press, New York.

Schwartz C.C. & Ellis J.E. (1981) Feeding ecology and niche separation in some native and domestic ungulates on the shortgrass prairie. *Journal of Applied Ecology* **18**, 343–53.

Scriber J.M. (1978) Cyanogenic glycosiders in *Lotus corniculatus*. *Oecologia* **34**, 143–55.

Scriber J.M. & Feeny P. (1979) Growth of herbivorous caterpillars in relation to feeding specialization and to the growth form of their food plants. *Ecology* **60**, 829–50.

Scriber J.M. & Slansky F. (1981) The nutritional ecology of immature insects. *Annual Review of Entomology* **26**, 183–211.

Seif el Din A. & Obeid M. (1971) Ecological studies of the vegetation of the Sudan. IV. The effects of simulated grazing on the growth of *Acacia senegal* (L.) Willd. seedlings. *Journal of Applied Ecology* **8**, 211–16.

Seigler D. & Price P.W. (1976) Secondary compounds in plants: primary functions. *American Naturalist* **110**, 101–5.

Seneviratne G. (1980) Can the panda survive in the wild? *New Scientist* **88**, 104–5.

Shahjahan M. & Streams F. (1973) Plant effects on host-finding by *Leiophron pseudopallipes* (Hymenoptera: Braconidae), a parasitoid of the tarnished plant bug. *Environmental Entomology* **2**, 921–5.

Shapiro A.M. (1981) The pierid red-egg syndrome. *American Naturalist* **117**, 276–94.

Shapiro A.M. & Carde R.T. (1970) Habitat selection and competition among sibling species of satyrid butterflies. *Evolution* **24**, 48–54.

Sharma R.D. (1971) Studies on the plant parasitic nematode *Tylenchorhynchus dubius*. *Mededelingen Landb. Hoogesch Wageningen* **71**, 1–9.

Shaver G.R. & Billings W.D. (1975) Root production and root turnover in a wet tundra ecosystem, Barrow, Alaska. *Ecology* **56**, 401–9.

Shaw M.W. (1968) Factors affecting the natural regeneration of sessile oak (*Quercus petraea*) in North Wales. *Journal of Ecology* **56**, 565–83; 647–60.

Shaw M.W. (1974) The reproductive characteristics of oak. In *The British Oak* (Eds M.G. Morris & F.H. Perring) pp. 162–81. Classey, Faringdon.

Sheard R.W. (1973) Organic reserves and plant regrowth. In *Chemistry and Biochemistry of Herbage* (Eds G.W. Butler & R.W. Bailey) vol. 2, pp. 353–78. Academic Press, New York.

Sheldon J.C. (1974) The behaviour of seeds in soil. III. The influence of seed morphology and the behaviour of seedlings on the establishment of plants from surface-lying seeds. *Journal of Ecology* **62**, 47–66.

Sheldrake A.R., Narayanan A. & Venkataratnam N. (1979) The effects of flower removal on the seed yield of pigeonpeas (*Cajanus cajan*). *Annals of Applied Biology* **91**, 383–90.

Shook R.S. & Baldwin P.H. (1970) Woodpecker predation on bark beetles in Englemann spruce logs as related to stand density. *Canadian Entomologist* **102**, 1345–54.

Shorten M. (1957) Damage caused by squirrels in Forestry Commission areas 1954–6. *Forestry* **30**, 151–72.

Shure D.J. (1971) Insecticide effects on early succession in an old-field ecosystem. *Ecology* **52**, 271–9.

Shutt D.A. (1976) The effect of plant oestrogens on animal reproduction. *Endeavour* **35**, 110–13.

Silen R.R. & Dimock E.J. (1978) Modeling feeding preferences by hare and deer among Douglas-fir genotypes. *Forest Science* **24**, 57–64.

Silvertown J.W. (1980a) The evolutionary ecology of mast seeding in trees. *Biological Journal of the Linnean Society, London* **14**, 235–250.

Silvertown J.W. (1980b) The dynamics of a grassland ecosystem: botanical equilibrium in the Park Grass Experiment. *Journal of Applied Ecology* **17**, 491–504.

Simenstad C.A., Estes J.A. & Kenyon K.W. (1978) Aleuts, sea otters, and alternate stable-state communities. *Science* **200**, 403–11.

Simmonds H.W. (1933) The biological control of the weed *Clidemia hirta* D. Don. in Fiji. *Bulletin of Entomological Research* **24**, 345–8.

Sinclair A.R.E. (1975) The resource limitation of trophic levels in tropical grassland ecosystems. *Journal of Animal Ecology* **44**, 497–520.

Sinclair A.R.E. (1977) *The African Buffalo*. University of Chicago Press, Chicago.

Sinclair A.R.E. (1979) The eruption of the ruminants. In *Serengeti: The Dynamics of an Ecosystem* (Eds A.R.E. Sinclair & M. Norton-Griffiths) pp. 82–103. University of Chicago Press, Chicago.

Sinclair A.R.E. & Norton-Griffiths M. (Eds) (1979) *Serengeti: Dynamics of an Ecosystem*. University of Chicago Press, Chicago.

Singer M.C. (1972) Complex components of habitat suitability within a butterfly colony. *Science* **176**, 75–7.

Singer M.C. & Ehrlich P.R. (1979) Population dynamics of the checkerspot butterfly *Euphydryas editha*. In *Population Ecology* (Eds U. Halbach & J. Jacobs). *Fortschritte der Zoologie* **25** (2/3): 29–52. Gustav Fischer Verlag, Stuttgart.

Singh B.B., Hadley H.H. & Bernard R.L. (1971) Morphology of pubescence in soybeans and its relationship to plant vigour. *Crop Science* **11**, 13–16.

Singh J.S. & Joshi M.C. (1979) Ecology of the semi-arid regions of India with emphasis on land use. In *Management of Semi-Arid Ecosystems* (Ed. B.H. Walker) pp. 243–75. Elsevier, Amsterdam.

Singh P. (1970) Host plant nutrition and composition: effects on agricultural pests. *Canadian Department of Agriculture Research Institute Information Bulletin* **6**, 1–102.

Skellam J.G. (1951) Random dispersal in theoretical populations. *Biometrika* **38**, 196–218.

Skinner G.J. & Whittaker J.B. (1981) An experimental investigation of interrelationships between the wood-ant (*Formica rufa*) and some tree-canopy herbivores. *Journal of Animal Ecology* **50**, 313–26.

Skogland T. (1980) Comparative summer feeding strategies of arctic and alpine *Rangifer*. *Journal of Animal Ecology* **49**, 81–98.

Slade N.A. & Balph D.F. (1974) Population ecology of Uinta ground squirrels. *Ecology* **55**, 989–1003.

Slansky F. & Feeny P. (1977) Stabilization of the rate of nitrogen accumulation by larvae of the cabbage butterfly on wild and cultivated food plants. *Ecological Monographs* **47**, 209–28.

Slatkin M. (1974) Competition and regional coexistence. *Ecology* **52**, 19–34.

Smalley A.E. (1959) The growth cycle of *Spartina* and its relation to the insect population in the marsh. *Proceedings of the Salt Marsh Conference* (Sapelo Island, Georgia, 1958) pp. 96–100.

Smigel B.W. & Rosenzweig M.L. (1974) Seed selection in *Dipodomys merriami* and *Perognathus penicillatus*. *Ecology* **55**, 329–39.

Smith A.E. & Leinweber C.L. (1971) Relationship of carbohydrate trend and morphological development of little bluestem tillers. *Ecology* **52**, 1052–7.

Smith C.C. (1970) The coevolution of pine squirrels (*Tamiasciurus*) and conifers. *Ecological Monographs* **40**, 349–71.

Smith C.C. (1975) The coevolution of plants and seed predators. In *Coevolution of Animals and Plants* (Eds L.E. Gilbert & P.H. Raven) pp. 53–77. University of Texas Press, Austin.

Smith C.C. & Follmer D. (1972) Food preferences of squirrels. *Ecology* **53**, 82–91.

Smith D. (1969) Removing and analysing total nonstructural carbohydrates from plant tissues. *Wisconsin Agricultural Experimental Station Research Report* **41**, 1–11.

Smith J.G. (1976a) Influence of crop background on aphids and other phytophagous insects on brussels sprouts. *Annals of Applied Biology* **83**, 1–13.

Smith J.G. (1976b) Influence of crop background on natural enemies of aphids on brussels sprouts. *Annals of Applied Biology* **83**, 15–29.

Smith J.N. (1962) Detoxification mechanisms. *Annual Review of Entomology* **7**, 465–80.

Smith R.H. & Bass M.H. (1972) Relation of artificial pod removal to soybean yields. *Journal of Economic Entomology* **65**, 606–8.

Smith T.J. & Odum W.E. (1981) The effects of grazing by snow geese on coastal salt marshes. *Ecology* **62**, 98–106.

Snow D.W. (1971) Evolutionary aspects of fruit-eating by birds. *Ibis* **113**, 194–202.

Soholt L.F. (1973) Consumption of primary production by a population of kangaroo rats (*Dipodomys merriami*) in the Mojave desert. *Ecological Monographs* **43**, 357–76.

Solomon M.E. (1949) The natural control of animal populations. *Journal of Animal Ecology* **18**, 1–35.

Soo Hoo C.F. & Fraenkel G. (1966) The selection of food plants in a polyphagous insect, *Prodenia eridania* (Cramer). *Journal of Insect Physiology* **12**, 693–709; 711–30.

Southward A.J. (1964) Limpet grazing and the control of vegetation on rocky shores. In *Grazing in Terrestrial and Marine Environments* (Ed. D.J. Crisp) pp. 265–73. Blackwell Scientific Publications, Oxford.

Southwood T.R.E. (1961) The number of species of insect associated with various trees. *Journal of Animal Ecology* **30**, 1–8.

Southwood T.R.E. (1973) The insect/plant relationship—an evolutionary perspective. *Symposium of the Royal Entomological Society of London* vol. 6, pp. 3–30.

Southwood T.R.E. (1976) *Ecological Methods*, 2e. Chapman & Hall, London.

Southwood T.R.E. (1977) Habitat, the templet for ecological strategies? *Journal of Animal Ecology* **46**, 337–65.

Southwood T.R.E., Brown V.K. & Reader P.M. (1979) The relationships of plant and insect diversities in succession. *Biological Journal of the Linnean Society of London* **12**, 327–48.

Southwood T.R.E. & Comins H.N. (1976) A synoptic population model. *Journal of Animal Ecology* **45**, 949–65.

Spatz G. & Mueller-Dombois D. (1973) The influence of feral goats on koa tree reproduction in Hawaii Volcanoes National Park. *Ecology* **54**, 870–6.

Spedding C.R.W. (1971) *Grassland Ecology*. Clarendon Press, Oxford.

Spencer D.A. (1964) Porcupine population fluctuations in past centuries revealed by dendrochronology. *Journal of Applied Ecology* **1**, 127–49.

Spencer K.A. (1973) *Agromyzidae (Diptera) of Economic Importance*. Junk, Hague.

Spinage C.A. & Guinness F.E. (1971) Tree survival in the absence of elephants in the Akagera National Park, Rwanda. *Journal of Applied Ecology* **8**, 723–8.

Sprague M.A. (1954) The effect of grazing management on forage and grain production from rye, wheat and oats. *Agronomy Journal* **46**, 29–33.

Staaland H., White R.G., Luick J.R. & Holleman D.F. (1980) Dietary influences on sodium and potassium metabolism of reindeer. *Canadian Journal of Zoology* **58**, 1728–34.

Stark R.W. & Dahlsten D.L. (Eds) (1970) *Studies on the Population Dynamics of the Western Pine Beetle*, Dendroctonus brevicomis Le Conte *(Coleptera: Scolytidae)*. University of California Division of Agricultural Science, Berkeley, California.

Steenbergh W.F. & Lowe C.H. (1969) Critical factors during the first years of life of the saguaro (*Cereus giganteus*) at Saguaro National Monument, Arizona. *Ecology* **50**, 825–34.

Stephens G.R. (1971) The relation of insect defoliation to mortality in Connecticut forests. *Connecticut Agricultural Experimental Station Bulletin* **723**, 1–16.

Stickler F.C. & Pauli A.W. (1961) Leaf removal in grain sorghum. *Agronomy Journal* **53**, 99–107.

Stiling P.D. (1980) Competition and coexistence among *Eupteryx* leafhoppers (Hemiptera: Cicadellidae) occurring on stinging nettles (*Urtica dioica*). *Journal of Animal Ecology* **49**, 793–805.

Stodart E. & Myers K. (1966) The effects of different foods on confined populations of wild rabbits *Oryctolagus cuniculus* (L.) *CSIRO Wildlife Research* **11**, 111–24.

Strobel G.A. & Lanier G.N. (1981) Dutch elm disease. *Scientific American* **245**(2), 40–50.

Strong D.R. (1974) Rapid asymptotic species accumulation in phytophagous insect communities: the pests of cacao. *Science* **185**, 1064–6.

Strong D.R., McCoy E.D. & Rey J.R. (1977) Time and the number of herbivore species: the pests of sugarcane. *Ecology* **58**, 167–75.

Strong F.E. & Sakamoto S.S. (1963) Some amino acid requirements of the green peach aphid, *Myzus persicae* (Sulzer), determined with glucose-U-C^{14}. *Journal of Insect Physiology* **9**, 875–9.

Stubbs M. (1977) Density dependence in the life cycles of animals and its importance in *K*- and *r*-strategies. *Journal of Animal Ecology* **46**, 677–88.

Stumpf P.K. & Conn E.E. (Eds) (1982) *Biochemistry of Plants: A Comprehensive Treatise: Volume 7 Secondary Plant Products*. Academic Press, London.

Sugimoto T. (1980) Models of the spatial pattern of egg population of *Ranunculus* leaf mining fly, *Phytomyza ranunculi* (Diptera: Agromyzidae), in host leaves. *Researches in Population Ecology* **22**, 13–32.

Sullivan T.P. (1979) The use of alternative foods to reduce conifer seed predation by the deermouse (*Peromyscus maniculatus*). *Journal of Applied Ecology* **16**, 475–95.

Summerhayes V.S. (1951) *Wild Orchids of Britain*. Collins, London.

Summers D.D.B. (1981) Bullfinch (*Pyrrhula pyrrhula*) damage in orchards in relation to woodland bud and seed feeding. In *Pests, Pathogens and Vegetation* (Ed. J.M. Thresh) pp. 385–91. Pitman, London.

Swank W.T., Waide J.B., Crossley D.A. & Todd R.L. (1981) Insect defoliation enhances nitrate export from forest ecosystems. *Oecologia* **51**, 297–9.

Sweet G.B. & Wareing P.F. (1966) Role of plant growth in regulating photosynthesis. *Nature* **210**, 77–9.

Swingland I.R. & Lessells C.M. (1979) The natural regulation of giant tortoise populations on Aldabra Atoll: movement polymorphism, reproductive success and mortality. *Journal of Animal Ecology* **48**, 639–54.

Symonides E. (1978) Effect of population density on the phenological development of individuals of annual plant species. *Ekologia Polska* **26**, 273–86.

Taber R.D. & Dasmann R.F. (1957) The dynamics of three natural populations of the deer *Odocoileus hemionus columbianus*. *Ecology* **38**, 233–46.

Tahvanainen J.O. (1972) Phenology and microhabitat selection of some flea beetles (Coleoptera: Chrysomelidae) on wild and cultivated crucifers in central New York. *Entomologica Scandinavica* **3**, 130–8.

Tahvanainen J.O. & Root R.B. (1972) The influence of vegetational diversity on the population ecology of a specialized herbivore *Phyllotreta cruciferae* (Coleoptera: Chrysomelidae). *Oecologia* **10**, 321–46.

Taitt M.J. & Krebs C.J. (1981) The effect of extra food on small rodent populations: II. voles (*Microtus townsendii*). *Journal of Animal Ecology* **50**, 125–37.

Tamm C.O. (1956) Further observations on the survival and flowering of some perennial herbs. *Oikos* **7**, 273–92.

Tamm C.O. (1972) Survival and flowering of perennial herbs. III. The behaviour of *Primula veris* on permanent plots. *Oikos* **23**, 159–66.

Tanner J.T. (1975) The stability and the intrinsic growth rates of prey and predator populations. *Ecology* **56**, 855–67.

Tapper S.C. (1976) Population fluctuations of field voles (*Microtus*): a background to the problems involved in predicting vole plagues. *Mammal Review* **6**, 93–117.

Tast J. & Kalela O. (1971) Comparisons between rodent cycles and plant production in Finnish Lapland. *Annales Academiae Scientiarum Fennicae Series A, IV Biologica* **186**, 1–14.

Taylor R.J. (1971) The value of clumping to prey: experiments with a mammalian predator. *Oecologia* **30**, 285–94.

Teal J.M. (1962) Energy flow in the salt marsh ecosystem of Georgia. *Ecology* **43**, 614–24.

Temple S.A. (1977) Plant–animal mutualism: coevolution with dodo leads to near extinction of plant. *Science* **197**, 885–6.

Thiegles B.A. (1968) Altered polyphenol metabolism in the foliage of *Pinus sylvestris* associated with European pine sawfly attack. *Canadian Journal of Botany* **46**, 724–5.

Thomas A.S. (1960) Changes in vegetation since the advent of myxomatosis. *Journal of Ecology* **48**, 287–306.

Thompson D.Q. (1955) The role of food and cover in population fluctuations of the brown lemming at Pt. Barrow, Alaska. *Transactions of the 20th North American Wildlife Conference*, pp. 166–76.

Thompson J.N. (1978) Within-patch structure and dynamics in *Pastinaca sativa* and resource availability to a specialized herbivore. *Ecology* **59**, 443–8.

Thompson J.N. & Price P.W. (1977) Plant plasticity, phenology, and herbivore dispersion: wild parsnip and the parsnip webworm. *Ecology* **58**, 1112–19.

Thompson K. & Grime J.P. (1979) Seasonal variation in the seed banks of herbaceous species in ten contrasting habitats. *Journal of Ecology* **67**, 893–921.

Thompson K., Grime J.P. & Mason G. (1977) Seed germination in response to diurnal fluctuations of temperature. *Nature* **67**, 147–9.

Thompson P. (1978) The oviposition sites of five leafhopper species (Hom. Auchenorrhyncha) on *Holcus mollis* and *H. lanatus*. *Ecological Entomology* **3**, 231–40.

Thomson A.I. (1926) *Problems of Birds Migration*. H.F. & G. Witherby, London.

Thorne G.N. & Evans A.F. (1964) Influence of tops and roots on net assimilation rate of sugar-beet and spinach beet and grafts between them. *Annals of Botany* **28**, 499–508.

Thornley J.H.M. (1976) *Mathematical Models in Plant Physiology*. Academic Press, London.

Thorsteinson A.J. (1960) Host selection in phytophagous insects. *Annual Review of Entomology* **5**, 193–218.

Thresh J.M. (Ed.) (1981) *Pests, Pathogens and Vegetation*. Pitman, London.

Tikhomiro B.A. (1959) *Interrelationships of the Animal World and the Vegetation Cover of the Tundra* (English summary). Academy of Science, USSR. Komarov Botanical Institute, Moscov–Leningrad.

Tillyard R.J. (1929) The biological control of noxious weeds. *Papers and Proceedings of the Royal Society of Tasmania* 1929 : 51–86.

Tilman D. (1978) Cherries, ants and tent caterpillars: timing of nectar production in relation to susceptibility of caterpillars to ant predation. *Ecology* **59**, 686–92.

Tingey W.M. & Pillemer E.A. (1977) Lygus bugs: crop resistance and physiological nature of feeding injury. *Bulletin of the Entomological Society of America* **23**, 277–87.

Tinker M.A.H. (1930) The effect of cutting to ground level upon the growth of established plants of *Dactylis glomerata* and *Phleum pratense*. *Welsh Journal of Agriculture* **6**, 182–98.

Tischler W. (1965) *Agraokologie*. Gustav Fishner, Jena.

Torrent J.A. (1955) Oak *Tortrix* and its control in Spain. *FAO Plant Protection Bulletin* **3**, 117–21.

Trenbath B.R. (1974) Biomass productivity in mixtures. *Advances in Agronomy* **26**, 177–210.

Trlica M.J. & Singh J.S. (1979) Translocation of assimilates and creation, distribution and utilization of reserves. In *Arid Land Ecosystems* (Eds D.W. Goodall & R.A. Perry) vol. 1, pp. 537–71. Cambridge University Press, Cambridge.

Troughton A. (1957) The underground organs of herbage grasses. *Commonwealth Advisory Bureau Bulletin* **44**, 1–10.

Trudell J. & White R.G. (1981) The effect of forage structure and availability on food intake, biting rate, bite size and daily eating time of reindeer. *Journal of Applied Ecology* **18**, 63–81.

Tucker J.J. & Fitter A.H. (1981) Ecological studies at Askham Bog nature reserve. *The Naturalist* **106**, 3–14.

Turcek F.J. (1960) Über Rotelmausschaden in den slowakischen Wäldern im Jahre 1959. *Zeitschrift für Angewandte Zoologie* **47**, 449–65.

Turner N. (Ed.) (1963) Effect of defoliation by the gypsy moth. *Connecticut Agricultural Experimental Station Bulletin* **658**, 1–30.

Turner R.M., Alcorn S.M. & Olin G. (1969) Mortality of transplanted saguaro seedlings. *Ecology* **50**, 835–44.

Ueckert D.N. (1979) Impact of white grub (*Phyllophaga crinita*) on a shortgrass community and evaluation of selected rehabilitation practices. *Journal of Range Management* **32**, 445–8.

Ueno M. & Smith D. (1970) Growth and carbohydrate changes in the root wood and bark of different sized alfalfa plants during regrowth after cutting. *Crop Science* **10**, 396–9.

Underwood A.J. (1978) An experimental evaluation of competition between three species of intertidal gastropods. *Oecologia* **33**, 185–208.

Underwood E.J. (1956) *Trace Elements in Human and Animal Nutrition.* Academic Press, New York.

Usher M.B. (1966) A matrix approach to the management of renewable resources with special reference to selection forests. *Journal of Applied Ecology* **3**, 355–67.

Usher M.B. (1972) Developments in the Leslie matrix model. In *Mathematical Models in Ecology* (Ed. J.N.R. Jeffers) pp. 29–60. Blackwell Scientific Publications, Oxford.

van den Bergh J.P. (1968) An analysis of yields of grasses in mixed and pure stands. *Versl. Landbouwk. Onderz.* **714**, 1–71.

van der Meijden E. (1976) Changes in the distribution patterns of *Tyria jacobaeae* during the larval period. *Netherlands Journal of Zoology* **26**, 136–61.

van der Meijden E. (1979) Herbivore exploitation of a fugitive plant species: local survival and extinction of cinnabar moth and ragwort in a heterogeneous environment. *Oecologia* **42**, 307–23.

van Dyne G.M., Brockington N.R., Szocs Z., Duek J. & Ribic C.A. (1980) Large herbivore subsystem. In *Grasslands, Systems Analysis and Man* (Eds A.I. Breymeyer & G.M. van Dyne) pp. 269–537. Cambridge University Press, Cambridge.

van Emden H.F. (1965) The role of uncultivated land in the biology of crop pests and beneficial insects. *Scientific Horticulture* **17**, 121–36.

van Emden H.F. (1966) Plant resistance to insects induced by environment. *Scientific Horticulture* **18**, 91–102.

van Emden H.F. (1978) Insects and secondary plant substances—an alternative viewpoint with special reference to aphids. In *Biochemical Aspects of Plant and Animal Coevolution* (Ed. J.B. Harborne) pp. 309–23. Academic Press, London.

van Emden H.F., Eastop V.F., Hughes R.D. & Way M.J. (1969) The ecology of *Myzus persicae. Annual Review of Entomology* **14**, 197–270.

van Emden H.F. & Way M.J. (1973) Host plants in the population dynamics of insects. In *Insect/Plant Relationships* (Ed. H.F. van Emden) pp. 181–99. Blackwell Scientific Publications, Oxford.

van Hook R.E., Reichle D.E. & Auerbach S.I. (1970) Energy and nutrient dynamics of predator and prey arthropod populations in a grassland ecosystem. Oak Ridge National Laboratory, ORNL-4509.

Van Poollen H.W. & Lacey J.R. (1978) Herbage responses to grazing systems and stocking intensities. *Journal of Range Management* **32**, 250–3.

Varley G.C. & Gradwell G.R. (1962) The effect of partial defoliation by caterpillars on the timber production of oak trees in England. *Proceedings of the 11th International Congress of Entomology*, Vienna 1960, vol. 2, pp. 211–14.

Varley G.C., Gradwell G.R. & Hassell M.P. (1973) *Insect Population Ecology.* Blackwell Scientific Publications, Oxford.

Vaughan M.R. & Keith L.B. (1981) Demographic response of experimental snowshoe hare populations to overwinter food shortage. *Journal of Wildlife Management* **45**, 354–80.

Vesey-Fitzgerald D.F. (1960) Grazing succession amongst East African game animals. *Journal of Mammalogy* **41**, 161–70.

Vickery P.J. (1972) Grazing and net primary production of a temperate grassland. *Journal of Applied Ecology* **9**, 307–14.

Volterra V. (1926) Variation and fluctuations of the number of individuals in animal species living together. In R.N. Chapman (Ed.) (1931) *Animal Ecology*, pp. 409–48. McGraw Hill, London.

von Bertalanffy L. (1957) Quantitative laws in metabolism and growth. *Quarterly Review of Biology* **32**, 217–31.

Vorontsov A.I. (1963) *The Biological Bases of Forest Production.* Vyssh Shkola, Moscow. (In Russian.)

Voute A.D. & Walenkamp J.F.G.M. (1946) De oorzaak van het optreden van gradaties van de dennenlotrups (*Evetria buoliana* Schff) en de mogelijkheid deze te voorkomen. *Modedelingen No 14. Comm. Best. en Bestr. Ins. pl. in bossen.*

Waage J.K. (1980) Sloth moths and other zoophilous Lepidoptera. *Proceedings of the British Entomological and Natural History Society* **1980**, 73–4.

Wagner F. & Ehrhardt R. (1961) Untersuchungen am Stickanal der Graswanzé *Miris dolorbratus* L., der Urheberin der totalen Weissahrigkeit der Rotschwingels (*Festuca rubra*). *Zeitschrift für Pflanzenkrankheiten und Pflanzenschutz* **68**, 615–20.

Waldbauer G.P. (1968) The consumption and utilization of food by insects. *Advances in Insect Physiology* **5**, 229–88.

Walker B.H., Ludwig D., Holling C.S. & Peterman R.M. (1981) Stability of semi-arid savanna grazing systems. *Journal of Ecology* **69**, 473–98.

Wallace J.W. & Mansell R.L. (Eds) (1976) *Biochemical Interactions Between Plants and Insects.* Recent Advances in Phytochemistry Volume 10. Plenum Press, New York.

Wallner W.E. & Walton G.S. (1979) Host defoliation: a possible determinant of gypsy moth population quality. *Annals of the Entomological Society of America* **72**, 62–7.

Waloff N. (1968) Studies on the insect fauna on scotch broom *Sarothamnus scoparius* (L.) Wimmer. *Advances in Ecological Research* **5**, 87–208.

Waloff N. & Richards O.W. (1977) The effect of insect fauna on growth, mortality and natality of broom, *Sarothamnus scoparius*. *Journal of Applied Ecology* **14**, 787–98.

Waloff Z. (1976) Some temporal characteristics of desert locust plagues. *Anti Locust Memoirs* vol. 13. COPR, London.

Walters R.J.K. & Evans E.M. (1979) Evaluation of a sward sampling technique for estimating herbage intake by grazing sheep. *Grass and Forage Science* **34**, 37–44.

Wang Y., Gutierrez A.P., Oster G. & Daxl R. (1977) A population model for plant growth and development: coupling cotton–herbivore interactions. *Canadian Entomologist* **109**, 1359–74.

Wapshere A.J. (1971) Selection and biological control organisms of weeds. *Commonwealth Institute of Biological Control Miscellaneous Publications* **6**, 56–61.

Ward L.K. & Lakhani K.H. (1977) The conservation of juniper: the fauna of food-plant island sites in southern England. *Journal of Applied Ecology* **14**, 121–35.

Ward P. (1965) Feeding ecology of the black-faced dioch *Quelea quelea* in Nigeria. *Ibis* **107**, 173–214.

Ward P. (1971) The migration patterns of *Quelea quelea* in Africa. *Ibis* **113**, 275–97.

Wardlaw I.F. (1968) The control and pattern of movement of carbohydrates in plants. *Botanical Review* **34**, 79–105.

Wareing P.F., Khalifa M.M. & Treharne K.J. (1968) Rate limiting processes in photosynthesis at saturating light intensities. *Nature* **220**, 453–7.

Wareing P.F. & Patrick J. (1975) Source-sink relations and the partition of assimilates in the plant. In *Photosynthesis and Productivity in Different Environments* (Ed. J.P. Cooper) pp. 481–99. Cambridge University Press, Cambridge.

Wasserman S.S. & Mitter C. (1978) The relationship of body size to breadth of diet in some Lepidoptera. *Ecological Entomology* **3**, 155–60.

Watkinson A.R. (1978) The demography of a sand dune annual: *Vulpia fasciculata* II. The dynamics of seed populations. *Journal of Ecology* **66**, 35–44.

Watkinson A.R. (1980) Density dependence in single species populations of plants. *Journal of Theoretical Biology* **83**, 345–57.

Watkinson A.R. & Harper J.L. (1978) The demography of a sand dune annual: *Vulpia fasciculata* I. The natural regulation of populations. *Journal of Ecology* **66**, 15–33.

Watkinson A.R., Huiskes A.H.L. & Noble J.C. (1979) The demography of sand dune species with contrasting life cycles. In *Ecological Processes in Coastal Environments* (Eds R.L. Jefferies & A.J. Davy) pp. 95–112. Blackwell Scientific Publications, Oxford.

Watson A. & Moss R. (1970) Dominance, spacing behaviour and aggression in relation to population limitation in vertebrates. In *Animal Populations in Relation to their Food Resources* (Ed. A. Watson) pp. 167–220. Blackwell Scientific Publications, Oxford.

Watt A.S. (1919) On the causes of failure of natural regeneration in British oakwoods. *Journal of Ecology* **7**, 173–203.

Watt A.S. (1955) Bracken versus heather, a study in plant sociology. *Journal of Ecology* **35**, 1–22.

Watt A.S. (1981) A comparison of grazed and ungrazed Grassland A in East Anglian Breckland. *Journal of Ecology* **69**, 499–508; 509–36.

Watt T.A. & Hagger R.J. (1980) The effect of defoliation upon yield, flowering and vegetative spread of *Holcus lanatus* growing with and without *Lolium perenne*. *Grass and Forage Science* **35**, 227–34.

Watts C.H.S. (1970a) Effect of supplementary food on breeding in woodland rodents. *Journal of Mammalogy* **51**, 169–71.

Watts C.H.S. (1970b) A field experiment on intraspecific interactions in the red-backed vole, *Clethrionomys gapperi*. *Journal of Mammalogy* **51**, 341–7.

Way M.J. (1968) Intra-specific mechanisms with special reference to aphid populations. In *Insect Abundance* (Ed. T.R.E. Southwood) pp. 18–36. Blackwell Scientific Publications, Oxford.

Way M.J. & Cammell M. (1970) Aggregation behaviour in relation to food utilization by aphids. In *Animal Populations in Relation to their Food Resources* (Ed. A. Watson) pp. 229–47. Blackwell Scientific Publications, Oxford.

Way M.J. & Cammell M.E. (1973) The problem of pest and disease forecasting—possibilities and limitations as exemplified by work on the bean aphid, *Aphis fabae*. *Proceedings of the 7th British Insecticide and Fungicide Conference*, pp. 933–54.

Way M.J., Cammell M.E., Alford D.V., Gould H.J., Graham C.W., Lane A., Light W.I. StG., Rayner J.M., Heathcote G.D., Fletcher K.E. & Seal K. (1977) Use of forecasting in chemical control of black bean aphid, *Aphis fabae* Scop., on spring-sown field beans, *Vicia faba* L. *Plant Pathology* **26**, 1–7.

Way M.J. & Heathcote G.D. (1966) Interactions of crop density of field beans, abundance of *Aphis fabae* Scop., virus incidence and aphid control by chemicals. *Annals of Applied Biology* **57**, 409–23.

Wearing C.H. (1972) Responses of *Myzus persicae* and *Brevicoryne brassicae* to leaf age and water stress in Brussels sprouts grown in pots. *Entomologia Experimentalis et Applicata* **15**, 61–80.

Weaver S.E. & Cavers P.B. (1980) Reproductive effort of two perennial weed species in different habitats. *Journal of Applied Ecology* **17**, 505–13.

Webb J.W. & Moran V.C. (1978) The influence of the host plant on the population dynamics of *Acizzia russellae* (Homoptera: Psyllidae). *Ecological Entomology* **3**, 313–21.

Webb W.L. & Karchesy J.J. (1977) Starch content of Douglas fir defoliated by the tussock moth. *Canadian Journal of Forest Research* **7**, 186–8.

Wellington W.G. (1960) Qualitative changes in natural populations during changes in abundance. *Canadian Journal of Zoology* **38**, 289–314.

Went F.W. (1973) Competition among plants. *Proceedings of the National Academy of Sciences of the United States of America* **70**, 585–90.

Werner P.A. (1977) Colonization success of a 'biennial' plant species: experimental field studies of species cohabitation and replacement. *Ecology* **58**, 840–9.

Werner P.A. & Caswell H. (1977) Population growth rates and age versus stage-distribution models for teasel (*Dipsacus sylvestris* Huds.). *Ecology* **58**, 1103–11.

Werner R.A. (1979) Influence of host foliage on development, survival, fecundity and oviposition of the spear-marked black moth, *Rheumaptera hastata* (Lepidoptera: Geometridae). *Canadian Entomologist* **111**, 317–22.

West N.E. (1968) Rodent-influenced establishment of ponderosa pine and bitter-brush seedlings in central Oregon. *Ecology* **49**, 1009–11.

West N.E., Rea K.H. & Harniss R.O. (1979) Plant demographic studies in sagebrushgrass communities of southeastern Idaho. *Ecology* **60**, 376–88.

Westoby M. (1974) An analysis of diet selection by large generalist herbivores. *American Naturalist* **108**, 290–304.

Westoby M. (1978) What are the biological bases of varied diets? *American Naturalist* **112**, 627–31.

Westoby M. (1980) Relations between genet and tiller population dynamics: survival of *Phalaris tuberosa* tillers after clipping. *Journal of Ecology* **68**, 863–9.

Westoby M. (1981) The place of the self thinning rule in population dynamics. *American Naturalist* **118**, 581–7.

White E.G. (1974) Grazing pressures of grasshoppers in an alpine tussock grassland. *New Zealand Journal of Agricultural Research* **17**, 357–72.

White J. (1980) Demographic factors in populations of plants. In *Demography and Evolution of Plant Populations* (Ed. O.T. Solbrig) pp. 21–48. Blackwell Scientific Publications, Oxford.

White J. (1981) The allometric interpretation of the self-thinning rule. *Journal of Theoretical Biology* **89**, 475–500.

White L.M. (1973) Carbohydrate reserves of grasses: a review. *Journal of Range Management* **26**, 13–18.

White R.G., Bunnell F.L., Gaare E., Skogland T. & Hubert B. (1981) Ungulates on arctic ranges. In *Tundra Ecosystems: A Comparative Analysis* (Ed. L.C. Bliss, O.W. Heal & J.J. Moore) pp. 397–483. Cambridge University Press, Cambridge.

White R.R. (1974) Food plant defoliation and larval starvation of *Euphydryas editha*. *Oecologia* **14**, 307–15.

White T.C.R. (1969) An index to measure weather-induced stress of trees associated with outbreaks of psyllids in Australia. *Ecology* **50**, 905–9.

White T.C.R. (1974) A hypothesis to explain outbreaks of looper caterpillars, with special reference to populations of *Selidosema suavis* in a plantation of *Pinus radiata* in New Zealand. *Oecologia* **16**, 279–301.

Whitford W.G. (1978) Foraging by seed-harvesting ants. In *Production Ecology of Ants and Termites* (Ed. M.V. Brian) pp. 107–10. Cambridge University Press, Cambridge.

Whitham T.C. (1978) Habitat selection by *Pemphigus* aphids in response to resource limitation and competition. *Ecology* **59**, 1164–76.

Whitham T.C. (1980) The theory of habitat selection: examined and extended using *Pemphigus* aphids. *American Naturalist* **115**, 449–66.

Whittaker J.B. (1973) Density regulation in a population of *Philaenus spumarius* (L.) (Homoptera: Cercopidae). *Journal of Animal Ecology* **42**, 163–72.

Whittaker R.H. (1975) *Communities and Ecosystems*, 2e. Macmillan, London.

Wickens G.E. & Whyte L.P. (1979) Land-use in the southern margins of the Sahara. In *Management of Semi-Arid Ecosystems* (Ed. B.H. Walker) pp. 205–42. Elsevier, Amsterdam.

Wiegert R.G. (1965) Energy dynamics of the grasshopper populations in old field and alfalfa field systems. *Oikos* **16**, 161–76.

Wiegert R.G. & Evans F.C. (1967) Investigations of secondary productivity in grasslands. In *Secondary Productivity in Terrestrial Ecosystems* (Ed. K. Petrusewicz) vol. II, pp. 499–518.

Wiklund C. (1975) The evolutionary relationship between adult oviposition preferences and larval host plant range in *Papilio machaon* L. *Oecologia* **18**, 185–97.

Wiklund C. (1981) Generalist vs. specialist oviposition behaviour in *Papilio machaon* (Lepidoptera), and functional aspects on the heirarchy of oviposition preferences. *Oikos* **36**, 163–70.

Wiklund C. & Ahrberg C. (1978) Host plants, nectar source plants, and habitat selection of males and females of *Anthocharis cardamines* (Lepidoptera). *Oikos* **31**, 169–83.

Willey R.W. & Heath S.B. (1969) The quantitative relationships between plant populations and crop yield. *Advances in Agronomy* **21**, 281–321.

Williamson M. (1972) *The Analysis of Biological Populations.* Edward Arnold, London.

Williamson P. (1976) Above-ground primary production of chalk grassland allowing for leaf death. *Journal of Ecology* **64**, 1059–75.

Willms W., Bailey A.W., McLean A. & Tucker R. (1981) The effects of fall defoliation on the utilization of blue bunchgrass and its influence on the distribution of deer in spring. *Journal of Range Management* **34**, 16–18.

Willoughby W.M. (1959) Limitations to animal production imposed by seasonal fluctuations in pasture and by management procedures. *Australian Journal of Agricultural Research* **10**, 248–68.

Willoughby W.M. & Davidson R.L. (1979) Use, management and conservation. In *Grassland Ecosystems of the World: Analysis of Grasslands and their Uses* (Ed. R.T. Coupland) pp. 287–98. Cambridge University Press, Cambridge.

Wilson A.G.L., Hughes R.D. & Gilbert N.E. (1972) The response of cotton to pest attack. *Bulletin of Entomological Research* **61**, 405–14.

Wilson D.E. & Janzen D.H. (1972) Predation on *Scheelea* palm seeds by bruchid beetles: seed density and distance from the parent palm. *Ecology* **53**, 954–9.

Wilson F. (1964) The biological control of weeds. *Annual Review of Entomology* **9**, 225–44.

Wilson L.F. (1966) Effects of different population levels of the European pine sawfly on young Scotch pine trees. *Journal of Economic Entomology* **59**, 1043–9.

Winter T.G. (1974) New host plant records of Lepidoptera associated with conifer aforestation in Britain. *Entomologist's Gazette* **25**, 247–58.

Witton S. (1981) The impact of herbivorous insects on acorn production in *Quercus robur*. Unpublished MSc thesis, University of London.

Woloff J.O. (1980) The role of habitat patchiness in the population dynamics of snowshoe hares. *Ecological Monographs* **50**, 111–30.

Wondolleck J.T. (1978) Forage-area separation and overlap in heteromyid rodents. *Journal of Mammalogy* **59**, 510–18.

Wood D.L. (1972) Selection and colonization of ponderosa pine by bark beetles. In *Insect/Plant Relationships* (Ed. H.F. Van Emden) pp. 101–17. Blackwell Scientific Publications, Oxford.

Wood D.L. (1980) Approach to research and forest management for western pine beetle control. In *New Technology of Pest Control* (Ed. C.B. Huffaker) pp. 417–48. John Wiley & Sons, New York.

Wooden A.N., Sellers K.C. & Tribe D.E. (1963) *Animal Health, Production and Pasture.* Longmans, London.

Woodwell G.M., Whittaker R.H. & Houghton R.A. (1975) Nutrient concentrations in plants in the Brookhaven oak-pine forests. *Ecology* **56**, 318–22.

Wratten S.D. (1974) Aggregation in the birch aphid *Euceraphis punctipennis* (Zett) in relation to food quality. *Journal of Animal Ecology* **43**, 191–8.

Yoda K., Kira T., Ogawa H. & Hozumi K. (1963) Self thinning in overcrowded pure stands under cultivated and natural conditions. *Journal of Biology of Osaka City University* **14**, 107–29.

Yodzis P. (1981) The structure of assembled communities. *Journal of Theoretical Biology* **92**, 103–17.

Young B.A. & Corbett J.L. (1972) Maintenance energy requirement of grazing sheep in relation to herbage availability. I. Calorimetric estimates. *Australian Journal of Agricultural Research* **23**, 57–76.

Youngner V.B. (1972) Physiology of defoliation and regrowth. In *The Biology and Utilization of Grasses* (Eds V.B. Youngner & C.M. McKell) pp. 292–303.

Youngner V.B. & McKell C.M. (Eds) (1972) *The Biology and Utilization of Grasses.* Academic Press, New York.

Zahirul Islam (1981) The influence of cinnabar moth on reproduction of ragwort. Unpublished MSc thesis, University of London.

Zelazny B. (1979) Loss in coconut yield due to *Oryctes rhinoceros* damage. *FAO Plant Protection Bulletin* **27**, 65–70.

Zimmermann H.G. (1979) Herbicidal control in relation to distribution of *Opuntia aurantiaca* Lindley and effects on cochineal populations. *Weed Research* **19**, 89–93.

Zimmerman R.H. (1972) Juvenility and flowering in woody plants: a review. *Horticultural Science* **7**, 447–55.

Zwolfer H. (1979) Strategies and counterstrategies in insect population systems competing for space and food in flowerheads and plant galls. *Fortschritte der Zoologie* **25(2/3)**, 331–53.

INDEX

Abies
  A. alba 203
  A. balsamea
    age structure and
      survivorship 36
    foliage thinning 23
    response to woolly aphid 203
    size distribution 51
    spruce budworm outbreaks 230
  A. concolor 39
Abortion
  herbivores
    and food 6
  plants
    of flowers 5
    of immature fruits 5
Abutilon theophrasti 59
Acacia spp. 193, 197, 203, 204
Acalymma
  A. thiemei 310
  A. vittata 307, 311
Acanthoscelides fraterculus 299
Acer
  A. campestre 318
  A. pseudoplatanus 46, 141, 318
  A. rubrum 99, 306
  A. saccharum 306
Acizzia russellae 193
Acmaea spp. 164, 321
Acorns 64, 205, 322
Acyrthosiphum pisum 56, 216, 328
Adelges
  A. abietis 49
  A. piceae 203
Aerial seed reserves *see* Seed bank
Age structure
  affected by emigration 147
  harvesting and herbivore 245
  and numerical responses 142
  and plant survivorship 34
  and susceptibility to pest
    attack 232
  of tree populations 36
Aggregation
  of aphids and improved
    feeding 124

  of herbivores at high plant
    densities 149
  host finding and pheromones 123,
    309
  of insect eggs 145, 170
  leading to pseudo-interference 126
  models of herbivore attack 270,
    273
  predator, prey refuge and 280
  of superior competitor leading to
    coexistence 324
  of tropical plants around
    parents 292
Aggression
  and emigration in vole
    populations 127
  see also interference
Agriolimax spp. 207
Agropyron spp. 45, 55, 92, 122, 159,
  300
Agrostemma githago 218
Agrostis stolonifera 159
Agrostis/Festuca grasslands 190, 291,
  294, 302
Air pollution
  and plant competition 39
  reduces plants defences against
    insects 39
Alaska 19, 44, 325
Alatae *see* Wing determination
Alauda arvensis 160
Alces alces see Moose
Alder, age structure 36
Alfalfa
  aphid attack leads to reduced sheep
    fertility 202
  defoliation and root growth 105
  regrowth and carbohydrate
    reserves 96
Algae *see* Marine algae
Alkaloids
  detoxification in sheep rumen 183
  discrimination by voles 120
  increase following ragwort
    defoliation 202
  and primary metabolites 196

403